T0301771

Mathematical Foundations of
Nonextensive Statistical Mechanics

Other World Scientific Titles by the Author

Beyond the Triangle: Brownian Motion, Ito Calculus, and Fokker–Planck Equation — Fractional Generalizations
ISBN: 978-981-3230-91-0

Mathematical Foundations of
Nonextensive Statistical
Mechanics

Sabir Umarov

University of New Haven, USA

Constantino Tsallis

Centro Brasileiro de Pesquisas Fisicas, Brazil

NEW JERSEY · LONDON · SINGAPORE · BEIJING · SHANGHAI · HONG KONG · TAIPEI · CHENNAI · TOKYO

Published by

World Scientific Publishing Co. Pte. Ltd.

5 Toh Tuck Link, Singapore 596224

USA office: 27 Warren Street, Suite 401-402, Hackensack, NJ 07601

UK office: 57 Shelton Street, Covent Garden, London WC2H 9HE

British Library Cataloguing-in-Publication Data

A catalogue record for this book is available from the British Library.

MATHEMATICAL FOUNDATIONS OF NONEXTENSIVE STATISTICAL MECHANICS

ISBN 978-981-124-515-2 (hardcover)
ISBN 978-981-124-516-9 (ebook for institutions)
ISBN 978-981-124-517-6 (ebook for individuals)

For any available supplementary material, please visit
https://www.worldscientific.com/worldscibooks/10.1142/12499#t=suppl

Typeset by Stallion Press
Email: enquiries@stallionpress.com

Printed in Singapore

To our families

Preface

The present book is dedicated to the mathematical foundations of a physical theory which emerged over three decades ago. This theory is currently referred to as nonextensive statistical mechanics, and it consists of a generalization of the celebrated Boltzmann–Gibbs statistical mechanics, one of the pillars of contemporary theoretical physics. In order to put this generalization into an appropriate context, let us briefly remind the history of mechanics since Newtons Principia Mathematica.

The systems involving masses which are neither too small nor too large compared to that of an apple, and whose velocities are small compared to c, the velocity of light, obey classical or Newtons mechanics (the second law of motion). The systems with masses neither too small nor too large but whose velocities are comparable to c obey Einsteins special relativity. The systems with very large masses (e.g., that of a star) and velocities comparable to c obey Einsteins general relativity. The systems with very small masses (e.g., that of an electron) and velocities small compared to c obey nonrelativistic quantum mechanics (Schroedinger equation). The systems with very small masses and velocities comparable to c obey relativistic quantum mechanics (Klein–Gordon or Dirac equations if they are spinless or have spin $1/2$, respectively).

The situation is very analogous to statistical mechanics — an expression coined by Gibbs, from New Haven — which adds to mechanics the theory of probabilities. The systems involving local space–time correlations obey Boltzmann–Gibbs statistical mechanics [constructed with the Boltzmann–Gibbs–von Neumann–Shannon (additive) entropy S_{BG}]. Notorious examples are the systems constituted by distinguishable particles with short-range interactions, indistinguishable particles with half-integer spin (Fermi–Dirac statistics), indistinguishable particles with integer spin (Bose–Einstein statistics with the celebrated Einstein condensation phenomenon), strongly chaotic nonlinear dynamical systems, specific random geometrical systems such as bond percolation (Kasteleyn and Fortuin theorem), self-avoiding random walk (de Gennes isomorphism).

In contrast, the systems with global space–time correlations obey non-Boltzmann–Gibbs statistical mechanics, e.g., nonextensive statistical mechanics (or q-statistics for short) [constructed with non-additive entropies such as S_q, with $S_1 = S_{\mathrm{BG}}$]. Notorious examples are the systems constituted by distinguishable particles with long-range interactions (e.g., gravitation), weakly chaotic nonlinear dynamical systems, specific random geometrical systems such as the ubiquitous (asymptotically) scale-invariant networks.

As just described, nonextensive statistical mechanics constitutes a generalization of the standard Boltzmann–Gibbs statistical mechanics since it recovers it for the particular instance where the entropic index q equals unity. This theory was formulated by one of us (CT) in 1985 and published in 1988. It took however quite a few years until the mathematical foundations of this theoretical physical theory were gradually understood and rigorously established. It is the aim of this book to develop systematically the mathematical foundations at their present stage.

The book contains seven chapters. Chapter 1 is introductory. Here, we briefly discuss the physical background of nonextensive statistical mechanics and probability theory establishing, in parallel, the notions and notations used throughout the book.

The next two chapters introduce key mathematical components of nonextensive statistical mechanics. Namely, in Chapter 2, we introduce a q-algebra with the corresponding operations, some fundamental functions consistent with q-algebra, and study in detail their properties. In this chapter, we also introduce the q-Gaussian distribution, which plays a profound role in the whole mathematical theory of nonextensive statistical mechanics. In Chapter 3, we introduce an important mathematical tool, named q-Fourier transform, and which represents a nonlinear generalization of the celebrated Fourier transform. The q-Fourier transform is useful in the derivation of many fundamental assertions of nonextensive statistical mechanics, including the q-central limit theorem. The q-Fourier transform, in contrast to the standard Fourier transform, is not a one-to-one mapping, a fact which creates some difficulties in analysis. In this chapter we also discuss how to overcome these difficulties and mention some other one-to-one versions of the q-Fourier transform.

Chapters 4 and 5 discuss the q-central limit theorem and q-generalizations of the classical α-stable distributions. The crucial difference of these objects from their classical counterparts is that the sequence of identically distributed random variables are not independent and not weakly dependent. In fact, they are strongly dependent no matter how distant are the elements of the sequence of random variables. From this point of view the q-central limit theorems and (q, α)-stable distributions significantly expand the domain of applications.

In Chapter 6, we will discuss applications of the theory developed in Chapters 1–5 to various processes arising in natural and social sciences. In particular, we present high-energy collisions of elementary particles, granular matter, financial operations, among others. In Chapter 7, we will discuss some interesting open mathematical questions of the present theory, the solution of which would significantly improve our knowledge along this avenue of modern statistical mechanics.

Finally, let us mention that each chapter contains a section which provides historical notes and additional information on related topics, as well as on other relevant sources for those readers who want further reading.

Acknowledgement

We are deeply indebted to colleagues and collaborators with whom various parts of the book were discussed in seminars, workshops, conferences, and in personal conversations. Our thanks go to Sumiyoshi Abe, Shavkat Alimov, Christian Beck, Ernesto P. Borges, Evaldo M.F. Curado, Shakir Formanov, Marjorie Hahn, Rudolf Hanel, Henrik J. Hilhorst, Max Jauregui, Xinxin Jiang, Renio S. Mendes, Kenric Nelson, Angelo Plastino, Angel R. Plastino, Alessandro Pluchino, Silvio M.D. Queiros, Andrea Rapisarda, Antonio Rodriguez, Guiomar Ruiz, Stanly Steinberg, Stefan Thurner, Ugur Tirnakli, and Christophe Vignat. Our deep gratitude goes also to Murray Gell-Mann and Reuben Hersh, in memoriam.

Sabir Umarov and Constantino Tsallis
June, 2021

Contents

4. q-Central Limit Theorems   **167**

5. (q, α)-Stable Distributions   **235**

Chapter 1

Background. q-Generalization of Boltzmann–Gibbs Entropy

1.1. Introduction

Boltzmann–Gibbs (BG) statistical mechanics (Boltzmann, 1872, 1877; Gibbs, 1902; von Neumann, 1927; Penrose, 1970; Huang, 1987) is one of the pillars of contemporary theoretical physics. It is based on electro-mechanics — namely, Newtonian, quantum and relativistic mechanics, as well as Maxwell electromagnetism — and theory of probabilities. Its goal is to provide, consistently with first principles, i.e., at a microscopic level, the analytical description of all thermodynamical quantities at thermal equilibrium, as well as various other related quantities. At the mesoscopic level, the BG theory is directly connected to many important descriptions, such as the Langevin equation, Fokker–Planck equation, master equation, to mention but a few. At the macroscopic level, it is consistent with classical thermodynamics.

We may say that Newtonian mechanics is not universal, meaning that, for high velocities, it must be generalized into relativistic mechanics, and also that, for small masses, it must be generalized into quantum mechanics. We may analogously say that BG statistical mechanics is not universal either. Indeed, it is well known that strongly mixing (hence ergodic) dynamical systems obey the BG theory (Eckman and Ruelle, 1985; Ruelle, 1986). But when such simplifying hypotheses are not satisfied, which is indeed the case for wide classes of complex systems (for example classical ones with *vanishing* largest Liapunov exponent), there is no legitimate reason for the BG theory to be valid. To be more precise, even in those cases where the usual thermostatistical quantities might remain computable within the BG prescriptions (e.g., some analytically solvable mean-field-like models), we do not expect those mathematical results to correspond to thermostatistical reality. Even much more so in those cases where the BG quantities are not even computable, e.g., the specific heat of a single non-ionized hydrogen atom at any finite temperature (and, strictly speaking, all similar systems within atomic physics). Indeed, the long-range Coulombian interaction between proton and electron makes the BG partition function to diverge unless the system is assumed to be confined in some *ad-hoc* finite box. More precisely, the non-ionized-atom BG partition function diverges, hence no physical quantity such as the specific heat can be computed. A preliminary discussion of this case is available at Lucena *et al.* (1995).

Since the BG theory is crucially based on the adoption of the well known Boltzmann–Gibbs-von Neuman–Shannon entropic functional (noted S_{BG} from now on) (see also Shannon, 1948), a natural manner to generalize it consists in generalizing the entropic functional itself. It is along this line that it was proposed in 1988 (Tsallis, 1988) the generalization of the BG theory. The BG entropic functional (or, occasionally, *BG entropy* for short) is defined, in its simplest discrete form, as follows:

$$S_{\mathrm{BG}} = -k \sum_{i=1}^{W} p_i \ln p_i \quad \left(\sum_{i=1}^{W} p_i = 1 \right), \qquad (1.1)$$

where k is a positive conventional constant chosen once for ever (physicists usually adopt $k = k_B$, k_B being Boltzmann constant; information theorists usually adopt $k = 1$), and W is the total number of possibilities whose probabilities are $\{p_i\}$. Among other important properties, let us mention that this entropic functional is *non-negative*, and also *additive*, meaning that, if A and B are two probabilistically independent arbitrary systems (i.e., $p_{ij}^{A+B} = p_i^A p_j^B$),

$$S_{\mathrm{BG}}(A + B) = S_{\mathrm{BG}}(A) + S_{\mathrm{BG}}(B), \qquad (1.2)$$

where $S_{\text{BG}}(A + B) \equiv S_{\text{BG}}(\{p_{ij}^{A+B}\})$, $S_{\text{BG}}(A) \equiv S_{\text{BG}}(\{p_i^A\})$ and $S_{\text{BG}}(B) \equiv S_{\text{BG}}(\{p_j^B\})$. The maximal value of S_{BG} occurs for equal probabilities, i.e., $p_i = 1/W$, and is given by the celebrated Boltzmann formula:

$$S_{\text{BG}}(\{p_i = 1/W\}) = k \ln W. \tag{1.3}$$

The generalized entropy S_q that was proposed in 1988 is defined as follows:

$$S_q = k \frac{1 - \sum_{i=1}^{W} p_i^q}{q - 1} \quad \left(q \in \mathcal{R}; \; S_1 = S_{\text{BG}}; \; \sum_{i=1}^{W} p_i = 1 \right). \tag{1.4}$$

Like BG entropy S_q is also non-negative. However, S_q is not additive, unless $q = 1$. In fact, for probabilistically independent systems A and B, it satisfies

$$\frac{S_q(A + B)}{k} = \frac{S_q(A)}{k} + \frac{S_q(B)}{k} + (1 - q) \frac{S_q(A)}{k} \frac{S_q(B)}{k}, \tag{1.5}$$

hence

$$S_q(A + B) = S_q(A) + S_q(B) + \frac{1 - q}{k} S_q(A) S_q(B). \tag{1.6}$$

It follows that when $q > 1$ the entropy S_q is *subadditive*:

$$S_q(A + B) < S_q(A) + S_q(B); \tag{1.7}$$

and when $q < 1$ it is *superadditive*:

$$S_q(A + B) > S_q(A) + S_q(B); \tag{1.8}$$

and when $q = 1$ we recover the additive BG entropy. The entropy S_q can also be written in the following forms:

$$S_q = -k \sum_{i=1}^{W} p_i^q \ln_q p_i = k \sum_{i=1}^{W} p_i \ln_q \frac{1}{p_i}$$

$$= -k \sum_{i=1}^{W} p_i \ln_{2-q} p_i, \tag{1.9}$$

where the q-logarithmic function is defined as follows:

$$\ln_q z \equiv \frac{z^{1-q} - 1}{1 - q}, \quad \ln_1 z = \ln z. \tag{1.10}$$

The entropy S_q is extremal at equal probabilities (maximal for $q > 0$, and minimal for $q < 0$), and its value is given by

$$S_q(\{p_i = 1/W\}) = k \ln_q W. \tag{1.11}$$

The BG statistical mechanics, based on the BG entropy has, since one century and a half, been fruitfully applied to uncountable physical systems. It fails however in a wide range of natural, artificial and social complex systems. In a large variety of those, the q-generalized statistical mechanics appears to satisfactorily apply. Analytical, numerical, experimental and observational studies exhibit that, through predictions, validations, and applications. Examples are found within physics (both in high- and low-energy systems), astrophysics, geophysics, chemistry, economics, engineering, medicine, computer science (including image and signal processing), linguistics, cognitive psychology, and others. A bibliography is available at the site tsallis.cat.cbpf.br/biblio.htm

The present book discusses mathematical foundations of the q-theory which emerged consistently with the generalization of the BG entropy. Since this theory is originated from BG entropy we first briefly mention some of its well-known properties. The dynamics described by BG entropy is closely related to the central limit theorem, the universal mathematical language for description of random limiting processes under certain restrictions. Before embarking into the properties of BG and q-entropy we briefly discuss some basic facts from probability theory.

1.2. Some Basic Facts from Probability Theory. Central Limit Theorem

In the present book, we discuss various types of random processes which require a probabilistic language. In this section, for the readers convenience and with the purpose of developing terminology, we briefly mention basic probabilistic notions and notations. The reader familiar with the probability basics can skip this section.

Let Ω be a set of simple outputs of physical experiments (observations). In probability theory Ω is called a *sample set* and its elements *elementary events*. An example is outputs of the coin tossing: $\Omega = \{tail, \; head\}$. Another example is random selection of a positive number less than 1: $\Omega = (0, 1)$. Denote by $\mathcal{F}$ the set of subsets of Ω. In the case of coin tossing example $\mathcal{F}$ is finite, containing only four elements: $\{\emptyset, \; tail, \; head, \; \{tail, \; head\}\}$. Here $\emptyset$ is the empty set, meaning an impossible event. In the case of random selection of numbers $\mathcal{F}$ is very reach. Therefore, a classification of the set of subsets is useful.

The set $\mathcal{F}$ is called an *algebra* if $A \in \mathcal{F}$ implies $\Omega \backslash A \in \mathcal{F}$, and $A, B \in \mathcal{F}$ implies $A \cup B \in \mathcal{F}$. Here $\Omega \backslash A$ is the set complementary to A, and $A \cup B$ is the union of sets A and B. If additionally, a countable number of subsets $A_k \in \mathcal{F}$, $k = 1, 2, \ldots$, imply $\cup_{k=1}^{\infty} A_k \in \mathcal{F}$, then $\mathcal{F}$ is called a Σ-*algebra*.

For a set Ω and a σ-algebra $\mathcal{F}$ of subsets of Ω a pair $(\Omega, \mathcal{F})$ is called a measurable space. For the same set Ω one can define different σ-algebras of subsets of Ω. For instance, for $\Omega = \mathbb{R}^n$ the minimal σ-algebra containing all open sets in Ω is called *Borel σ-algebra*. The minimal σ-algebra containing all Borel sets and sets of zero Lebesgue measure[1] is the σ-algebra of Lebesgue measurable sets.

Further, a set function $\mu : \mathcal{F} \to [0, \infty]$ is called a measure if $\mu(\emptyset) = 0$ and for countable disjoint sets $A_k \in \mathcal{F}$, $k = 1, 2, \ldots$, one has

$$\mu(\cup_{k=1}^{\infty} A_k) = \sum_{k=1}^{\infty} \mu(A_k).$$

The triplet $(\Omega, \mathcal{F}, \mu)$, where $(\Omega, \mathcal{F})$ is measurable and μ is a measure in it, is called a *measure space*. Finally, if μ possesses the additional property $\mu(\Omega) = 1$, then it is called a *probability space* and the corresponding measure space a *probability space*. We note that the

---

[1] We assume that the reader is familiar with the Lebesgue measure.

probability measure is traditionally denoted by $\mathbb{P}$. Thus, in this case the measure $\mu = \mathbb{P}$ is bounded and one has $\mathbb{P} : \mathcal{F} \to [0,1]$.

Let $(\Omega, \mathcal{F}, \mathbb{P})$ be a probability space. A mapping $X : \Omega \to \mathbb{R}$ such that $X^{-1}(B) \in \mathcal{F}$ for any open interval $B \in \mathbb{R}$, is called a *random variable*. We use capital Latin letters for random variables: X, Y, Z, etc. A function $F_X(x) = \mathbb{P}(X \leq x)$, $x \in \mathbb{R}$, is called a *distribution function* of the random variable X. Here $\mathbb{P}(X \in A)$ for a Borel set A means the probability of the random variable X being in the set A. In the case when $F_X(x)$ is absolute continuous, one can differentiate $F_X(x)$ (almost everywhere in the sense of Lebesgue measure), and the derivative $f_X(x) = F'_X(x)$ is called a *density function*. We denote the density function of X by $f_X(x)$. If a random variable takes values on some interval $[a, b] \subseteq \mathbb{R}$, we call this interval a range, or a support of X. For a random variable X with a range $[a, b]$ its distribution function can be written as

$$F_X(x) = \int_a^x f_X(y) dy,$$

through the corresponding density function $f_X(x)$. For example, the density function of the *standard normal random variable* is given by

$$f_X(x) = \frac{1}{\sqrt{2\pi}} e^{-\frac{x^2}{2}}, \quad x \in \mathbb{R}, \tag{1.12}$$

and the corresponding *standard normal distribution* is

$$F_X(x) = \frac{1}{\sqrt{2\pi}} \int_{-\infty}^{x} e^{-\frac{t^2}{2}} dt.$$

It follows from the definition of the distribution function, that it possesses the following properties: $F_X(-\infty) = 0$, non-decreasing, and $F_X(\infty) = 1$.

In the book we also face with distributions not being absolute continuous. A random variable X is called *discrete* if there is a countable (at most) set $S \subset \mathbb{R}$, such that $\mathbb{P}(X \in S) = 1$. The corresponding distribution function is piecewise constant and the density function is defined only at points $s_k \in S$, $k = 1, 2, \ldots$, called *atoms.* In this case the density function is also called a *probability mass function.*

For a given random variable X the integral

$$\mathbb{E}[X] = \int_\Omega X \mathbb{P}(d\omega)$$

is called a *mathematical expectation* or *mean* of X, and $\mathbb{E}[(X - \mathbb{E}(X))^2]$ is called a *variance* of X. If X has a density function $f_X(x)$, then $\mathbb{E}[X]$ can be written as

$$\mathbb{E}[X] = \int_\mathbb{R} x f_X(x) dx.$$

In general, if $g(\cdot)$ is a $\mathbb{P}$-measurable function then the mean of the random variable $g(X)$ is

$$\mathbb{E}[g(X)] = \int_\Omega g(X) \mathbb{P}(d\omega) = \int_\mathbb{R} g(x) f_X(x) dx.$$

In particular, the function $\varphi_X(\xi) = \mathbb{E}[e^{i\xi X}]$, $\xi \in \mathbb{R}$, is called a *characteristic function* of the random variable X. It is not hard to verify that the characteristic function of the standard normal random variable X is

$$\varphi_X(\xi) = e^{-\frac{\xi^2}{2}}, \quad \xi \in \mathbb{R}.$$

The standard normal distribution has zero mean and the variance equal to 1. The normal distribution N with mean μ and variance σ^2 has the density function

$$f_N(x) = \frac{1}{\sigma\sqrt{2\pi}} e^{-\frac{(x-\mu)^2}{2\sigma^2}}.$$

The random variable N can be reduced to the standard normal by rescaling and shifting. Namely, one can easily verify that the random variable

$$X = \frac{1}{\sigma}N - \frac{\mu}{\sigma}$$

is the standard normal. Inverting the latter we also have $N = \sigma X + \mu$. Hence, the characteristic function of N is

$$\varphi_N(\xi) = \mathbb{E}[e^{i\xi\sigma X + i\xi\mu}] = e^{i\xi\mu}e^{-\frac{\sigma^2\xi^2}{2}}.$$

Random variables X and Y, defined on the same probability space, are said to be *independent* if $\mathbb{P}(X \in A | Y \in B) = \mathbb{P}(X \in A)$ for any sets $A, B \in \mathcal{F}$, and the symbol $X \in A | Y \in B$ means that the event $X \in A$ has occurred under the condition that $Y \in B$. In other words the probability of the event $X \in A$ does not depend on the event $Y \in B$ for any sets in the sigma-algebra $\mathcal{F}$. The independence of random variables X and Y can be expressed through their density functions or characteristic functions, as well. For instance, the distribution function

$$F_{(X,Y)}(x, y) = \mathbb{P}(X \leq x, \ Y \leq y)$$

of the *joint random variable* (X, Y) with independent X and Y is

$$F_{(X,Y)}(x, y) = \mathbb{P}(X \leq x, Y \leq y)$$
$$= \mathbb{P}(X \leq x)\mathbb{P}(Y \leq y) = F_X(x)F_Y(y). \tag{1.13}$$

This immediately implies the following relation for density functions

$$f_{(X,Y)}(x, y) = f_X(x)f_Y(x), \tag{1.14}$$

and for the characteristic functions

$$\varphi_{(X,Y)}(\xi_1, \xi_2) = \mathbb{E}[e^{i(X\xi_1 + Y\xi_2)}]$$
$$= \int_{\mathbb{R}^2} f_{(X,Y)}(x, y)e^{i(x\xi_1 + y\xi_2)}dxdy$$
$$= \int_{\mathbb{R}^2} f_X(x)f_Y(y)e^{i(x\xi_1 + y\xi_2)}dxdy$$
$$= \int_{\mathbb{R}} f_X(x)e^{ix\xi_1}dx \int_{\mathbb{R}} f_Y(y)e^{y\xi_2}dy$$
$$= \mathbb{E}[e^{iX\xi_1}]\mathbb{E}[e^{iY\xi_2}] = \varphi_X(\xi_1)\varphi_Y(\xi_2). \tag{1.15}$$

Similarly for the linear combination $aX + bY$ of independent random variables X and Y the characteristic function is

$$\varphi_{aX+bY}(\xi) = \mathbb{E}[e^{i(aX+bY)\xi}] = \varphi_X(a\xi)\varphi_Y(b\xi). \qquad (1.16)$$

In particular for the sum $X + Y$ of two independent random variables X and Y,

$$\varphi_{X+Y}(\xi) = \varphi_X(\xi)\varphi_Y(\xi).$$

The latter implies that

$$f_{X+Y}(x) = (f_X * f_Y)(x) = \int_{\mathbb{R}} f_X(x - y)f_Y(y)dx, \qquad (1.17)$$

meaning that the density function of $X + Y$ is the convolution of density functions of X and Y.

A sequence of random variables X_n, $n = 1, 2, \ldots$, defined on the probability space $(\Omega, \mathcal{F}, \mathbb{P})$ is called *weakly convergent* to a random variable X, if $\mathbb{P}(X_n \leq x) \to \mathbb{P}(X \leq x)$ for all $x \in \mathbb{R}$, for which $\mathbb{P}(\{x\}) = 0$. The sequence X_n is called *convergent in distribution* to X, if the sequence of distribution functions $F_{X_n}(x)$ converges to the distribution function $F_X(x)$ of X as $n \to \infty$ at every discontinuity point x of $F_X(x)$.

Proposition 1.1 (Billingsley, 1995). *The following statements are equivalent:*

(1) *X_n weakly converges to X;*
(2) *X_n converges to X in distribution;*
(3) *X_n converges to X if and only if the sequence of characteristic functions $\varphi_{X_n}(\xi)$ converges to $\varphi_X(\xi)$ at every point $\xi \in \mathbb{R}$.*

Let $X_1, X_2, \ldots, X_n, \ldots$, be a sequence of independent random variables with a same distribution. Such sequences of random variables are called *independent identically distributed* (i.i.d. for short) random variables. Denote by S_n the sum of n copies of the i.i.d.

sequence $\{X_n\}$:

$$S_n = \sum_{k=1}^{n} X_k. \tag{1.18}$$

Since $\{X_n\}$ is i.i.d. all members of the sequence have the same density function, and hence, the same characteristic function. Denote the corresponding density and characteristic functions by $f(x)$ and $\varphi(\xi)$, respectively. Then it follows from (1.16) and (1.17) that

$$f_{S_n}(x) = (f * \cdots * f)(x) \quad (n \text{ times convolution}), \tag{1.19}$$

and

$$\varphi_{S_n}(\xi) = [\varphi(\xi)]^n. \tag{1.20}$$

Theorem 1.1 (Central limit theorem). *Let $\{X_n\}$ be a sequence of i.i.d. random variables with a mean μ and a finite variance $Var(X_1) = \sigma^2$. Then the sequence*

$$Z_n = \frac{S_n - n\mu}{\sigma\sqrt{n}}$$

weakly converges to the standard normal random variable as $n \to \infty$.

Proof. It follows from (1.16) and (1.20) that

$$\varphi_{Z_n}(\xi) = \varphi_{\frac{S_n}{\sigma\sqrt{n}} - \frac{\mu\sqrt{n}}{\sigma}}(\xi)$$

$$= e^{-i\mu\xi\sqrt{n}/\sigma} \varphi_{S_n}\left(\frac{\xi}{\sigma\sqrt{n}}\right)$$

$$= e^{-i\mu\xi\sqrt{n}/\sigma} \left[\varphi_{X_1}\left(\frac{\xi}{\sigma\sqrt{n}}\right)\right]^n.$$

This implies

$$\ln \varphi_{Z_n}(\xi) = -\frac{\mu\xi\sqrt{n}}{\sigma}i + n \ln\left(1 + i\frac{\mu\xi}{\sigma\sqrt{n}} - \frac{\sigma\xi^2}{2\sigma n} + o(|\xi^2|/n)\right)$$

$$= -\frac{\xi^2}{2} + o(1), \quad n \to \infty,$$

for all $\xi \in \mathbb{R}$. Thus $\varphi_{Z_n}(\xi) \to e^{-\xi^2/2}$ as $n \to \infty$ at every point $\xi \in \mathbb{R}$, and therefore, due to Proposition 1.1 the sequence Z_n weakly converges to the standard normal distribution. $\qquad\square$

Remark 1.1.

(1) The sketched proof can be found in probability textbooks. We provided it here only by the reason that this simple idea of proof based on the characteristic functions will be implemented for the proof of the q-central limit theorem, as well. However, we will see in that case that the proof becomes significantly harder.

(2) The central limit theorem (CLT) (in the sense that the limit distribution is normal) is valid for some weakly dependent or not identically distributed random variables as well, under certain conditions. For corresponding citations and some historical facts on the central limit theorem see Section 1.5.

The limiting random variable in the CLT is normal with the exponential density in equation (1.12). In the physics literature frequently this density is also named Gaussian. As we will see below the CLT plays an important role in Statistical Mechanics and it is closely related to the BG entropy.

It follows from (1.16) that for two independent copies X_1, X_2 of the standard normal random variable X

$$\varphi_{aX_1+bX_2}(\xi) = e^{-\frac{(a^2+b^2)\xi^2}{2}},$$

where a, b are arbitrary real numbers. That is, $aX_1 + bX_2$ is again a normal random variable with zero mean and standard deviation $\sqrt{a^2 + b^2}$. Hence, it can be reduced to the standard normal by rescaling and shifting. This property of the standard normal random variable is an implication of its stability. In fact, any normal random variable is stable. By definition, a random variable X is called *stable* if for arbitrary numbers a, b and two independent copies X_1, X_2 of X there exist numbers c and d, such that the equality

$$aX_1 + bX_2 = cX + d$$

holds in the sense of distributions. That is, distributions of the right-
and left-hand sides are the same.

Stable random variables (or distributions) were first studied by
Paul Lévy in the 1930s. Therefore, they are also called Lévy's α-stable
distributions. We will discuss stable distributions in Chapter 5 in
detail. But here we formulate a theorem which describes Lévy's α-
stable distributions in general.

**Theorem 1.2 (Generalized central limit theorem (Nolan,
2002)).** *A random variable Z is α-stable for some $\alpha \in (0, 2]$ if and
only if there exist an independent, identically distributed sequence of
random variables $X_1, X_2, \ldots$, and constants $a_n > 0$, $b_n \in \mathbb{R}$, such that*

$$a_n(x_1 + \cdots + X_n) - b_n \to Z$$

in the sense of distributions.

1.3. Boltzmann–Gibbs Entropy. Its Properties and Connections with the Central Limit Theorem and the Large Deviation Theory

The BG entropy defined in equation (1.1) can be associated with
a random variable X with a finite number of outcomes/states with
probabilities p_i, $i = 1, \ldots, W$. In the general case when X has a
probability density function $f_X(x)$, $x \in \Omega \subseteq \mathbb{R}^d$, BG entropy can be
defined as

$$S_{\mathrm{BG}}(f_X) = -k \int_\Omega f_X(x) \ln f_X(x) dx. \tag{1.21}$$

The functional $S_{\mathrm{BG}}(f_X)$, defined in the set of density functions, is
typically non-negative unless the density is excessively thin around
some average value. Physically speaking, this is an evidence of the
failure of classical mechanics for a system in the very low temperature regime. Under those circumstances, it is unavoidable to use
quantum instead of classical mechanics. Indeed, at low temperature,
the population of the ground state becomes predominant and the
entropy approaches zero, in conformity with the third principle of
thermodynamics.

As one can see easily, the functional $S_{\mathrm{BG}}(f_X)$ is additive. Indeed, if $f_Z(x, y)$ is the joint probability density function of $Z = (X, Y)$ for independent random variables X and Y with the corresponding densities $f_X(x)$, $x \in \Omega_X$, and $f_Y(y)$, $y \in \Omega_Y$, then

$$
\begin{aligned}
S_{\mathrm{BG}}(f_Z) &= -k \iint_{\Omega_X \times \Omega_Y} f_Z(x, y) \ln f_X(x) f_Y(y) dx dy \\
&= -k \iint_{\Omega_X \times \Omega_Y} f_Z(x, y) \ln f_X(x) dx dy \qquad (1.22) \\
&\quad - k \iint_{\Omega_X \times \Omega_Y} f_Z(x, y) \ln f_Y(y) dx dy \\
&= -k \int_{\Omega_X} f_X(x) \ln f_X(x) dx - k \iint_{\Omega_Y} f_Y(y) \ln f_Y(y) dy \\
&= S_{\mathrm{BG}}(f_X) + S_{\mathrm{BG}}(f_Y). \qquad (1.23)
\end{aligned}
$$

In the physics terminology, the latter can be written as

$$
S_{\mathrm{BG}}(A + B) = S_{\mathrm{BG}}(A) + S_{\mathrm{BG}}(B), \qquad (1.24)
$$

as we did in equation (1.2), where A and B correspond to the systems associated with random variables X and Y respectively, and $A + B$ corresponds to the system associated with the joint variable $Z = (X, Y)$. The relation (1.24) expresses the additivity of the BG entropy S_{BG} in the general setting.

By definition, a functional $F(f)$, defined on a Banach space $\mathcal{D}$ is said to be *concave* if for elements $f, g \in \mathcal{D}$, one has

$$
F(\lambda f + (1 - \lambda)g) \geq \lambda F(f) + (1 - \lambda)F(g), \qquad (1.25)
$$

for any $\lambda \in [0, 1]$. If the inequality sign in equation (1.25) appears to be opposite, then the functional F is called *convex*.

Below we show that S_{BG} is concave. Indeed, consider the function $\varphi(u) = -u \ln u$, $u > 0$. One can easily verify that $\varphi''(u) < 0$ for all $u > 0$ implying concavity of $\varphi(u)$:

$$
\varphi(\lambda u + (1 - \lambda)v) > \lambda \varphi(u) + (1 - \lambda)\varphi(v) \qquad (1.26)
$$

for any $\lambda \in [0,1]$. Now, let f and g be density functions defined on $\Omega \subseteq \mathbb{R}^d$ and $\lambda \in [0,1]$. Then, making use of inequality (1.26), we have

$$S_{\mathrm{BG}}(\lambda f + (1-\lambda)g)$$

$$= -k \int_\Omega (\lambda f + (1-\lambda)g) \ln(\lambda f + (1-\lambda)g) dx$$

$$= k \int_\Omega \varphi(\lambda f + (1-\lambda)g) dx$$

$$> k\lambda \int_\Omega \varphi(f(x)) dx + k(1-\lambda) \int_\Omega \varphi(g(x)) dx$$

$$= -k\lambda \int_\Omega f(x) \ln f(x) dx - k(1-\lambda) \int_\Omega g(x) \ln g(x) dx$$

$$= \lambda S_{\mathrm{BG}}(f) + (1-\lambda)S_{\mathrm{BG}}(g), \qquad (1.27)$$

proving the concavity of S_{BG}.

Another notion related to BG and other entropic forms is the composability. In general, an entropic form S/k is said to be *composable*, if for arbitrary two independent subsystems A and B of a generic system, it can be expressed in the form $S(A+B) = F(A, B, \eta)$, where $F(x, y; \eta)$ is symmetric in variables x and y, associative, and satisfies $F(x, 0, \eta) = x$. Here η is a parameter; see details in Tsallis (2009a). We call $F(x, y; \eta)$ a *composability function* associated with the entropy S. In this sense BS entropy is obviously composable with the associated composability function $F(x, y; \eta) = x + y$.

The optimization of the BG entropy under appropriate constraints, namely, normalization and energy mean value of the extremizing distribution, yields the celebrated BG exponential weight $P_i \propto e^{-\beta E_i(N)}$, operational heart of BG statistical mechanics. This distribution is connected in various manners to the CLT and to the large deviation theory (LDT). Let us briefly mention these connections.

In what concerns the CLT, the thermal equilibrium distribution of momenta (proportional to velocities) is the Maxwellian distribution, whose Gaussian form corresponds to the CLT attractor in the space of distributions. Moreover, in the context of the Kolmogorov–Sinai entropy production rate (basically per unit time), the Pesin identity establishes an important equality between the BG entropic functional and the set of positive Lyapunov exponents of quite general nonlinear dynamical systems with *strong* chaos. For various strongly chaotic low-dimensional conservative and dissipative dynamical systems, Gaussian distributions are observed for time-averaged quantities (see Ruiz *et al.*, 2017a,b; Tirnakli and Borges, 2016).

In what concerns the LDT, the mathematical correspondence of the BG factor $P_i(N) \propto e^{-\beta E_i(N)} = e^{-\left[\beta \frac{E_i(N)}{N}\right]N}$, where the quantity $\left[\beta \frac{E_i(N)}{N}\right]$ is *intensive* for all *extensive* Hamiltonian systems, particularly those involving short-range interactions. In other words, the BG weight decays exponentially with N in the large N limit. This fact precisely corresponds (see Touchette, 2009) to the asymptotic behavior of the large deviation probability $P(x; N) \sim e^{r(x)N}$, where the intensive *rate function* $r(x)$ equals a relative BG entropy per particle, x characterizing the deviation out from the $N \to \infty$ distribution.

1.4. *q*-Entropy and Its Properties

The *q*-entropy for a system with probabilities p_i, $i = 1, \ldots, W$, is defined in (1.4). In the general case, for a system with a continuous probability density function $f(x)$, $x \in \Omega \subseteq \mathbb{R}^d$, the *q*-entropy $S_q(f)$ is defined by

$$S_q(f) = \frac{k}{1-q}\left(1 - \int_\Omega [f(x)]^q dx\right), \quad q \in (-\infty, \infty), \qquad (1.28)$$

subject to the integral on the right exists. If $f(x)$ is discrete, then the latter reduces to (1.4), and if $q \to 1$, then to the BG entropy.

Note that S_q also can be written as

$$S_q(f) = -k \int_\Omega [f(x)]^q \ln_q f(x) dx$$

$$= k \int_\Omega f(x) \ln_q [f(x)]^{-1} dx$$

$$= -k \int_\Omega f(x) \ln_{2-q} f(x) dx, \tag{1.29}$$

using the q-logarithmic function in (1.10) and its properties, to be established in Chapter 2.

The q-entropy defined in equation (1.28) or (1.4) is non-additive for $q \neq 1$, concave for $q > 0$, convex for $q < 0$, and composable for all $q \in \mathbb{R}$. For a discrete density function the non-additivity property of the q-entropy is shown in (1.6)–(1.8). For a continuous density, one can derive this fact as the limit case of the discrete approximation of the density function. The concavity (convexity) can be proved similarly to the concavity of the BG entropy, making use of the fact that the function

$$\varphi_q(u) = -u \ln_q(u), \quad u > 0,$$

has negative second derivative for $q > 0$ and positive second derivative for $q < 0$. This, like the BG entropy, implies the concavity of S_q for $q > 0$, and the convexity of S_q for $q < 0$. Finally, the composability holds with the associated composability function $F(x, y; \eta) = x + y + (1 - \eta)xy$.

The optimization of the q-entropy under appropriate constraints, namely once again normalization and energy mean value of the extremizing distribution, yields the q-exponential weight $P_i \propto e_q^{-\beta_q E_i(N)}$. Analogously to the BG case, this distribution appears to be connected in various manners to a q-generalized CLT (discussed in Chapter 4). Moreover, it is possibly connected to a q-generalized LDT, still to be developed within a mathematical frame. Various numerical evidences are already available (Ruiz and Tsallis, 2012,

2013; Touchette, 2013; Tirnakli *et al.*, 2021) in the literature, but their analytical progress still is in its infancy.

In what concerns the q-CLT theorem its mathematical proof was presented in paper Umarov *et al.* (2008), and the quasi-equilibrium distribution of one-particle momenta (proportional to the velocities) has been numerically shown to have the q-Gaussian form for various many-body classical Hamiltonian d-dimensional systems (XY and Heisenberg rotators, Fermi–Pasta–Ulam model) involving long-range two-body interactions (Cirto *et al.*, 2014, 2018; Christodoulidi *et al.*, 2014, 2016; Carati *et al.*, 2019; Bagchi and Tsallis, 2016, 2017, 2018; Rodriguez *et al.*, 2019).

In these same systems, numerical evidence strongly suggests the q-exponential form for the one-particle energy distributions. Moreover, the q-generalized form of a Pesin-like identity has been exhibited (see Tsallis *et al.*, 1997; Borges *et al.*, 2002; Ananos and Tsallis, 2004; Celikoglu and Tirnakli, 2006; Boldovin and Robledo, 2002, 2004; Ruiz and Tsallis, 2009; Mayoral and Robledo, 2004, 2005) in various nonlinear dynamical systems at the edge of chaos (in the sense that the maximal Lyapunov exponent vanishes, which is a necessary condition for *weak* chaos).

For various weakly chaotic low-dimensional conservative and dissipative dynamical systems, q-Gaussian distributions are observed for time-averaged quantities (Tirnakli and Borges, 2016; Ruiz *et al.*, 2017a,b). Also, in restricted random walks q-Gaussians are numerically observed; see, e.g., Tirnakli *et al.* (2011) .

In what concerns the q-LDT, the extension of the q-exponential factor $P_i(N) \propto e^{-\beta_q E_i(N)}$ can be rewritten, for long-range-interacting classical d-dimensional Hamiltonian systems, as

$$P_i(N) \propto e_q^{-\left[(\beta_q \tilde{N})\frac{E_i(N)}{N\tilde{N}}\right]N},$$

where $\tilde{N} \propto N^{\alpha-d}$ for $0 \leq \alpha/d < 1$ and $\tilde{N}$ is a constant for $\alpha/d > 1$, α being the exponent which characterizes the power-law decay with distance for the two body interactions (Tsallis, 2009a,b; Tirnakli *et al.*, 2021). Notice that $\left[(\beta_q \tilde{N})\frac{E_i(N)}{N\tilde{N}}\right]$ is an intensive quantity for all values of $\alpha/d \geq 0$, which implies that $P_i(N)$ q-exponentially decays with N.

The mathematical counterpart of this behavior would therefore be expected to be $P(x; N) \sim e_q^{r_q(x)N}$, where the intensive *rate function* $r_q(x)$ could correspond to some sort of relative q-entropy per particle, x characterizing, as for the $q = 1$ case, the deviation out from the $N \to \infty$ distribution. This behavior has indeed been numerically exhibited in various models (Ruiz and Tsallis, 2012, 2013; Touchette, 2013; Tirnakli *et al.*, 2021), but its analytical approach still remains elusive. Its proof would signify a formidable step forward in the mathematical foundations of non-extensive statistical mechanics. Among other implications, such a result would be consistent with the thermodynamical extensivity of the entropy in all cases, since $r_q(x)N$ would essentially correspond to the *total* entropy of the system.

The theorem below summarizes the properties of the q-entropy mentioned above.

Theorem 1.3. *The q-entropy S_q defined in (1.28) (or in (1.4) in the discrete case) possesses the following properties:*

(1) **Recovery of BG entropy:** $\lim_{q \to 1} S_q(f) = S_{\mathrm{BG}}(f)$ *for any probability density function f;*
(2) **Non-negativity:** $S_q(f) \geq 0$ *for all $q \in \mathbb{R}$ and discrete probability mass functions $f \equiv \{p_j, \ j = 1, \ldots, W\}$;*
(3) **Non-additivity for $q \neq 1$:** *For arbitrary two independent subsystems A and B in the state space the relation*

$$S_q(A + B) = S_q(A) + S_q(B) + \frac{1 - q}{k} S_q(A) S_q(B) \qquad (1.30)$$

holds;
(4) **Subadditivity for $q < 1$:** $S_q(A + B) > S_q(A) + S_q(B)$;
(5) **Superadditivity for $q > 1$:** $S_q(A + B) < S_q(A) + S_q(B)$;
(6) **Concavity for $q > 0$:** *For arbitrary two densities f and g the concavity relation*

$$S_q(\lambda f + (1 - \lambda)g) \geq \lambda S_q(f) + (1 - \lambda)S_q(g)$$

holds;

(7) **Convexity for $q < 0$:** *For arbitrary two densities f and g the convexity relation*

$$S_q(\lambda f + (1 - \lambda)g) \leq \lambda S_q(f) + (1 - \lambda)S_q(g)$$

holds;

(8) **Composability:** S_q/k *is composable with the associated composabilty function*

$$F(x, y; \eta) = x + y + (1 - \eta)xy. \tag{1.31}$$

1.5. Additional Notes

(1) On central limit theorems

Limit theorems and, particularly, the central limit theorems, surely are among the most important theorems in probability theory and statistics. They play an essential role in various applied sciences, including statistical mechanics.

Historically A. de Moivre, P.S. de Laplace, S.D. Poisson and C.F. Gauss originated the classical CLT, the fact that the standard normal distribution is a limit in the sense of weak convergence of the properly rescaled sum of i.i.d random variables with a finite second moment. A sketch of a proof of this theorem is provided in Section 1.2. Chebyshev, Markov, Lyapunov, Feller, Lindeberg, Lévy, and many others contributed to the development of the CLT. We refer the reader to the following sources for reading on the history and modern state of the theory related to various limit theorems, including CLTs: Feller (1945), Fischer (2011), Adams (2009), Dudley (1999), Jacod and Shiryaev (2003) and Shiryaev (2016).

The CLT extends to weakly dependent random variables as well with the Gaussian limiting distribution. An introduction to this area can be found in Yoshihara (1992), Peligrad (1986), Rio (2000), Doukhan (1994), Dehling *et al.* (1986), Bradley (2003) (see also references therein), where different types of weak dependences are considered, as well as the history of the developments. The CLT does not hold (in the sense that the limit distribution is normal), if correlation between far-ranging random variables is not neglectable (see Dehling *et al.*, 2002).

(2) Levy's α-stable distributions

The classic theory of α-stable distributions was originated by Paul Lévy and developed by Lévy, Kolmogorov, Gnedenko, Feller, and many others; for details and history of the topic we refer the reader to books Gnedenko and Kolmogorov (1954), Feller (1966), Samorodnitsky and Taqqu (1994), Meerschaert and Scheffler (2001), Nolan (2002), Metzler and Klafter (2000), Uchaykin and Zolotarev (1999) and references therein.

By definition, the α-*stable distribution* is a distribution of a random variable X with the characteristic function (see e.g., Nolan, 2002)

$$\varphi_X(\xi; \alpha, \beta, \gamma, \mu) = \mathbb{E}[e^{iX\xi}]$$

$$= \begin{cases} e^{i\mu\xi - \gamma^\alpha |\xi|^\alpha \left[1 + i\beta \mathrm{sign}(x) \tan \frac{\pi\alpha}{2}((\gamma|\xi|)^{1-\alpha} - 1)\right]}, & \text{if } \alpha \neq 1, \\ e^{i\mu\xi - \gamma|\xi|\left[1 + i\beta \mathrm{sign}(x) \frac{2}{\pi} \ln(\gamma|\xi|)\right]}, & \text{if } \alpha = 1, \end{cases}$$

where α is the stability index (describes the thickness of the tail), β is the skewness, γ is the scale parameter (describes the width of the density), and μ is the location parameter. The distribution is symmetric if $\beta = 0$, otherwise it is skewed to the right if $\beta > 0$, and to the left if $\beta < 0$.

α-Stable distributions asymptotically behave as power laws. Such behavior emerges in a many applications in various fields. Just to name a few examples — anomalous processes in polymer physics, rheology, biophysics, thermodynamics, movements of cultured monkeys or sharks in food searching, price fluctuations in stock markets, financial transactions in insurance companies, and protein movements in cell biology (see, e.g., McCulloch, 1996; Meerschaert and Scheffler, 2001; Metzler and Klafter, 2000; Montroll and Shlesinger, 1982; Montroll and Bendler, 1984; Schmitt and Seuront, 2001; Tsallis and Bukman, 1996; Uchaykin and Zolotarev, 1999; Umarov *et al.*, 2018).

Figure 1.1 shows probability density and cumulative distribution functions of symmetric α-stable distributions for some values of α.

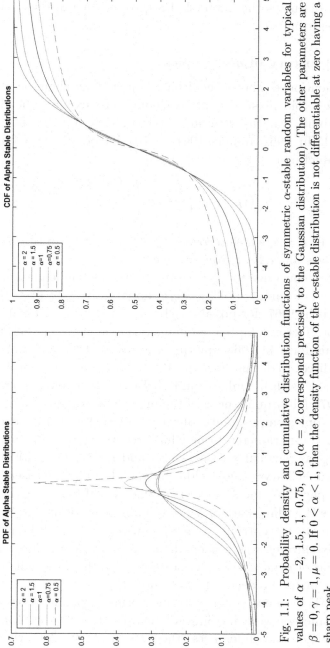

Fig. 1.1: Probability density and cumulative distribution functions of symmetric α-stable random variables for typical values of $\alpha = 2$, 1.5, 1, 0.75, 0.5 ($\alpha = 2$ corresponds precisely to the Gaussian distribution). The other parameters are $\beta = 0, \gamma = 1, \mu = 0$. If $0 < \alpha < 1$, then the density function of the α-stable distribution is not differentiable at zero having a sharp peak.

Chapter 5 discusses the detailed theory of q-generalized versions of α-stable distributions.

(3) Brief history of q-entropies

As was mentioned above the Boltzmann–Gibbs–Shannon entropy plays a key role in statistical mechanics. To expand the domain of applicability various generalizations of this entropic form were introduced. In 1961 Renyi (1961, 1970) introduced the following entropic functional:

$$S_q^R(\{p_j\}) = \frac{k}{1-q} \ln \left(\sum_{j=1}^{W} p_j^q \right),$$

where $q \geq 0$, $q \neq 1$, which recovers S_{BG} when $q \to 1$. In the literature this form was named a *Renyi entropy*.

In what concerns the q-entropy S_q defined in equation (1.4), Havrda and Charvat (1967) were apparently the first to ever introduce it in 1967, for possible cybernetic purposes, though with a different pre-factor, adapted to binary variables. Vajda (1968) further studied this form in 1968, quoting Havrda and Charvat. Daroczy (1970) rediscovered this form in 1970 (he quotes neither Havrda-Charvat nor Vajda). Linhard and Nielsen (1971), rediscovered this form in 1971 (they quote none of the predecessors) through the property of entropic composability. Sharma and Mittal (1975) introduced in 1975 a two-parameter form which recovers both S_q and Renyi entropy S_q^R as particular cases. Aczel and Daroczy (1975) quote in 1975 Havrda–Charvat and Vajda, but not Lindhard-Nielsen. Wehrl (1978) (in page 247) mentions in 1978 the form of S_q, quotes Daroczy, but ignores Havrda–Charvat, Vajda, Lindhard-Nielsen, and Sharma–Mittal. Finally, the entropic form S_q, now referred in the literature as the *Tsallis entropy*, was independently introduced in 1988 with the specific aim of generalizing BG statistical mechanics itself (Tsallis, 1988; none of the predecessors was quoted).

Since then many other entropic forms dependent on one or more parameters were introduced. See Tsallis (2009b), Lima and Tsallis (2020) and references therein for their definitions and detailed properties. Among these entropies is the *Kaniadakis entropy* (Kaniadakis, 2002) defined by

$$S_\kappa^K(\{p_j\}) = \frac{1}{2\kappa} \sum_{j=1}^{W} \left(p_j^{1-\kappa} - p_j^{1+\kappa} \right),$$

where $-1 < \kappa < 1$, $\kappa \neq 0$, which recovers the BG-entropy in the limit case $\kappa \to 0$. Since $S_\kappa^K(\{p_j\}) = S_{-\kappa}^K(\{p_j\})$, it is enough to consider $0 \leq \kappa < 1$.

Summarizing we would like to mention some basic properties of these most popular entropic functionals. S_{BG} is additive, concave, composable, and is of the trace-form (meaning that the entropic functional can be written in the form $S = \sum_i g(p_i)$, where $g(z)$ is some smooth function). Its extremal value is $k \ln W$. The Renyi entropy S_q^R is additive, concave (convex) for $0 < q \leq 1$ ($q < 0$), and neither concave nor convex for $q > 1$. Its extremal value is $k \ln W$. It is in fact the most general additive entropy: it is not of the trace-form if $q \neq 1$. S_q is non-additive for $q \neq 1$, concave (convex) for $q > 0$ ($q < 0$), composable, and trace-form. Its extremal value is $k \ln_q W$. It is in fact the most general composable and trace-form entropy. The Kaniadakis entropy S_κ^K is non-additive for $\kappa \neq 0$, concave for $0 \leq \kappa \leq 1$, convex or neither concave nor convex for $\kappa > 1$ (depending on W), non-composable for $\kappa \neq 0$, and trace-form. Its extremal value is $k \ln_\kappa W$, with $\ln_\kappa^K z \equiv \frac{z^\kappa - z^{-\kappa}}{2\kappa}$ ($\ln_0^K z = \ln z$).

(4) *q*-Entropy and related functions in Wolfram Mathematica

The readers who are familiar with Wolfram Mathematica may use the following functions related to the *q*-entropy, which are implemented therein:

- Resource Function ["TsallisEntropy"];
- Resource Function ["TsallisQExponential"];
- Resource Function ["TsallisQGaussian"].

The *q*-exponential and *q*-Gaussian functions will be introduced in Chapter 2.

Chapter 2

q-Algebra and q-Generalizations of Some Elementary Functions

2.1. q-Algebra. Introduction

In this chapter, we introduce an algebraic structure consistent with non-extensive statistical mechanics and called a q-algebra. We define basic operations of this algebraic structure and present their key properties. As we will see, the summation and multiplication operations of the q-algebra are commutative, associative, and for $q = 1$ they recover the standard summation and multiplication operations of the ring of real (or complex) numbers. However, if $q \neq 1$, then unlike many abstract algebraic structures, the q-sum and q-product defined below do not possess the distributive property.

In this chapter, we also introduce q-generalizations of some elementary functions like exponential, logarithmic, and trigonometric functions adjusted to the q-algebra and study their main properties. Moreover, the univariate and multivariate analogs of the Gaussian distribution are presented as well.

2.2. q-Sum and q-subtraction

Relation (1.2) exhibited in Chapter 1 states that if A and B are two independent subsystems, then the Boltzmann–Gibbs (BG) entropy of the total system $A + B$ satisfies the additivity property

$$S_{\text{BG}}(A + B) = S_{\text{BG}}(A) + S_{\text{BG}}(B). \tag{2.1}$$

The additivity property of S_{BG} fails to hold for q-entropy S_q, unless $q = 1$. Indeed, Theorem 1.3 says that for two independent subsystems A and B the total q-entropy $S_q(A + B)$ can be computed via the formula

$$\frac{S_q(A + B)}{k} = \frac{S_q(A)}{k} + \frac{S_q(B)}{k} + (1 - q)\frac{S_q(A)}{k}\frac{S_q(B)}{k}. \qquad (2.2)$$

We denote by $\mathbb{C}$ and $\mathbb{R}$ the sets of complex and real numbers, respectively. The addition operation in the q-algebra is defined consistently with the right-hand side of (2.2). Namely, we define the q-addition in the following form.

Definition 2.1. For a fixed real number q define the q-sum $x \oplus_q y$ of $x \in \mathbb{C}$ and $y \in \mathbb{C}$ as a binary operation

$$x \oplus_q y = x + y + (1 - q)xy. \qquad (2.3)$$

In accordance with this definition, Part (3) of Theorem 1.3 can be reformulated as follows.

Theorem 2.1. *Let A and B be independent systems with q-entropies $S_q(A)$ and $S_q(B)$ defined in (1.28) (or in (1.4) in the discrete case). Then*

$$\frac{S_q(A + B)}{k} = \frac{S_q(A)}{k} \oplus_q \frac{S_q(B)}{k}. \qquad (2.4)$$

A functional S satisfying condition (2.4) is called *q-additive*. Hence, S_q is a q-additive entropy.

Proposition 2.1. *The q-sum possesses the following properties:*

(i) *commutative:* $x \oplus_q y = y \oplus_q x$ *for all* $x, y \in \mathbb{C}$;
(ii) *associative:* $(x \oplus_q y) \oplus_q z = x \oplus_q (y \oplus_q z) = x \oplus_q y \oplus_q z$ *for all* $x, y, z \in \mathbb{C}$;
(iii) *recovers the usual summation operation of $\mathbb{C}$ if $q = 1$, i.e.,* $x \oplus_1 y = x + y$;
(iv) *preserves $0 \in \mathbb{C}$ as the neutral element, i.e.,* $x \oplus_q 0 = x$.

Proof. The commutativity of the q-sum immediately follows from the commutativity of the summation and multiplication operations of $\mathbb{C}$:

$$x \oplus_q y = x + y + (1-q)xy = y + x + (1-q)yx = y \oplus_q x.$$

Let us show the associativity of $\oplus_q$. We have

$$(x \oplus_q y) \oplus_q z$$
$$= [x + y + (1-q)xy] + z + (1-q)[x + y + (1-q)xy]z$$
$$= x + y + z + (1-q)(xz + yz + xy) + (1-q)^2 xyz. \qquad (2.5)$$

Similarly,

$$x \oplus_q (y \oplus_q z)$$
$$= x + [y + z + (1-q)yz] + (1-q)x[y + z + (1-q)yz]$$
$$= x + y + z + (1-q)(yz + xy + xz) + (1-q)^2 xyz. \qquad (2.6)$$

Equations (2.5) and (2.6) imply

$$(x \oplus_q y) \oplus_q z = x \oplus_q (y \oplus_q z) = x \oplus_q y \oplus_q z.$$

$\square$

By inversion, we can define the *q-subtraction* as

$$x \ominus_q y = \frac{x - y}{1 + (1-q)y}. \qquad (2.7)$$

By inversion we mean $(x \oplus_q y) \ominus_q y = (x \ominus_q y) \oplus_q y = x$, valid for all $x, y \in \mathbb{C}$. This operation preserves some properties of standard subtraction. For instance, $x \ominus_q 0 = x$. However, $0 \ominus_q y = \frac{-y}{1+(1-q)y}$, which differs from $(-y)$ unless $q = 1$. Consistently, one can verify that $0 \ominus_q \frac{y}{1-(1-q)y} = -y$.

2.3. *q*-Exponential and *q*-Logarithmic Functions

The *q*-analysis relies essentially on the analogs of exponential and logarithmic functions, which are called *q*-exponential and *q*-logarithmic functions. The exponential function $y = e^t$, obviously satisfies the differential equation $y'(t) = y(t)$ with the initial condition $y(0) = 1$.

To introduce a q-generalization of the exponential function consider the following nonlinear differential equation ($q \neq 1$):

$$\frac{dy(t)}{dt} = [y(t)]^q, \quad q \in \mathbb{R}, \tag{2.8}$$

with the initial condition $y(0) = 1$. To solve this initial-value problem we rewrite the equation in (2.8) in the form $y^{-q}dy = dt$ and integrate it over the interval $(0, t)$. Then, we have

$$\frac{y^{1-q}}{1-q} - \frac{1}{1-q} = t,$$

or

$$y(t) = [1 + (1-q)t]^{\frac{1}{1-q}}. \tag{2.9}$$

This function generalizes the exponential function $y = e^t$, recovering it when $q \to 1$. Indeed, using L'Hopital's rule, we have

$$\lim_{q \to 1} \frac{\ln\left(1 + (1-q)t\right)}{1-q} = t,$$

which implies

$$\lim_{q \to 1}[1 + (1-q)t]^{\frac{1}{1-q}} = e^t. \tag{2.10}$$

Therefore, the function in (2.9) is called a q-exponential function and denoted by e_q^t. Obviously, this function is well defined for all t real satisfying the condition $t > -(1-q)^{-1}$. Extending by zero for $t \leq -(1-q)^{-1}$, we can assume that e_q^t is defined for all real t. Thus, we have the following definition of the q-exponential function:

Definition 2.2. The function

$$e_q^t = [1 + (1-q)t]_+^{\frac{1}{1-q}}, \quad t \in \mathbb{R}, \tag{2.11}$$

is called a q-exponential function. Here the symbol $(\cdot)_+$ means

$$(a)_+ = \begin{cases} a & \text{if } a \geq 0 \\ 0 & \text{if } a < 0. \end{cases}$$

Extension of the q-exponential function for complex arguments will be discussed in Section 2.5.

We note that in the mathematical literature there are other generalizations of the classic exponential function distinct from the q-exponential defined in (2.11). These generalizations were introduced by Euler (1748), Jackson (1908), and others. See Ernst (2003) for details.

It is easy to verify that the inverse function to (2.9) is

$$f(t) = \frac{t^{1-q} - 1}{1 - q}. \tag{2.12}$$

As the inverse to the q-exponential, this function is naturally called a q-logarithmic function and denoted by $\ln_q t$. This function is defined for all $t > 0$ and recovers the log-function $y = \ln t$, as $t \to 1$. The latter can easily be verified using L'Hopital's rule. Thus, we have the following q-generalization of the logarithmic function.

Definition 2.3. The function.

$$\ln_q t = \frac{t^{1-q} - 1}{1 - q}, \quad t > 0, \tag{2.13}$$

is called a q-logarithmic function.

2.4. *q*-Product and *q*-Division

Consider how transforms the property $e^t e^s = e^{t+s}$ of exponentials in the case of q-exponentials. We have

$$e_q^t \, e_q^s = [1 + (1-q)t]^{\frac{1}{1-q}} \, [1 + (1-q)s]^{\frac{1}{1-q}}$$

$$= \{[1 + (1-q)t][1 + (1-q)s]\}^{\frac{1}{1-q}}$$

$$= \{[1 + (1-q)[t + s + (1-q)(ts)]]\}^{\frac{1}{1-q}}$$

$$= \{[1 + (1-q)(t \oplus_q s)]\}^{\frac{1}{1-q}}$$

$$= e_q^{t \oplus_q s}.$$

Hence, the above mentioned property of exponential functions reduces to the relation

$$e_q^t \, e_q^s = e_q^{t \oplus_q s} \tag{2.14}$$

in the case of q-exponentials. In other words, instead of the usual addition operation we have the q-plus operation.

It follows from relation (2.14) that

$$\ln_q(xy) = \ln_q x \oplus_q \ln_q y \tag{2.15}$$

valid for all $x > 0$, $y > 0$. Indeed, taking q-logarithm of both sides of (2.14),

$$\ln_q(e_q^{t \oplus_q s}) = t \oplus_q s = \ln_q(e_q^t \, e_q^s).$$

Setting $t = \ln_q x$ and $s = \ln_q y$, we obtain (2.15), generalizing the well-known property $\ln(xy) = \ln x + \ln y$ of logarithms.

Similarly, one can ask the natural question: into what relation transforms the expression e_q^{t+s} with the usual addition operation. Due to equality (2.14) we cannot expect that this expression equals to the product $e_q^t \, e_q^s$ with the usual product operation. We have

$$
\begin{aligned}
e_q^{t+s} &= [1 + (1-q)(t+s)]_+^{\frac{1}{1-q}} \\
&= [[1 + (1-q)t] + [1 + (1-q)s] - 1]_+^{\frac{1}{1-q}} \\
&= \left\{ [[1 + (1-q)t]^{\frac{1}{1-q}}]^{1-q} + [[1 + (1-q)s]^{\frac{1}{1-q}}]^{1-q} - 1 \right\}_+^{\frac{1}{1-q}} \\
&= [(e_q^t)^{1-q} + (e_q^s)^{1-q} - 1]_+^{\frac{1}{1-q}}. \tag{2.16}
\end{aligned}
$$

The right-hand side, as we expected, is not the standard product of exponentials. However, one can accept the expression on the right-hand side of (2.16) as a product $e_q^t \otimes_q e_q^s$ with a new product operation $\otimes_q$ of q-exponentials e_q^t and e_q^s. Namely, we introduce the q-product as follows.

Definition 2.4. Let $q \in \mathbb{R}$. For real numbers $x, y \in \mathbb{R}$ we define the *q-product* by the binary relation

$$x \otimes_q y = \text{sign}(x)\,\text{sign}(y)\,[|x|^{1-q} + |y|^{1-q} - 1]_+^{\frac{1}{1-q}}, \qquad (2.17)$$

where $\text{sign}(a)$ stands for the sign of a number $a \in \mathbb{R}$. If $x > 0$ and $y > 0$, then (2.17) takes the form

$$x \otimes_q y = [x^{1-q} + y^{1-q} - 1]_+^{\frac{1}{1-q}}, \qquad (2.18)$$

By virtue of this definition the right-hand side of (2.16) takes the form $e_q^t \otimes_q e_q^s$. Hence we have the relation

$$e_q^{t+s} = e_q^t \otimes_q e_q^s, \qquad (2.19)$$

obtaining another generalization of the equality $e^{t+s} = e^t s^s$, since the q-product reduces to the usual product for $q = 1$ (see Proposition 2.2).

The q-product introduced above has the following properties.

Proposition 2.2. *The q-product possesses the following properties:*

(i) *commutative:* $x \otimes_q y = y \otimes_q x$ *for all* $x, y \in \mathbb{R}$;
(ii) *associative:* $(x \otimes_q y) \otimes_q z = x \otimes_q (y \otimes_q z) = x \otimes_q y \otimes_q z$ *for all* $x, y, z \in \mathbb{R}$;
(iii) *recovers the usual product operation if* $q = 1$, *i.e.,* $x \otimes_1 y = xy$;
(iv) *preserves* 1 *as the unit element, i.e.,* $x \otimes_q 1 = x$.

Proof. Without loss of generality and to avoid manipulations with cumbersome expressions, we can assume that $x > 0$, $y > 0$. The commutativity of the q-product immediately follows from its definition (2.17):

$$x \otimes_q y = [x^{1-q} + y^{1-q} - 1]_+^{\frac{1}{1-q}} = [y^{1-q} + x^{1-q} - 1]_+^{\frac{1}{1-q}} = y \otimes_q x.$$

Let us show the associativity of $\oplus_q$. We have

$$(x \otimes_q y) \otimes_q z = \{[[x^{1-q} + y^{1-q} - 1]_+^{\frac{1}{1-q}}]^{1-q} + z^{1-q} - 1\}_+^{\frac{1}{1-q}}$$

$$= \{[x^{1-q} + y^{1-q} - 1]_+ + z^{1-q} - 1\}_+^{\frac{1}{1-q}}$$

$$= \{x^{1-q} + [y^{1-q} + z^{1-q} - 1]_+ - 1\}_+^{\frac{1}{1-q}}$$

$$= \{x^{1-q} + [[y^{1-q} + z^{1-q} - 1]_+^{\frac{1}{1-q}}]^{1-q} - 1\}_+^{\frac{1}{1-q}}$$

$$= x \otimes_q (y \otimes_q z).$$

Thus, the result does not depend in which order the q-product is performed:

$$x \otimes_q y \otimes_q z = (x \otimes_q y) \otimes_q z = x \otimes_q (y \otimes_q z).$$

To show that for $q = 1$ the q-product coincides with the ordinary product, we evaluate the limit $\lim_{q \to 1} \ln(x \otimes_q y)$. Using the L'Hopital's rule, we have

$$\lim_{q \to 1} \ln(x \otimes_q y) = \lim_{q \to 1} \frac{\ln(x^{1-q} + y^{1-q} - 1)_+}{1 - q}$$

$$= \lim_{q \to 1} \frac{-x^{1-q} \ln x - y^{1-q} \ln y}{-(x^{1-q} + y^{1-q} - 1)}$$

$$= \ln x + \ln y = \ln(xy).$$

The latter implies that $\lim_{q \to 1}(x \otimes_q y) = xy$.

The property $x \otimes_q 1 = x$ for all q immediately follows from the definition of the q-product. $\qquad\square$

Remark 2.1. Definition 2.4 of the q-product that came from the attempt to establish a q-version of the equality $e^{t+s} = e^t e^s$, is restrictive. In particular, for $q < 1$ the q-product in (2.17) vanishes for all x and y if $x^{1-q} + y^{1-q} \le 1$. For our further considerations we need to extend the notion of the q-product for the complex plane $\mathbb{C}$, as well. For this purpose, we recall that the complex function $f(z) = z^q$, $0 \ne z \in \mathbb{C}$, for $q \in \mathbb{R}$, is defined as $z^q = \exp(q \mathrm{Ln}(z))$, where $\exp(\cdot)$ is the complex exponential function and $\mathrm{Ln}(z)$ is the principal value of the complex logarithm $\ln(z)$ cut alone the negative real axis $Re(z) < 0$ (see details, e.g., Conway, 1978). We also assume conventions: $0^q = 0$ for all $q > 0$ and $0^q = \infty$ for $q < 0$.

Definition 2.5. Let $z, \zeta \in \mathbb{C}$. Then the q-product of z and ζ is defined by

$$z \otimes_q \zeta = (z^{1-q} + \zeta^{1-q} - 1)^{\frac{1}{1-q}}.$$

All the statements in Proposition 2.2 extend to the complex q-product, as well. Namely, for all $z, \zeta, \xi \in \mathbb{C}$, the properties

(1) (commutativity) $z \otimes_q \zeta = \zeta \otimes_q z$;
(2) (associativity) $(z \otimes_q \zeta) \otimes_q \xi = z \otimes_q (\zeta \otimes_q \xi) = z \otimes_q \zeta \otimes_q \xi$;
(3) (standard product for $q = 1$) $z \otimes_1 \zeta = z\zeta$;
(4) (unity preserving) $z \otimes_q 1 = z$,

hold. Below we present some other properties.

Proposition 2.3. *For $z, \zeta \in \mathbb{C}$ the q-product possesses the following properties*:

(1) $(z \otimes_q \zeta)^{-1} = z^{-1} \otimes_{2-q} \zeta^{-1}$;
(2) $z^q \otimes_{1/q} \zeta^q = (z \otimes_{2-q} \zeta)^q$;
(3) *for n-th q-product power $(\otimes_q z)^n = z \otimes_q \cdots \otimes_q z$ (n times) of z the following equality holds*:

$$(\otimes_q z)^n = [nz^{1-q} - (n-1)]^{\frac{1}{1-q}};$$

(4) *If $q \geq 1$, then $z \otimes_q 0 = 0$*;
(5) *If $q < 1$, then $z \otimes_q 0 = (z^{1-q} - 1)^{\frac{1}{1-q}}$*;
(6) *if x and y are positive real numbers such that $x \leq a$ and $y \leq b$, then $x \otimes_q y \leq a \otimes_q b$.*

Proof.

(1) Notice that $1 - (2 - q) = -(1 - q)$. Taking this into account, we have

$$z^{-1} \otimes_{2-q} \zeta^{-1} = (z^{-[1-(2-q)]} + \zeta^{-[1-(2-q)]} - 1)^{\frac{1}{1-(2-q)}}$$

$$= (z^{1-q} + \zeta^{1-q} - 1)^{\frac{-1}{1-q}}$$

$$= (z \otimes_q \zeta)^{-1}.$$

(2) Again using the relation $1 - (2 - q) = -(1 - q)$, we have

$$z^q \otimes_{1/q} \zeta^q = \left(z^{q(1-\frac{1}{q})} + \zeta^{q(1-\frac{1}{q})} - 1\right)^{\frac{1}{1-\frac{1}{q}}}$$

$$= \left(z^{-(1-q)} + \zeta^{-(1-q)} - 1\right)^{\frac{1}{-(1-q)}}$$

$$= [(z^{1-(2-q)} + \zeta^{1-(2-q)} - 1)^{\frac{1}{1-(2-q)}}]^q$$

$$= (z \otimes_{2-q} \zeta)^q.$$

(3) We prove this statement by induction. If $n = 2$, then

$$z \otimes_q z = (2z^{1-q} - 1)^{1/(1-q)}.$$

Assume that the statement is true for $n = k$:

$$(\otimes_q z)^k = [kz^{1-q} - (k-1)]^{\frac{1}{1-q}}.$$

Then for $n = k + 1$ we have

$$(\otimes_q z)^{k+1} = (\otimes_q z)^k \otimes_q z$$

$$= [kz^{1-q} - (k-1) + z^{1-q} - 1]^{\frac{1}{1-q}}$$

$$= [(k+1)z^{1-q} - k]^{\frac{1}{1-q}}.$$

(4),(5) Properties (4) and (5) follow immediately from the definition of the complex q-product.

(6) Let $q < 1$. Then $x^{1-q} \leq a^{1-q}$ and $y^{1-q} \leq b^{1-q}$. This immediately implies the desired result. If $q > 1$, then $x^{1-q} \geq a^{1-q}$ and $y^{1-q} \geq b^{1-q}$. We have

$$x \otimes_q y = \frac{1}{[x^{1-q} + y^{1-q} - 1]^{\frac{1}{q-1}}} \leq \frac{1}{[a^{1-q} + b^{1-q} - 1]^{\frac{1}{q-1}}} = a \otimes_q b,$$

again obtaining the desired result. The case $q = 1$ is obvious.

$\square$

Example 2.1. Let $z = i = \sqrt{-1}$. The equality $i^2 = -1$ in the q-product is not verified if $q \neq 1$. In fact, as follows from Proposition 2.3, part (3), that $(\otimes_q i)^2 = (2i^{1-q} - 1)^{1/(1-q)}$. For example,

$(\otimes_{-1} i)^2 = \sqrt{3}i$, if $q = -1$; $(\otimes_0 i)^2 = 2i - 1$, if $q = 0$; $(\otimes_2 i)^2 = (2i - 1)/5$, if $q = 2$. However, $(\otimes_q i)^2 = 1$, if $q = 1 - 4m$, where $m = 1, 2, \ldots$. Moreover, $(\otimes_q i)^n = 1$, for any natural n, if $q = 1 - 4m$, $m = 1, 2, \ldots$.

We define the *q-division*, denoted by the symbol $\oslash_q$, as the inverse operation to the q-product.

Definition 2.6. For $z, \zeta \in \mathbb{C}$ define the q-division by

$$z \oslash_q \zeta = (z^{1-q} - \zeta^{1-q} + 1)^{\frac{1}{1-q}}.$$

The q-division, similar to the q-product, inherits some properties of the usual division. Here are some properties of the q-division.

Proposition 2.4.

(1) $z \oslash_q 1 = z$;
(2) $z \otimes_q (1 \oslash_q z) = 1$. *In general, the equality*

$$(z \oslash_q \zeta) \otimes_q (\zeta \oslash_q z) = 1$$

holds, representing a q-version of the equality $\frac{z}{\zeta} \cdot \frac{\zeta}{z} = 1$;
(3) $z \oslash_1 \zeta = \frac{z}{\zeta}$;
(4) *If* $q \neq 1$, *then* $z \oslash_q 0 = (1 + z^{1-q})^{\frac{1}{1-q}}$;
(5) *For* $q \neq 1$,

$$z \oslash_q \left((1 + z^{1-q})^{\frac{1}{1-q}} \right) = 0^{\frac{1}{1-q}} = \begin{cases} 0 & \text{if } q < 1 \\ \infty & \text{if } q > 1. \end{cases}$$

Remark 2.2. Notice that, for $q \neq 1$, q-division by zero is allowed (see part (4) of Proposition 2.4).

Proof. Parts (1), (4), and (5) can be verified directly. Part (3) can be shown similarly to the q-product. We prove part (2).

$$(z \oslash_q \zeta) \otimes_q (\zeta \oslash_q z)$$

$$= \left(z^{1-q} - \zeta^{1-q} + 1 \right)^{\frac{1}{1-q}} \otimes_q \left(\zeta^{1-q} - z^{1-q} + 1 \right)^{\frac{1}{1-q}}$$

$$= \left[(z^{1-q} - \zeta^{1-q} + 1) + (\zeta^{1-q} - z^{1-q} + 1) - 1 \right]^{\frac{1}{1-q}} = 1. \qquad \square$$

2.5. Properties of q-exponential and q-logarithmic functions

In Section 2.3, we introduced the *q-exponential* and *q-logarithmic* functions respectively defined by

$$e_q^x = [1 + (1 - q)x]_+^{\frac{1}{1-q}}, \quad x \in \mathbb{R},$$

and

$$\ln_q x = \frac{x^{1-q} - 1}{1 - q}, \quad x > 0.$$

Here x is a real variable. However, for our further analysis we will need the q-exponential and q-logarithmic functions extended to the whole complex plane. Replacing x by $z \in \mathbb{C}$ in the above expression e^x, we can write

$$e_q^z = e^{\text{Ln}[1+(1-q)z]^{\frac{1}{1-q}}} = e^{\frac{\text{Ln}[1+(1-q)z]}{1-q}},$$

where $\text{Ln}(\cdot)$ is the principal value of the complex logarithmic function $\ln(\cdot)$ (see Remark 2.1). Similarly, we can rewrite $\ln_q z$ also through the function $\text{Ln}(\cdot)$. Thus, we have the following definitions of the complex q-exponential and q-logarithmic functions.

Definition 2.7. Let $q \neq 1$ and $z \in \mathbb{C}$. Then the complex q-exponential function $\exp_q(z)$ is defined by

$$\exp_q(z) = \exp\left[\frac{\text{Ln}(1 + (1 - q)z)}{1 - q}\right], \quad z \neq -\frac{1}{1 - q}, \tag{2.20}$$

and the complex q-logarithmic function is defined by

$$\text{Ln}_q(z) = \frac{\exp\left[(1 - q)\text{Ln}(z)\right] - 1}{1 - q}, \quad z \neq 0. \tag{2.21}$$

Remark 2.3.

(1) For functions $\exp_q(z)$ and $\text{Ln}_q(z)$ the numbers $z = -\frac{1}{1-q}$ and $z = 0$, respectively, are branching points, and therefore these functions are not defined at these points.

(2) Evidently, if $z = x$ is real, then the complex q-exponential and q-logarithmic functions defined in (2.20) and (2.21) reduce to e^x and $\ln_q x$ defined in (2.11) and (2.13), respectively.

(3) The definitions of the q-exponential and q-logarithmic functions given in (2.20) and (2.21) are valid for complex $1 \neq q \in \mathbb{C}$, as well. We note that complex q is not only of theoretical interest, it also arises in applications (Rybczynski *et al.*, 2015; Wilk and Wlodarczyk, 2015).

Proposition 2.5. *The complex q-logarithm satisfies the equation*

$$\text{Ln}_q(z) \oplus_q \text{Ln}_q(\zeta) = \text{Ln}_q(z\zeta). \tag{2.22}$$

Proof. Let $z, \zeta \neq 0$. The proof follows from the following chain of equations:

$$\text{Ln}_q(z) \oplus_q \text{Ln}_q(\zeta) = \frac{\exp[(1-q)\text{Ln}(z)] - 1}{1-q} \oplus_q \frac{\exp[(1-q)\text{Ln}(\zeta)] - 1}{1-q}$$

$$= \frac{\exp[(1-q)\text{Ln}(z)] - 1}{1-q} + \frac{\exp[(1-q)\text{Ln}(\zeta)] - 1}{1-q}$$

$$+ (1-q)\frac{\exp[(1-q)\text{Ln}(z)] - 1}{1-q} \times \frac{\exp[(1-q)\text{Ln}(\zeta)] - 1}{1-q}$$

$$= \frac{\exp[(1-q)\text{Ln}(z)] \, \exp[(1-q)\text{Ln}(z)] - 1}{1-q}$$

$$= \frac{\exp[(1-q)(\text{Ln}(z) + \text{Ln}(\zeta)] - 1}{1-q}$$

$$= \frac{\exp[(1-q)\text{Ln}(z\,\zeta)] - 1}{1-q} = \text{Ln}_q(z\zeta). \qquad \square$$

Corollary 2.1. *The complex q-exponential function satisfies the following equality*

$$\exp_q(z)\exp_q(\zeta) = \exp_q(z \oplus_q \zeta). \tag{2.23}$$

Proof. Let $0 \neq u, w \in \mathbb{C}$. Then, due to Proposition 2.5,

$$\text{Ln}_q(u) \oplus_q \text{Ln}_q(w) = \text{Ln}_q(uw). \tag{2.24}$$

Let $\text{Ln}_q(u) = z$ and $\text{Ln}_q(w) = \zeta$. Then we have $u = \exp_q(z)$ and $w = \exp_q(\zeta)$. Now equation (2.24) takes the form

$$z \oplus_q \zeta = \text{Ln}_q(\exp_q(z)\exp_q(\zeta)),$$

which is equivalent to

$$\exp_q(z \oplus_q \zeta) = \exp_q(z)\exp_q(\zeta). \qquad \square$$

Proposition 2.6. *The complex q-logarithm satisfies the equation*

$$\text{Ln}_q(z) + \text{Ln}_q(\zeta) = \text{Ln}_q(z \otimes_q \zeta). \tag{2.25}$$

Proof. We have

$$
\begin{aligned}
\text{Ln}_q(z \otimes_q \zeta) &= \frac{\exp[(1-q)\text{Ln}(z \otimes_q \zeta)] - 1}{1-q} \\[2mm]
&= \frac{\exp\left[(1-q)\text{Ln}[z^{1-q} + \zeta^{1-q} - 1]^{\frac{1}{1-q}}\right] - 1}{1-q} \\[2mm]
&= \frac{\exp\left[\text{Ln}(z^{1-q} + \zeta^{1-q} - 1)\right] - 1}{1-q} \\[2mm]
&= \frac{z^{1-q} + \zeta^{1-q} - 2}{1-q} \\[2mm]
&= \frac{z^{1-q} - 1}{1-q} + \frac{\zeta^{1-q} - 1}{1-q} \\[2mm]
&= \frac{\exp[(1-q)\text{Ln}(z)] - 1}{1-q} + \frac{\exp[(1-q)\text{Ln}(\zeta)] - 1}{1-q} \\[2mm]
&= \text{Ln}_q(z) + \text{Ln}_q(\zeta). \qquad \square
\end{aligned}
$$

Corollary 2.2. *The complex q-exponential function satisfies the following equality*

$$\exp_q(z + \zeta) = \exp_q(z) \otimes_q \exp_q(\zeta). \tag{2.26}$$

The proof is similar to the proof of Corollary 2.1. Formula (2.26) generalizes to the case of complex arguments the relation (2.19) obtained in Section 2.4 for real numbers.

Next proposition uses the following q-sum and q-product symbols for $z_1, \ldots, z_d \in \mathbb{C}^d$:

$$\sum_{j=1}^{d}{}_q z_j = z_1 \oplus_q \cdots \oplus_q z_d,$$

$$\prod_{j=1}^{d}{}_q z_j = z_1 \otimes_q \cdots \otimes_q z_d.$$

Proposition 2.7. *For numbers* $z_1, \ldots, z_d \in \mathbb{C}^d$ *the following relations hold:*

$$\mathrm{Ln}_q \left(\prod_{j=1}^{d} z_j \right) = \sum_{j=1}^{d}{}_q \mathrm{Ln}_q(z_j), \qquad (2.27)$$

$$\mathrm{Ln}_q \left(\prod_{j=1}^{d}{}_q z_j \right) = \sum_{j=1}^{d} \mathrm{Ln}_q(z_j), \qquad (2.28)$$

$$\exp_q \left(\sum_{j=1}^{d}{}_q z_j \right) = \prod_{j=1}^{d} \exp_q z_j, \qquad (2.29)$$

$$\exp_q \left(\sum_{j=1}^{d} z_j \right) = \prod_{j=1}^{d}{}_q \exp_q z_j. \qquad (2.30)$$

Proof. The proof follows directly from Propositions 2.5, 2.6, and Corollaries 2.1 and 2.2. $\qquad \square$

Proposition 2.8. *For the derivative of order* n *of the* q-*exponential function* $\exp_q(z)$ *the following formula holds:*

$$\frac{d^n \exp_q(z)}{dz^n} = \prod_{k=0}^{n-1} a_k(q) \, [\exp_q(z)]^{n(q-1)+1}, \quad n = 1, 2, \ldots, \qquad (2.31)$$

where

$$a_k(q) = k(q-1) + 1, \quad k = 0, 1, \ldots. \qquad (2.32)$$

Proof. We recall that the q-exponential function is defined as a solution to the differential equation $\frac{dy}{dz} = [y(z)]^q$. Therefore, for the first derivative of the q-exponential function we have

$$\frac{d \exp_q(z)}{dz} = [\exp_q(z)]^q, \tag{2.33}$$

which is consistent with (2.31) for $n = 1$. Now we prove (2.31) by induction. Assume that formula (2.31) is valid for $n = m$, that is

$$\frac{d^m \exp_q(z)}{dz^m} = \prod_{k=0}^{m-1} a_k(q) \, [\exp_q(z)]^{m(q-1)+1}. \tag{2.34}$$

Then using (2.33) and (2.34), we have

$$\frac{d^{m+1} \exp_q(z)}{dz^{m+1}} = \frac{d}{dz} \left(\frac{d^m \exp_q(z)}{dz^m} \right)$$

$$= \prod_{k=0}^{m-1} a_k(q) \, [m(q-1)+1] \, [\exp_q(z)]^{m(q-1)} \frac{d \exp_q(z)}{dz}$$

$$= \prod_{k=0}^{m} a_k(q) \, [\exp_q(z)]^{m(q-1)+q}.$$

The latter implies (2.31) for $n = m + 1$, since $m(q-1) + q = (m+1)(q-1) + 1$. □

It follows immediately from Proposition 2.8 the following corollary due to the condition $\exp_q(0) = 1$.

Corollary 2.3.

$$\frac{d^n \exp_q(z)}{dz^n}\Big|_{z=0} = \prod_{k=0}^{n-1} a_k(q), \quad n = 1, 2, \ldots. \tag{2.35}$$

Proposition 2.9. *The q-exponential function $\exp_q(z)$ has the following power series representation centered at z_0:*

$$\exp_q(z) = \exp_q(z_0) + \sum_{n=1}^{\infty} \frac{\prod_{k=0}^{n-1} a_k(q)[\exp_q(z_0)]^{n(q-1)+1}}{n!} (z - z_0)^n, \tag{2.36}$$

which is absolutely and uniformly convergent in $|z - z_0| < \frac{1}{|1-q||\exp_q(z_0)|^{q-1}}$.

In particular, if $z_0 = 0$, then $\exp_q(z)$ has the power series representation

$$\exp_q(z) = 1 + \sum_{n=1}^{\infty} \frac{\prod_{k=0}^{n-1} a_k(q)}{n!} z^n, \qquad (2.37)$$

which absolutely and uniformly converges in the disk

$$|z| < \frac{1}{|1-q|}.$$

Proof. The power series representation (2.37) immediately follows from Proposition 2.8. We show that the radius of convergence of this power series is $r = (|1-q||\exp_q(z_0)|^{q-1})^{-1}$. Consider the sequence

$$A_0 = 1, A_n = \prod_{k=0}^{n-1} a_k(q), \quad n = 1, 2, \ldots, \qquad (2.38)$$

where the sequence $a_k(q), k = 0, 1, \ldots$, is defined in (2.32). It follows from the definition of $a_k(q)$ that it is asymptotically equivalent to $(q-1)k$ for large k. Therefore, A_n is asymptotically equivalent to

$$\left[(q-1)(\exp_q(z_0))^{q-1}\right]^n n!$$

for large n. It follows that the power series in (2.37) converges (or diverges) simultaneously with the power series

$$P(z) = \sum_{n=1}^{\infty} \left[(q-1)(\exp_q(z_0))^{q-1}\right]^n (z - z_0)^n.$$

The latter converges absolutely and uniformly (on any compact set) in the disk $|z - z_0| < (|1-q||\exp_q(z_0)|^{q-1})^{-1}$.

The second part of the proposition follows from Corollary 2.3 and the fact that $\exp_q(0) = 1$. $\qquad \square$

Remark 2.4.

(1) The power series in (2.36) and (2.37) coincide with the power series representations of $\exp(z)$ if $q = 1$, since in this case $a_k = 1$ for all $k = 0, 1, \ldots$, and therefore $A_n = 1, n = 0, 1, \ldots$.

(2) It follows from the definition of the q-exponential function that $\exp_q(z)$ is a polynomial of order $n + 1$ if $q = n/(n + 1), n = 1, 2, \ldots$. We notice that if $q = n/(n + 1)$, then $a_n = 0$. Hence, $A_{n+1} = 0, A_{n+2} = 0, \ldots$. This fact implies that the power series representations (2.36) and (2.37), in fact, are polynomials of order $n + 1$.

Corollary 2.4. *The q-exponential function $\exp_q(z)$ has the following asymptotic relation*

$$\exp_q(z) = 1 + z + \frac{q}{2}z^2 + o(z^2), \quad |z| \to 0. \tag{2.39}$$

There is another approach to the definition of the complex q-exponential function convenient for our further analysis. Let $q \neq 1$. We define $\exp_q(iy)$, where i is the imaginary unit and y is a real number, as the principal value of $[1 + i(1-q)y]^{\frac{1}{1-q}}$, namely

$$\exp_q(iy) = [1 + (1-q)^2 y^2]^{\frac{1}{2(1-q)}} e^{\frac{i \arctan[(q-1)y]}{1-q}}, \quad y \in \mathbb{R}, \ q \neq 1. \tag{2.40}$$

Then, for $z = x + iy \in \mathbb{C}$ we define $\exp_q(z) = \exp_q(x + iy)$ using the relation (2.26). Namely,

$$\exp_q(z) = \exp_q(x + iy) = e_q^x \otimes_q \exp_q(iy).$$

It follows from (2.40) that

$$|\exp_q(iy)| = [1 + (1-q)^2 y^2]^{\frac{1}{2(1-q)}}, \ y \in \mathbb{R}. \tag{2.41}$$

The latter provides a behavior of $\exp_q(iy)$ for $q > 1$ and $q < 1$. Namely, the following proposition holds. In the limit case $q \to 1$ Eq. (2.41) implies $|\exp_1(iy)| = |\exp(iy)| = 1$.

Proposition 2.10. *If $q < 1$, then*

(1) $|\exp_q(iy)| \geq 1$, $y \in \mathbb{R}$;

(2) $|\exp_q(iy)| = O(|y|^{\frac{1}{1-q}})$, $|y| \to \infty$.

    Similarly, if $q > 1$, then

(3) $0 < |\exp_q(iy)| \leq 1$, $y \in \mathbb{R}$; *and*

(4) $|\exp_q(iy)| = O\left(\dfrac{1}{|y|^{\frac{1}{q-1}}}\right)$, $|y| \to \infty$.

    Finally, if $q = 1$, then

(5) $|\exp_1(iy)| = 1$ *for all $y \in \mathbb{R}$.*

Remark 2.5. Proposition 2.10 and equation (2.40) imply that the function $\exp_q(iy)$, unlike e^{iy}, is increasing oscillating if $q < 1$, and decreasing oscillating, if $q > 1$. That is, the pure oscillatory behavior occurs only in the case $q = 1$.

Proposition 2.11. *For $0 \neq z$, $\zeta \in \mathbb{C}$ the following equality holds:*

$$z \otimes_q \exp_q(\zeta) = z \exp_q(\zeta \, z^{q-1}). \qquad (2.42)$$

Proof. We have

$$
\begin{aligned}
z \otimes_q \exp_q(\zeta) &= \left[z^{1-q} + 1 + (1-q)\zeta - 1\right]^{\frac{1}{1-q}} \\
&= \left[z^{1-q} + (1-q)\zeta\right]^{\frac{1}{1-q}} \\
&= z\left[1 + (1-q)\zeta \, z^{q-1}\right]^{\frac{1}{1-q}} \\
&= z \exp_q(\zeta \, z^{q-1}). \qquad \square
\end{aligned}
$$

Corollary 2.5. *Let $q > 1$. Then for $z \in \mathbb{C}$ and $y \in \mathbb{R}$ the following estimate holds:*

$$|\exp_q(z) \otimes_q \exp_q(iy)| \leq |\exp_q(z)|.$$

Proof. The proof immediately follows from Proposition 2.11 if one takes into account Proposition 2.10, Part (3). $\square$

Proposition 2.12. *For the derivative of order n of the q-logarithmic function $Ln_q(z)$, $z \neq 0$, the following formula holds:*

$$\frac{d Ln_q(z)}{dz} = z^{-q}, \quad (n = 1), \tag{2.43}$$

$$\frac{d^n Ln_q(z)}{dz^n} = (-1)^{n-1} \prod_{k=0}^{n-2} b_k(q) \, z^{-(n+q-1)}, \quad n = 2, 3, \ldots, \tag{2.44}$$

where

$$b_k(q) = k + q, \quad k = 0, 1, \ldots . \tag{2.45}$$

Remark 2.6. Notice that if $q = -m$, where $m = 0, 1, 2, \ldots$, then $b_m = 0$. This implies that $\prod_{k=0}^{n-2} b_k(q) = 0$ for all $n \geq m + 2$. Hence all the derivatives of orders $m + 2, m + 3, \ldots$, are identically zero, implying that $Ln_{-m}(z)$ is a polynomial of order $m + 1$.

Corollary 2.6.

$$\frac{d Ln_q(z)}{dz}\Big|_{z=1} = 1,$$

$$\frac{d^n Ln_q(z)}{dz^n}\Big|_{z=1} = (-1)^{n-1} \prod_{k=0}^{n-2} b_k(q), \quad n = 2, 3, \ldots, \tag{2.46}$$

with $b_k(q), k = 0, 1, \ldots$, defined in (2.45).

Proposition 2.13. *The function $Ln_q(z + 1)$ has the power series representation*

$$Ln_q(1 + z) = z + \sum_{n=2}^{\infty} \frac{(-1)^{n-1} \prod_{k=0}^{n-2} b_k(q)}{n!} z^n, \tag{2.47}$$

where $b_k(q), k = 0, 1, \ldots$, are defined in (2.45). If $q = 0, -1, -2, \ldots$, then the right-hand side of (2.47) is a polynomial of order $1 - q$. In contrast, if $q \neq 0, -1, -2, \ldots$, then the power series in (2.47) converges absolutely and uniformly only in the disk $|z| < 1$.

Proof. It follows from Corollary 2.6 that the power series of the function $\mathrm{Ln}_q(z)$ centered at $z = 1$ has the form

$$\mathrm{Ln}_q(z) = z - 1 + \sum_{n=2}^{\infty} \frac{(-1)^{n-1} \prod_{k=0}^{n-2} b_k(q)}{n!} (z-1)^n.$$

Changing z to $1 + z$ we obtain (2.47).

Further, if $q = -m$, $m = 0, 1, \ldots$, then $b_m = 0$, and hence

$$B_n = \prod_{k=0}^{n-2} b_k = 0, \quad n = m+2, m+3, \ldots.$$

In this case power series (2.47) contains a finite number of terms and becomes a polynomial of order $m + 1$ (see Remark 2.6), which is convergent everywhere. Now assume $q \neq 0, -1, -2, \ldots$. We distinguish cases $q < 0$ and $q > 0$. Suppose, first that $q > 0$ and an integer number: $q = \ell = 1, 2, \ldots$. Then

$$B_n = \ell(\ell+1)\ldots(\ell+n-2) = \frac{(n+\ell-2)!}{(\ell-1)!}.$$

It follows that

$$\frac{B_n}{n!} = \frac{(n+1)\ldots(n+\ell-2)}{(\ell-1)!}.$$

The latter implies the following estimate for $B_n/(n!)$:

$$\frac{(n+1)^{\ell-2}}{(\ell-1)!} \leq \frac{B_n}{n!} \leq \frac{(n+\ell-2)^{\ell-2}}{(\ell-1)!}, \quad n = 2, 3, \ldots.$$

It is easy to verify that both power series

$$\sum_{n=2}^{\infty} \frac{(n+1)^{\ell-2}}{(\ell-1)!} z^n \quad \text{and} \quad \sum_{n=2}^{\infty} \frac{(n+\ell-2)^{\ell-2}}{(\ell-1)!} z^n$$

are absolutely and uniformly convergent in $|z| < 1$. This fact implies that the power series in (2.47) is also absolutely and uniformly convergent in $|z| < 1$. Now, let $\ell < q < \ell + 1$, where $\ell = 1, 2, \ldots$.

Then, we have

$$\frac{(n + \ell - 2)!}{(\ell - 1)!} \leq B_n \leq \frac{(n + \ell - 1)!}{\ell!}.$$

This implies the estimate

$$\frac{(n + 1)^{\ell-2}}{(\ell - 1)!} \leq \frac{B_n}{n!} \leq \frac{(n + \ell - 1)^{\ell-1}}{\ell!}, \quad n = 2, 3, \dots.$$

Since the power series

$$\sum_{n=2}^{\infty} \frac{(n + 1)^{\ell-2}}{(\ell - 1)!} z^n \quad \text{and} \quad \sum_{n=2}^{\infty} \frac{(n + \ell - 1)^{\ell-1}}{\ell!} z^n$$

are absolutely and uniformly convergent in $|z| < 1$, it follows that the power series in (2.47) is also absolutely and uniformly convergent in $|z| < 1$.

Now let $-m < q < -m + 1$, where $m = 0, 1, 2, \dots$. In this case $B_n \neq 0$ for all $n = 2, 3, \dots$. Further, we have

$$\lim_{n \to \infty} \frac{B_{n+1}}{(n + 1)!} \times \frac{n!}{B_n} = \lim_{n \to \infty} \left(1 + \frac{q - 2}{n + 1}\right) = 1,$$

which means that the radius of convergence of power series in (2.47) equals 1, proving the proposition. □

Corollary 2.7. *The function* $\mathrm{Ln}_q(z+1)$ *has the following asymptotic relation*

$$\mathrm{Ln}_q(1 + z) = z - \frac{q}{2}z^2 + o(z^2), \quad |z| \to 0.$$

All the propositions proved above are valid for the q-exponential and q-logarithmic functions of a real variable, as well. However, in the real case one can obtain specific properties of these functions valid only for real variable. Consider the function $\exp_q(x)$, which is the restriction of the complex q-exponential function $\exp(z)$ to the real axis $-\infty < x < \infty$. This function has singularity at $x = -(1 - q)^{-1}$. Recall that this point is a branching point for the complex q-exponential function $\exp_q(z)$ (see Remark 2.3). The function $\exp_q(x)$ takes positive real values on the semi-infinite interval $(-\infty, -(1 - q)^{-1})$ if $q > 1$, and on the semi-infinite interval $(-(1 - q)^{-1}, \infty)$

if $q < 1$. In what concerns the values of $\exp_q(x)$ on the complements of these intervals, they can be complex or negative depending on q. The restriction of $\exp_q(z)$ to the interval $(-\infty, (q-1)^{-1})$, when $q > 1$, and to the interval $(-(1-q)^{-1}, \infty)$, when $q < 1$, extended by zero to the complements of these intervals respectively, was denoted by e_q^x; see Definition 2.2. Thus, e_q^x is defined on the whole real axis $\mathbb{R} \equiv (-\infty, \infty)$ and we use this definition in the future, when the argument x is real.

Proposition 2.14. *For any $\beta \in \mathbb{C}$ and real $a \neq 0$, the relation*

$$(\exp_q(\beta x))^a = \exp_{(a-1+q)/a}(a\beta x)$$

holds.

Proof. We have

$$\exp_{(a-1+q)/a}(a\beta x) = \left[1 + \left(1 - \frac{a-1+q}{a}\right)a\beta x\right]^{\frac{1}{1-\frac{a-1+q}{a}}}$$

$$= [1 + (1-q)\beta x]^{\frac{a}{1-q}}$$

$$= (\exp_q(\beta x))^a. \qquad \square$$

In particular, if $\beta = 1$, $a = q \neq 0$, then we obtain the equality

$$(\exp_q(x))^q = \exp_{2-1/q}(qx). \tag{2.48}$$

The following proposition provides a power series of e_q^{-x} for $q > 1$ through the degrees of $1/x$, thus describing in detail its behavior at infinity.

Proposition 2.15. *Let $q > 1$. Then for $x > (q-1)^{-1}$ the following representation holds:*

$$e_q^{-x} = [(q-1)x]^{-\frac{1}{q-1}} \sum_{n=0}^{\infty} \frac{(-1)^n A_n}{n!(q-1)^{2n}} \left(\frac{1}{x}\right)^n \tag{2.49}$$

where $A_n, n = 0, 1, \ldots,$ are defined in (2.38).

Proof. To prove this statement we use

$$\frac{1}{(1-z)^a} = 1 + \frac{a}{1!}z + \frac{a(a+1)}{2!}z^2 + \frac{a(a+1)(a+2)}{3!}z^3 + \dots, \quad |z| < 1.$$
$$(2.50)$$

It is easy to check that

$$e_q^{-x} = [1 - (1-q)x]^{\frac{1}{1-q}}$$
$$= [(q-1)x]^{-\frac{1}{q-1}} \frac{1}{[1 - \frac{-1}{(q-1)x}]^{\frac{1}{q-1}}}.$$

Now using (2.50) with

$$a = \frac{1}{q-1} \quad \text{and} \quad z = \frac{-1}{(q-1)x},$$

we obtain (2.49). $\qquad\qquad\square$

Similarly, one can prove the following statement on a representation of e_q^{-x} for $q < 1$, describing its behavior at minus infinity.

Proposition 2.16. *Let $q < 1$. Then for $x < -(1-q)^{-1}$ the representation*

$$e_q^{-x} = [-(1-q)x]^{\frac{1}{1-q}} \sum_{n=0}^{\infty} \frac{(-1)^n A_n}{n!(1-q)^{2n}} \left(\frac{1}{x}\right)^n \qquad (2.51)$$

holds.

Remark 2.7.

(1) It follows from Proposition 2.16 that if $q < 1$ then e_q^{-x} increases at minus infinity with the power $1/(1-q)$, and hence, is not integrable over any interval $(-\infty, a)$, but is integrable over any finite interval or intervals of the form (b, ∞). Here a and b are some real numbers;

(2) It follows from Proposition 2.15 that if $q > 1$ then the function e_q^{-x} decreases at infinity with the power $-1/(q-1)$, and therefore, is integrable over $(0, \infty)$ if $q < 2$. Combining with Part (1) of this remark, we can conclude that e_q^{-x} is integrable over $(0, \infty)$ for all

$q < 2$. Similarly, the function xe_q^{-x} is integrable over the interval $(0, \infty)$ for all $q < 3/2$.

Proposition 2.17. *Let $q > 1$. Then $\ln_q x$ is a strictly increasing function on the interval $(0, \infty)$ with the asymptote*

$$\lim_{x \to \infty} \ln_q x = \frac{1}{q-1}. \tag{2.52}$$

Proof. One can easily verify that the derivative of $\ln_q(x)$ is $(\ln_q x)' = x^{-q} > 0$ if $x > 0$, and hence $\ln_q x$ is strictly increasing. The asymptote (2.52) can be verified directly. □

Proposition 2.18. *Let $1 \leq q_1 < q_2$. Then*

(1) $\ln_{q_1} x \geq \ln_{q_2} x$ *for all $x > 0$, being equal only at $x = 1$,*
(2) $e_{q_1}^x \leq e_{q_2}^x$ *for all $x \in (-\infty, \frac{1}{q_2-1})$, being equal only at $x = 0$.*

Proof. Consider the function $g(x) = \ln_{q_1} x - \ln_{q_2} x$. Its derivative is

$$g'(x) = \frac{1}{x^{q_1}} - \frac{1}{x^{q_2}} = \frac{x^{q_2} - x^{q_1}}{x^{q_1} x^{q_2}}.$$

Let $x > 1$. Since $q_2 > q_1$, obviously $g'(x) > 0$ for $x > 1$. Hence, $g(x)$ is strictly increasing function on the interval $(1, \infty)$. Moreover, $g(1) = 0$. Therefore, $g(x) > 0$ for all $x > 1$, implying $\ln_{q_1} x > \ln_{q_2} x$. If $0 < x < 1$, then $g'(x) < 0$, and hence $g(x)$ is decreasing on the interval $(0, 1)$. Moreover, $g(1) = 0$, and therefore, $g(x) > 0$ on the interval $(0, 1)$. This implies that $\ln_{q_1} x > \ln_{q_2} x$ on the interval $(0, 1)$, as well, proving Part (1). Part (2) follows from Part (1) as inverse functions. □

2.6. q-Generalizations of Trigonometric Functions

In this section, we introduce and study some useful properties of q-generalizations of trigonometric functions. Setting $z = iy$, where $y \in \mathbb{R}$ and $i = \sqrt{-1}$, the imaginary unit, in the power series representation (2.37) of the complex q-exponential function $\exp_q(z)$, we have

$$\exp_q(iy) = \sum_{n=0}^{\infty} \frac{A_n(q)i^n}{n!} y^n, \quad |y| < \frac{1}{|1-q|}, \qquad (2.53)$$

where the sequence A_n is defined as follows:

$$A_0(q) = 1, \quad A_n(q) = \prod_{k=0}^{n-1} a_k(q), \quad n = 1, 2, \ldots, \qquad (2.54)$$

$$a_k(q) = k(q-1) + 1, \quad k = 0, 1, \ldots. \qquad (2.55)$$

It follows from (2.53) that

$$\exp_q(iy) = \sum_{n=0}^{\infty} \frac{(-1)^n A_{2n}(q)}{(2n)!} y^{2n}$$

$$+ i \sum_{n=0}^{\infty} \frac{(-1)^n A_{2n+1}(q)}{(2n+1)!} y^{2n+1}. \qquad (2.56)$$

By analogy of Euler's formula $e^{iy} = \cos y + i \sin y$, the latter leads us to the following definition of q-generalizations of the cosine and sine functions.

Definition 2.8. Define q-cosine and q-sine functions by

$$\cos_q(x) = \sum_{n=0}^{\infty} \frac{(-1)^n A_{2n}(q)}{(2n)!} x^{2n}, \qquad (2.57)$$

and

$$\sin_q(x) = \sum_{n=0}^{\infty} \frac{(-1)^n A_{2n+1}(q)}{(2n+1)!} x^{2n+1}. \qquad (2.58)$$

respectively.

Power series representing $\cos_q(x)$ and $\sin_q(x)$ in (2.57) and (2.58) converge absolutely and uniformly in the interval $|x| < \frac{1}{|1-q|}$. However, these functions can be extended to the whole real axis making use of representation (2.40).

Representations (2.57) and (2.58) immediately imply asymptotic behaviors of theses functions near the origin.

Proposition 2.19.

(1) $\cos_q(x) = 1 - \frac{q}{2}x^2 + o(x^3)$, $x \to 0$,
(2) $\sin_q(x) = x - \frac{q}{6}x^3 + o(x^4)$, $x \to 0$.

The relation (2.56) and definitions (2.57) and (2.58) of q-cosine and q-sine imply the follwing proposition generalizing Euler's formulas.

Proposition 2.20. *The functions* $\cos_q(x)$ *and* $\sin_q(x)$ *satisfy the following relations:*

$$e_q^{ix} = \cos_q(x) + i \, \sin_q(x), \tag{2.59}$$

and

$$\cos_q(x) = \frac{e_q^{ix} + e_q^{-ix}}{2}, \quad \sin_q(x) = \frac{e_q^{ix} - e_q^{-ix}}{2i}. \tag{2.60}$$

Definition 2.9. The functions q-tangent and q-cotangent are defined as follows:

$$\tan_q(x) = \frac{\sin_q(x)}{\cos_q(x)}, \quad \cot_q(x) = \frac{\cos_q(x)}{\sin_q(x)}.$$

Definition 2.10. The functions q-secant and q-cosecant are defined as follows:

$$\sec_q(x) = \frac{1}{\cos_q(x)}, \quad \csc_q(x) = \frac{1}{\sin_q(x)}.$$

Similarly, one can define q-hyperbolic functions: $\sinh_q(x)$, $\cosh_q(x)$, $\tanh_q(x)$, etc. Some properties of $\sin_q$, $\cos_q$, and corresponding q-hyperbolic functions, were studied in Borges (1998). Below we provide some examples of q-trigonometric identities.

Proposition 2.21. *The following q-trigonometric identities hold:*

(1) $\sin_q(x)\csc_q(x) = 1$, $\cos_q(x)\sec_q(x) = 1$, $\tan_q(x)\cot_q(x) = 1$,
(2) $(\cos_q(x))^2 + (\sin_q(x))^2 = e_q^{(1-q)x^2}$,
(3) $(\cos_q(x))^2 - (\sin_q(x))^2 = \cos_{(1+q)/2}(2x)$,

(4) $1 + (\tan_q(x))^2 = e_q^{(1-q)x^2}(\sec_q(x))^2,$

(5) $1 + (\cot_q(x))^2 = e_q^{(1-q)x^2}(\csc_q(x))^2,$

(6) $\cos_q(2x) = e_{2q-1}^{2(1-q)x^2} - 2\,\sin_{2q-1}^2(x).$

Proof.

(1) The proof of identities in Part (1) immediately follows from the definitions of q-trigonometric functions.

(2) Using formulas in (2.60) and Corollary 2.1 we have

$$(\cos_q(x))^2 + (\sin_q(x))^2 = \exp_q(ix)\,\exp_q(-ix)$$
$$= \exp_q(ix \oplus_q (-ix))$$
$$= e_q^{(1-q)x^2}.$$

(3) First, we note the equality $(\exp_q(ix))^2 = \exp_{(1+q)/2}(2ix)$, which follows from Proposition 2.14 with $\beta = 1$ and $a = 2$. Now using this relation and (2.60), we have

$$(\cos_q(x))^2 - (\sin_q(x))^2 = \frac{(\exp_q(ix))^2 + (\exp_q(-ix))^2}{2}$$
$$= \frac{\exp_{(1+q)/2}(2ix) + \exp_{(1+q)/2}(-2ix)}{2}$$
$$= \cos_{(1+q)/2}(2x).$$

(4),(5) Proofs of these two properties follow from identity (2) dividing by $(\cos_q(x))^2$ and $(\sin_q(x))^2$, respectively.

(6) It follows from properties (2), (3) of this proposition that

$$\cos_{(q_1+1)/2}(2x) = e_{q_1}^{2(1-q_1)x^2} - 2(\sin_{q_1}(x))^2.$$

Setting $q_1 = 2q - 1$ in the latter equation we obtain the identity in Part (6). $\square$

Proposition 2.22. *Let $q > 1$. Then*

(1) $|\sin_q(x)| \le 1$ *for all* $x \in \mathbb{R}$;

(2) $\lim_{x \to 0} \sin_q(x)/x = 1$;

(3) $\left|\sin_q(x)/x\right| \le 1$ *for all* $x \in \mathbb{R}$.

Proof. Property (1) immediately follows from (2.60) and Proposition 2.10, Part (3). Property (2) is obvious implication of Proposition 2.19. This and Property (1) imply (3). □

Further, consider the function $h_q(x) = \cos_q 2x - 1$. Due to identity (6) of Proposition 2.21 this function can be expressed in the form

$$h_q(x) = (e_{2q-1}^{2(1-q)x^2} - 1) - 2\,\sin_{2q-1}^2(x). \tag{2.61}$$

Proposition 2.23. *For* $h_q(x)$ *the following asymptotic behavior holds near zero*:

$$h_q(x) = -2\,q\,x^2 + o(x^3), \quad x \to 0.$$

Proof. Using the asymptotic relation (2.39), we have

$$e_{2q-1}^{2(1-q)x^2} - 1 = 2(1-q)x^2 + o(x^3),\ x \to 0. \tag{2.62}$$

In turn, it follows from Proposition 2.19 that

$$-2\,\sin_{2q-1}^2(x) = -2\,x^2 + o(x^3),\ x \to 0. \tag{2.63}$$

Now (2.61), (2.62) and (2.63) imply the statement. □

Proposition 2.24. *Let* $q \geq 1$. *Then the function* $h_q(x)$ *defined in* (2.61) *satisfies the estimate* $-3 \leq h_q(x) \leq 0$ *for all* $x \in \mathbb{R}$.

Proof. Assume $q \geq 1$. Then it is readily seen from the explicit form

$$e_{2q-1}^{2(1-q)x^2} = \frac{1}{[1 + 4(1-q)^2 x^2]^{\frac{1}{2(q-1)}}}$$

of the function $e_{2q-1}^{2(1-q)x^2}$, that it satisfies the two-sided estimate

$$0 \leq e_{2q-1}^{2(1-q)x^2} \leq 1.$$

or, equivalently,

$$-1 \leq e_{2q-1}^{2(1-q)x^2} - 1 \leq 0. \tag{2.64}$$

On the other hand it follows from Corollary 2.19, Part 2, that

$$-2 \leq -2(\sin_q(x))^2 \leq 0. \tag{2.65}$$

Adding inequalities (2.64) and (2.65) and using the form of $h_q(x)$ in equation (2.61) one obtains $-3 \leq h_q(x) \leq 0$.   □

2.7. *q*-Gaussian Distribution

We recall that the mathematical expectation $\mathbb{E}(X)$ or the mean of a random variable X with a range $\mathcal{X}$ is defined by

$$\mathbb{E}[X] = \int_{\mathcal{X}} x f_X(x) dx,$$

and the variance $\mathrm{Var}(X)$ is defined by

$$\mathrm{Var}(X) = E[(X - E[X])^2] = \int_{\mathcal{X}} (x - E[X])^2 f_X(x) dx.$$

We also use notations μ_X or $\langle X \rangle$ for the mathematical expectation of X and σ_X^2 for the variance, assuming that σ_X is the standard deviation of X.

The Gaussian (or normal) distribution plays an important role in modern probability theory and its various applications in science and engineering. Let N be a random variable associated to the Gaussian distribution. The density function of the standard Gaussian distribution[1] N is given by

$$f_N(x) = G(x) = C e^{-\frac{1}{2}|x|^2}, \quad x \in \mathbb{R},$$

where $C = (2\pi)^{-1/2}$ is the normalizing constant, so that the integral of $G(x)$ over $\mathbb{R}$ equals 1. The density $G(x)$ describes the normal random variable N with mean zero and variance 1. The normal distribution $N(\mu, \sigma)$ with mean μ and variance σ^2 has the density given by

$$f_{N(\mu,\sigma)}(x) = G(x; \mu, \sigma) = \frac{1}{\sigma \sqrt{2\pi}} e^{-\frac{(x-\mu)^2}{2\sigma}}.$$

The relation between the normal random variable $N(\mu, \sigma)$ and its standard version N is $N(\mu, \sigma) = \sigma N + \mu$.

---

[1]Sometimes, loosely speaking, by the word "distribution" we mean the corresponding random variable.

Below we discuss a q-generalization of the Gaussian distribution based on the q-exponential function e_q^x, introduced in Section 2.3, and study some of its useful properties.

Definition 2.11. Let $q \in (-\infty, 3)$ and $\beta \in (0, \infty)$. The q-Gaussian random variable N_q is a random variable whose density function is given by

$$f_{N_q}(x) = G_q(\beta; x) = \frac{\sqrt{\beta}}{C_q} e_q^{-\beta x^2} \qquad (2.66)$$

with domains, depending on q, given below:

(1) if $q < 1$, then $G_q(\beta; x)$ is defined on the compact set $[-K_\beta, K_\beta]$, where $K_\beta = (\beta(1 - q))^{-1/2}$;
(2) if $1 \leq q < 3$, then $G_q(\beta; x)$ is defined on the whole real axis $\mathbb{R} = (-\infty, \infty)$.

In expression (2.66) C_q is the normalizing constant, i.e.,

$$C_q = \sqrt{\beta} \int_{-\infty}^{\infty} e_q^{-\beta x^2} dx = 2 \int_0^{\infty} e_q^{-x^2} dx. \qquad (2.67)$$

Remark 2.8.

(1) In Definition 2.11 it is assumed $-\infty < q < 3$. We note that if $q \geq 3$, then the q-Gaussian density $G_q(\beta; x)$ is not normalizable. That is, its integral over $\mathbb{R}$ diverges due to a non-integrable singularity at infinity. Thus, in this case the q-Gaussian random variable N_q is meaningless as a random variable.
(2) The range of the N_q in the case $q < 1$ is the finite closed interval $[-K_\beta, K_\beta]$. However, if $q \geq 1$, then the q-Gaussian random variable can take any value on the real axis $\mathbb{R}$. To unify domains in these two cases we also use the convention:

$$K_\beta = \begin{cases} [\beta(1 - q)]^{-1/2}, & \text{if } q < 1, \\ \infty, & \text{if } q \geq 1. \end{cases}$$

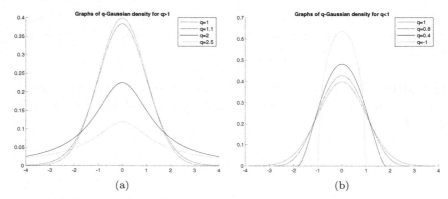

Fig. 2.1:   Graphs of $G_q(\beta; x)$ for $\beta = 0.5$. (a) $q \geq 1$, namely, $q = 1, 1.1, 2, 2.5$. (b) $q \leq 1$, namely, $q = 1, 0.8, 0.4, -1$. The support is the whole axis if $q \geq 1$, and is compact if $q < 1$. The red curve represents the standard Gaussian.

For example, using this convention we can express the distribution function of N_q in the form

$$F_{N_q}(x) = \int_{-K_\beta}^{x} G_q(\beta; y)dy.$$

(3) The q-Gaussian distribution is connected with Student's t-distribution (if $q > 1$) and with the r-distribution (if $q < 1$). See details in Section 2.9.

The graphs of $G_q(\beta; x)$ for some specific values of q are given in Fig. 2.1.

Proposition 2.25. *The explicit form of the normalizing constant C_q of the q-Gaussian density is given by*

$$C_q = \begin{cases} \dfrac{2\sqrt{\pi}\,\Gamma\left(\frac{1}{1-q}\right)}{(3-q)\sqrt{1-q}\,\Gamma\left(\frac{3-q}{2(1-q)}\right)}, & -\infty < q < 1, \\[4mm] \sqrt{\pi}, & q = 1, \\[4mm] \dfrac{\sqrt{\pi}\,\Gamma\left(\frac{3-q}{2(q-1)}\right)}{\sqrt{q-1}\,\Gamma\left(\frac{1}{q-1}\right)}, & 1 < q < 3, \end{cases} \qquad (2.68)$$

where $\Gamma(\cdot)$ is Euler's gamma-function: $\Gamma(s) = \int_0^\infty x^{s-1}e^{-x}dx$.

Proof. Let $q \in (-\infty, 1)$. Changing the integration variable $x = x(s)$, where

$$x(s) = \sqrt{\frac{s}{1-q}}, \quad 0 \leq s \leq 1,$$

reduces the integral on the right of (2.67) to

$$C_q = 2 \int_0^{K_\beta} e_q^{-x^2} dx$$

$$= 2 \int_0^{1/\sqrt{1-q}} [1 - (1-q)x^2]^{\frac{1}{1-q}} dx$$

$$= \frac{1}{\sqrt{1-q}} \int_0^1 s^{-\frac{1}{2}} (1-s)^{\frac{1}{1-q}} ds = \frac{B\left(\frac{1}{2}, \frac{2-q}{1-q}\right)}{\sqrt{1-q}}, \tag{2.69}$$

where $B(\alpha, \beta)$ is Euler's beta function:

$$B(\alpha, \beta) = \int_0^1 s^{\alpha-1}(1-s)^{\beta-1} ds.$$

Using the relation

$$B(\alpha, \beta) = \frac{\Gamma(\alpha)\,\Gamma(\beta)}{\Gamma(\alpha+\beta)} \tag{2.70}$$

between Euler's beta- and gamma-functions and the property $\Gamma(\alpha + 1) = \alpha\,\Gamma(\alpha)$ of the gamma-function, we obtain (2.68) in the case $q < 1$.

Now assume $1 < q < 3$. Similar to the previous case the change of variable $(q-1)x^2 = s/(1-s)$ reduces the integral on the right of (2.67) to

$$C_q = \frac{1}{\sqrt{q-1}} \int_0^1 (1-s)^{\frac{1}{q-1} - \frac{1}{2} - 1} s^{\frac{1}{2}-1} ds$$

$$= \frac{1}{\sqrt{q-1}} B\left(\frac{1}{q-1} - \frac{1}{2}, \frac{1}{2}\right),$$

which implies (2.68) for $q > 1$.

The case $q = 1$ is obvious. □

Proposition 2.26. *For the moment of order m of the q-Gaussian random variable N_q with the density function defined in (2.66) the following formulas hold:*

(1) *If $q < 1$, then*

$$\mathbb{E}[N_q^m] = \int_{-K_\beta}^{K_\beta} x^m G_q(\beta; x)\, dx$$

$$= \begin{cases} \dfrac{(3-q)\Gamma\left(\frac{3-q}{2(1-q)}\right)\Gamma\left(k+\frac{1}{2}\right)}{2\sqrt{\pi}\beta^k(1-q)^{k+1}\Gamma\left(k+\frac{1}{1-q}+\frac{3}{2}\right)}, & \text{if } m = 2k, \\[4mm] 0, & \text{if } m = 2k+1, \end{cases}$$

$$(2.71)$$

for all $k = 0, 1, \ldots$.

(2) *If $1 < q < \frac{m+3}{m+1}$, then*

$$\mathbb{E}[N_q^m] = \int_{-\infty}^{\infty} x^m G_q(\beta; x)\, dx$$

$$= \begin{cases} \dfrac{\Gamma\left(k+\frac{1}{2}\Gamma\left(\frac{1}{q-1}-k-\frac{1}{2}\right)\right)}{\pi\beta^k(q-1)^k\Gamma\left(\frac{3-q}{2(q-1)}\right)}, & \text{if } m = 2k, \\[4mm] 0, & \text{if } m = 2k+1, \end{cases}$$

$$(2.72)$$

for all $k = 0, 1, \ldots$.

Proof. The proof is similar to the proof of Proposition 2.25. Namely, if $q < 1$, then we use the change $y = \beta(1-q)\,x^2$, and if $1 < q < (m+3)/(m+1)$, then the change $y/(1-y) = \beta(1-q)\,x^2$. The fact that the moments vanish for odd m follows from the symmetry of $G_q(\beta; x)$ about the origin. If $(m+3)/(m+1) \le q < 3$, then the integral in (2.72) diverges due to a non-integrable slow decay of the integrand at infinity.   □

Remark 2.9. The mth moment of the q-Gaussian diverges if $q > \frac{m+3}{m+1}$.

Corollary 2.8. *The q-Gaussian random variable N_q possesses the following properties:*

(1) *it has a mean $\mu_{N_q} = 0$;*
(2) *its variance*

$$
\text{Var}(N_q) =
\begin{cases}
\dfrac{(3-q)\Gamma\left(\frac{3-q}{2(1-q)}\right)}{2\beta(1-q)(2-q)\Gamma\left(\frac{2-q}{2(1-q)}\right)}, & \text{if } q < 1, \\[4mm]
\dfrac{\Gamma\left(\frac{5-3q}{2(q-1)}\right)}{2\beta(q-1)\Gamma\left(\frac{3-q}{2(q-1)}\right)}, & \text{if } 1 < q < \frac{5}{3}.
\end{cases}
$$

(3) *its variance diverges if $q \geq \frac{5}{3}$;*
(4) *For $q < 1$ its moment generating function is*

$$
M_{N_q}(z) = \frac{(3-q)\Gamma\left(\frac{3-q}{2(1-q)}\right)}{2\sqrt{\pi}} \sum_{k=0}^{\infty} \frac{\Gamma(k+\frac{1}{2})}{\beta^k(1-q)^{k+1}\Gamma(k+\frac{1}{1-q}+\frac{3}{2})} z^{2k},
$$

$$(2.73)$$

which converges in the interval $\left(-\sqrt{\beta(1-q)}, \sqrt{\beta(1-q)}\right)$.

Proof. We only need to show that the convergence of the moment generating function for $q < 1$ on the interval $[-K_\beta, K_\beta]$. All other parts follow directly from Proposition 2.26. Let

$$
a_k(q, \beta) = \frac{\Gamma(k+\frac{1}{2})}{\beta^k(1-q)^{k+1}\Gamma(k+\frac{1}{1-q}+\frac{3}{2})}, \quad k = 0, 1, \ldots.
$$

It is not hard to see that

$$
\lim_{k\to\infty} \frac{a_{k+1}(q, \beta)}{a_k(q, \beta)} = \frac{1}{\beta(1-q)}.
$$

Hence, the series in (2.73) converges if $|z|^2 < \beta(1-q)$, proving the statement. $\qquad\square$

Remark 2.10. If $q > 1$, then the q-Gaussian random variable does not have a moment generating function. The reason is in this case the moment of order m exists only if $q < (m+3)/(m+1)$. For all the moments to exist one needs the condition $q \leq \lim_{m\to\infty}(m+3)/(m+1) = 1$, which contradicts to $q > 1$.

It follows from the latter remark that standard moments of the q-Gaussian random variable are not effective in the case $q > 1$. Therefore, below we introduce the notion of q-moments, which play an important role in the study of q-Gaussian random variables in the case $q > 1$. For this purpose we introduce the notion of *escort density function* (denoted by $f_q(x)$) associated with a random variable X whose density function is $f_X(x)$. Namely, by definition, the escort density function of X is defined by

$$f_q(x) = \frac{[f_X(x)]^q}{\int_{\mathcal{X}}[f_X(x)]^q dx}, \qquad (2.74)$$

where $\mathcal{X}$ is the range of X.

Definition 2.12. Let a probability density function $f(x)$, $x \in \mathcal{X} \subseteq \mathbb{R}$, of a random variable X be given. The integral

$$\mathbb{E}_{m,q}[X^m] = \int_{\mathcal{X}} x^m f_{mq-(m-1)}(x)dx,$$

is called a q-moment of order m of the random variable X. In particular, the q-moment of first order

$$\mu_q = \mathbb{E}_{1,q}[X] = \int_{\mathcal{X}} x f_q(x)dx$$

is called a q-mean of X, and the expression

$$\sigma_{2q-1}^2 = E_{2,q}[(X - \mu_q)^2] = \int_{\mathcal{X}} (x - \mu_q)^2 f_{2q-1}(x)dx,$$

is called a $(2q-1)$-variance of X.

The q-moment of order m of the random variable X can be written in the form

$$\mathbb{E}_{m,q}[X^m] = \frac{\int_{\mathcal{X}} x^m [f(x)]^{mq-(m-1)} dx}{\nu_{m,q}[X]}, \qquad (2.75)$$

where

$$\nu_{m,q}[X] = \int_{\mathcal{X}} [f(x)]^{mq-(m-1)} dx. \tag{2.76}$$

Proposition 2.27. *For the q-moment of order m of the q-Gaussian random variable N_q the following formulas hold:*

(1) *If $q < 1$, then $\mu_q = \mathbb{E}_{1,q}[N_q]$ exists end equals 0,*
(2) *If $m \geq 2$ and $1 - \frac{1}{m-1} < q < 1$, then*

$$\mathbb{E}_{m,q}[N_q^m] = \begin{cases} \dfrac{\Gamma(k+\frac{1}{2})\Gamma\left(\frac{1}{1-q} - 2k + \frac{3}{2}\right)}{\sqrt{\pi}q^k(1-q)^k\Gamma\left(\frac{1}{1-q} - k + \frac{3}{2}\right)}, & \text{if } m = 2k, \\[4mm] 0, & \text{if } m = 2k+1, \end{cases} \tag{2.77}$$

for all $k = 1, 2, \dots$.
(3) *If $1 < q < 3$, then*

$$\mathbb{E}_{m,q}[N_q^m] = \begin{cases} \dfrac{\Gamma(k+\frac{1}{2})\Gamma\left(k + \frac{1}{q-1} - \frac{1}{2}\right)}{\sqrt{\pi}\beta^k(q-1)^k\Gamma\left(2k + \frac{1}{q-1} - \frac{1}{2}\right)}, & \text{if } m = 2k, \\[4mm] 0, & \text{if } m = 2k+1. \end{cases} \tag{2.78}$$

for all $k = 0, 1, \dots$.

Proof.

(1) Consider the q-moment of first order of the q-Gaussian random variable

$$\mathbb{E}_{1,q}[N_q] = \frac{\beta^{q/2}}{\nu_{1,q}[N_q]C_q^q} \int_{-K_\beta}^{K_\beta} x\left[1 - (1-q)\beta x^2\right]^{\frac{q}{1-q}} dx.$$

This integral obviously converges if $0 \leq q < 1$. If $q < 0$, then $q/(1-q) < 0$. Therefore, the integral converges if $-q/(1-q) < 1$. But this inequality is valid for all $q < 0$.

(2) Let $m \geq 2$ and $1 - (m-1)^{-1} < q < 1$. Then, due to (2.75) and (2.76), the q-moment of order m of N_q has the form

$$\mathbb{E}_{m,q}[N_q] = \frac{1}{\nu_{m,q}[N_q]} \left(\frac{\sqrt{\beta}}{C_q}\right)^{mq-m+1} \int_{-K_\beta}^{K_\beta} \frac{x^m}{[1-(1-q)\beta x^2]^{m-\frac{1}{1-q}}} dx,$$

and converges if $m - (1-q)^{-1} < 1$. The letter implies $q > 1 - (m-1)^{-1}$. Obviously, the integral vanishes if m is an odd number. Let $m = 2k, k = 1, 2, \ldots$. Then we have

$$\mathbb{E}_{2k,q}[N_q] = \frac{\int_0^{K_\beta} \frac{x^m}{[1-(1-q)\beta x^2]^{m-\frac{1}{1-q}}} dx}{\int_0^{K_\beta} \frac{dx}{[1-(1-q)\beta x^2]^{m-\frac{1}{1-q}}}}.$$

Further, using the change of variable $\beta(1-q)x^2 = s$ in both integrals, we obtain formula (2.77).

(3) Let $1 < q < 3$. Then the q-moment of order m of N_q, due to formulas (2.75) and (2.76), has the form

$$\mathbb{E}_{m,q}[N_q] = \frac{1}{\nu_{m,q}[N_q]} \left(\frac{\sqrt{\beta}}{C_q}\right)^{mq-m+1}$$
$$\times \int_{-\infty}^{\infty} \frac{x^m}{[1+(q-1)\beta x^2]^{m+\frac{1}{q-1}}} dx,$$

and converges if $m + 2/(q-1) > 1$. But this inequality is valid for all $1 < q < 3$ and $m = 0, 1, \ldots$. If $m = 2k+1, \ k = 0, 1, \ldots$, then the q-moment of order m of N_q vanishes. Let $m = 2k, k = 0, 1, \ldots$. Then the q-moment of order $m = 2k$ takes the form

$$\mathbb{E}_{m,q}[N_q] = \frac{\int_0^{\infty} \frac{x^m}{[1+(q-1)\beta x^2]^{m+\frac{1}{q-1}}} dx}{\int_0^{\infty} \frac{dx}{[1+(q-1)\beta x^2]^{m+\frac{1}{q-1}}}}.$$

Further, using the change of variable $\beta(q-1)x^2 = s/(1-s)$ in both integrals, we obtain formula (2.78). $\qquad\square$

Corollary 2.9. *The q-Gaussian random variable N_q defined by the density function $G_q(\beta; x)$ in (2.66) possesses the following properties:*

(1) *it has the q-mean $\mu_q = 0$;*

(2) *its $(2q-1)$-variance is*

$$\mathbb{E}_{2,q}[N_q] = \sigma^2_{2q-1} = \frac{1}{\beta(1+q)},$$

if $0 < q < 3$;

(3) *its $(2q-1)$-variance diverges if $q \leq 0$;*

(4) *For $1 < q < 3$ its q-moment generating function is*

$$\mathcal{M}_{N_q}(z) = \frac{1}{\sqrt{\pi}} \sum_{k=0}^{\infty} \frac{A_{2k}\Gamma(k+\frac{1}{2})\Gamma(k+\frac{1}{q-1}-\frac{1}{2})}{\beta^k(q-1)^k(2k)!\Gamma(2k+\frac{1}{q-1}-\frac{1}{2})} z^{2k},$$

which converges in the interval $|z| < \sqrt{\beta(q-1)}$. Here $A_{2k}, k = 0, 1, \ldots$, are numbers defined in (2.38).

Proof. The proof is similar to the proof of Corollary 2.8. $\qquad\square$

Remark 2.11. If $q < 1$, then the q-Gaussian random variable does not have a q-moment generating function. The reason is in this case the moment of order m exists only if $q > 1 - 1/(m-1), m = 2, 3, \ldots$. For all the moments to exist one needs the condition $q \geq \lim_{m\to\infty}(1 - 1/(m-1)) = 1$, which contradicts to $q < 1$.

Finally, the proposition below follows directly from the definition of the q-Gaussian.

Proposition 2.28.

(1) *Let $q > 1$. Then*

$$G_q(x) = O\left(\frac{1}{|x|^{\frac{2}{q-1}}}\right), \quad |x| \to \infty;$$

(2) *$q < 1$. Then $G_q(x)$ has the compact support*

$$\operatorname{supp}\,[G_q] \equiv \left[-\frac{1}{\sqrt{1-q}}, \frac{1}{\sqrt{1-q}}\right].$$

2.8. Multivariate q-Gaussian Distribution

Consider a d-dimensional random vector $\mathcal{N}_q = (N_{1,q}, \ldots, N_{d,q})$ of independent q-Gaussian random variables $N_{j,q}, j = 1, \ldots, d$.

It follows from the fact that the density of joint independent random variables is the product of densities and Proposition 2.7 that the density function of $\mathcal{N}_q$ is

$$
\begin{aligned}
f_{\mathcal{N}_q}(x_1, \dots, x_d) &= \prod_{j=1}^{d} f_{N_{j,q}}(x_j) \\
&= \left(\frac{\sqrt{\beta}}{C_q} \right)^d \prod_{j=1}^{d} e_q^{-\beta x_j^2} \\
&= \frac{\beta^{d/2}}{(C_q)^d} \exp_q \left(\sum_{j=1}^{d} {}_q(-\beta x_j^2) \right),
\end{aligned}
\tag{2.79}
$$

with the q-sum in the exponent. It is easy to verify that the mean of $\mathcal{N}_q$ is $\mu = (0, \dots, 0)$, and the covariance matrix is diagonal with all diagonal entries equal to the variance of $N_{1,q}$.

As is seen from (2.79), the density function of the multivariate q-Gaussian distribution with independent components can be written through a q-exponential function. Below we introduce a d-dimensional q-Gaussian random vector, components of which depend in a special way. Let Σ be a positive definite symmetric $(d \times d)$-matrix, which means that all the eigenvalues $\lambda_1, \dots, \lambda_d$ are positive.

Definition 2.13. Let $-\infty < q < 1 + 2/d$. A d-dimensional q-Gaussian random vector $\mathbb{N}_q$ is a random vector whose probability density function is defined as

$$
G_q(\Sigma; x) = \frac{\sqrt{|\Sigma|}}{C_{d,q}} e_q^{-(\Sigma x, x)}, \quad x \in \mathbb{R}^d,
\tag{2.80}
$$

where Σ is a positive definite $d \times d$ matrix, $|\Sigma|$ is the determinant of the matrix Σ, the symbol (ξ, x) for $\xi, x \in \mathbb{R}^d$ means $(\xi, x) = \xi_1 x_1 + \cdots + \xi_d x_d$, and

$$
C_{d,q} = \sqrt{|\Sigma|} \int_{\mathbb{R}^d} e_q^{-(\Sigma x, x)} dx
$$

is the normalizing constant.

Proposition 2.29. *The normalizing constant $C_{d,q}$ for the q-Gaussian density function defined in (2.80) can be expressed as*

$$
C_{d,q} = \begin{cases}
\dfrac{\pi^{\frac{d}{2}}\Gamma\left(\frac{1}{1-q}\right)}{(1-q)^{\frac{d}{2}+1}\Gamma\left(\frac{d}{2}+\frac{2-q}{1-q}\right)}, & \text{if } q < 1, \\[4mm]
\pi^{d/2}, & \text{if } q = 1, \\[3mm]
\dfrac{\pi^{d/2}\Gamma\left(\frac{1}{q-1}-\frac{d}{2}\right)}{(q-1)^{d/2}\Gamma(\frac{1}{q-1})}, & \text{if } 1 < q < 1+\frac{2}{d}.
\end{cases} \tag{2.81}
$$

Proof. Let $q < 1$. We have

$$
C_{d,q} = |\Sigma| \int_{(\Sigma x, x) \le \frac{1}{1-q}} \left[1 - (1-q)(\Sigma x, x)\right]^{\frac{1}{1-q}} dx.
$$

It is known that for any positive definite matrix Σ there exists an orthogonal matrix U such that $\Sigma = U^T \Lambda U$, where U^T is the transpose of U and Λ is a diagonal matrix. Moreover, all the diagonal entries are positive and, in fact, eigenvalues of Σ. Let $y = Ux$. Then

$$
(\Sigma x, x) = (U^T \Lambda U x, x) = (\Lambda U x, U x) = (\Lambda y, y).
$$

Therefore,

$$
C_{d,q} = \sqrt{|\Sigma|} \int_{(\Lambda y, y) \le \frac{1}{1-q}} \left[1 - (1-q)(\Lambda y, y)\right]^{\frac{1}{1-q}} d\, U^T y.
$$

Further, since $\Lambda y = (\lambda_1 y_1, \dots, \lambda_d y_d)$, where $\lambda_1, \dots, \lambda_d$ are diagonal entries of Λ, the change of variables $z_k = \sqrt{\lambda_k(1-q)}y_k$, $k = 1, \dots, d$, gives

$$
\begin{aligned}
C_{d,q} &= \sqrt{|\Sigma|} \int_{|z^2| \le 1} [1 - |z|^2]^{\frac{1}{1-q}} \frac{d\, U^T z}{\sqrt{|\Sigma|}(1-q)^{d/2}} \\
&= \frac{A_d}{(1-q)^{d/2}} \int_0^1 r^{d-1}(1-r^2)^{\frac{1}{1-q}} dr, \tag{2.82}
\end{aligned}
$$

where

$$
A_d = \frac{2\pi^{d/2}}{\Gamma(d/2)}, \tag{2.83}
$$

is the area of the $(d-1)$-dimensional sphere in $\mathbb{R}^d$. The expression $U^T z$ in the first integral in (2.82) means a rotation of the unit ball in $\mathbb{R}^d$, which does not effect in the last integral in (2.82). Now making use of the change of variable $r^2 = s$ in the integral in (2.82), we have

$$
\begin{aligned}
C_{d,q} &= \frac{\pi^{d/2}}{(1-q)^{d/2}\Gamma(d/2)} \int_0^1 s^{\frac{d}{2}-1}(1-s)^{\frac{1}{1-q}} ds \\
&= \frac{\pi^{d/2}}{(1-q)^{d/2}\Gamma(d/2)} B\left(\frac{d}{2}, \frac{1}{1-q}+1\right) \\
&= \frac{\pi^{\frac{d}{2}}\Gamma(\frac{1}{1-q})}{(1-q)^{\frac{d}{2}+1}\Gamma(\frac{d}{2}+\frac{2-q}{1-q})},
\end{aligned}
$$

proving (2.81) in the case $q < 1$. Here $B(\cdot, \cdot)$ is Euler's beta-function and we used the relationship (2.70) in the calculation.

In the case $q = 1$, as is known, $C_{d,q} = \pi^{d/2}$.

Now assume $1 < q < 1 + 2/d$. We have

$$
C_{d,q} = \sqrt{|\Sigma|} \int_{\mathbb{R}^d} \frac{dx}{[1+(q-1)(\Sigma x, x)]^{\frac{1}{q-1}}}.
$$

Again we use the change of variables $x = Uy$, where U is an orthogonal matrix such that $U^T \Sigma U = \Lambda$, the diagonal matrix with entries $\lambda_k, k = 1, \dots, d$, being eigenvalues of Σ. Then the latter integral takes the form

$$
C_{d,q} = \sqrt{|\Sigma|} \int_{\mathbb{R}^d} \frac{d\,Uy}{[1+(q-1)(\sum_{k=1}^d \lambda_k y_k^2)]^{\frac{1}{q-1}}}.
$$

Now setting $\sqrt{(q-1)\lambda_k} y_k = z_k, k = 1, \dots, d$, and taking into account $|\Sigma| = |\Lambda| = \lambda_1 \dots \lambda_d$, we have

$$
C_{d,q} = \sqrt{|\Sigma|} \int_{\mathbb{R}^d} \frac{1}{[1+|z|^2]^{\frac{1}{q-1}}} \frac{d\,Uz}{\sqrt{|\Sigma|}(q-1)^{d/2}}.
$$

Then using spherical coordinates reduces the latter intergral to

$$
C_{d,q} = \frac{A_d}{(q-1)^{d/2}} \int_0^\infty \frac{r^{d-1}}{(1+r^2)^{\frac{1}{q-1}}} dr,
$$

where A_d is the surface area of the $(d-1)$-dimensional unit sphere given in (2.83). Further, changing $r^2 = s/(1-s)$, we have

$$
\begin{aligned}
C_{d,q} &= \frac{\pi^{d/2}}{(q-1)^{d/2}\Gamma(d/2)} \int_0^1 s^{\frac{d}{2}-1}(1-s)^{\frac{1}{q-1}-\frac{d}{2}-1}ds \\
&= \frac{\pi^{d/2}}{(q-1)^{d/2}\Gamma(d/2)} B\left(\frac{d}{2}, \frac{1}{q-1}-\frac{d}{2}\right),
\end{aligned}
\tag{2.84}
$$

which is meaningful, since $\frac{1}{q-1} - \frac{d}{2} > 0$ due to the condition $q < 1 + 2/d$. Finally, using formula (2.70) in (2.84), we obtain

$$
C_{d,q} = \frac{\pi^{d/2}\Gamma(\frac{1}{q-1}-\frac{d}{2})}{(q-1)^{d/2}\Gamma(\frac{1}{q-1})}.
$$
$\square$

Remark 2.12. Note that the d-dimensional q-Gaussian density function $G_q(\Sigma; x)$ is not normalizable if $q \geq 1 + 2/d$.

2.9. Additional Notes

(1) q-Algebra with operations $\otimes_q$, $\oplus_q$

As noted above, the q-algebra based on the operations $\oplus_q$ and $\otimes_q$ was introduced relatively recently. The operation $\otimes_q$ was introduced simultaneously and independently in papers Nivanen *et al.* (2003) and Borges (2004). In Section 2.5, we studied properties of these operations in the field $\mathbb{C}$ of complex numbers. However, in most applications one needs to stay in the field of real numbers $\mathbb{R}$. Obviously, $x \oplus_q y = x + y + (1-q)xy$ is defined for any $x, y \in \mathbb{R}$. However,

$$
x \otimes_q y = [x^{1-q} + y^{1-q} - 1]^{1/(1-q)},
$$

is real for not all $x, y \in \mathbb{R}$, depending on the value of q. There are some exclusive values of q, for which this product is real for all $x, y \in \mathbb{R}$. For example, $q = 1$, $q = 0$, or $q = 2/3$ are such values. The q-product of $x, y \in \mathbb{R}$ is real for all q, only if x and y satisfy the inequality

$$
x^{1-q} + y^{1-q} \geq 1.
\tag{2.85}
$$

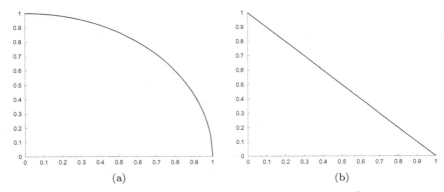

Fig. 2.2: The real valued q-product is defined for all $(x, y) \in \mathbb{R}^2$ on, above and to the right of the boundary curves. The boundary curve on (a) corresponds to $q = -1$ and is the graph of $y = \sqrt{1 - x^2}$. The line on (b) corresponds to $q = 0$ and is the graph of $y = 1 - x$. Both graphs join points $(0, 1)$ and $(1, 0)$ in $\mathbb{R}^2$.

The boundary of the set $(x, y) \in \mathbb{R}_+$ satisfying the latter inequality is a curve joining the points $(0, 1)$ in the y-axis and $(1, 0)$ in the x-axis and defined by the function $y = [1 - x^{1-q}]^{\frac{1}{1-q}}$. For example, if $q = -1$, this is a part of the unit circle in the first quadrant; if $q = 0$ it is a line joining points $(0, 1)$ and $(1, 0)$ (see Fig. 2.2); if $q = 1$, then the inequality (2.85) reduces to $2 \geq 1$, implying validity of the q-product for all $x, y \in \mathbb{R}$.

Note also that if $q > 1$, then this curve is located in the region $x > 0, y > 0$, having the vertical line $x = 1$ and the horizontal line $y = 1$ as asymptotes; see Fig. 2.3(b).

(2) Other forms of q-exponential and q-logarithmic functions

There are many other forms of the q-exponential and q-logarithmic functions. Attempts to generalize *logarithmic* and *exponential* functions goes back to Leonhard Euler[2] and Carl Friedrich Gauss.[3] L. Euler introduced a q-generalized logarithmic and exponential functions. The idea used by him to generalize q-exponential is as

---

[2]L. Euler, 1707–1783.
[3]C.F. Gauss, 1777–1855.

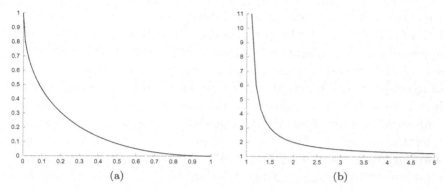

Fig. 2.3: The boundary curve on (a) corresponds to $q = 1/2$ and is the graph of $y = (1 - \sqrt{x})^2$. It joins points $(0, 1)$ and $(1, 0)$. The curve on (b) corresponds to $q = 2$, and is the graph of $y = x/(x - 1)$. This curve has asymptotes $x = 1$ and $y = 1$.

follows. Let the numbers $[k]_q$, $[\tilde{k}]_q$ be defined by

$$[k]_q = \frac{1 - q^k}{1 - q} = 1 + q + \cdots + q^{k-1},$$

$$[\tilde{k}]_q = \frac{1 - \frac{1}{q}}{1 - \frac{1}{q}} = 1 + \frac{1}{q} + \cdots + \frac{1}{q^{k+1}}.$$

Notice that if $q = 1$, then $[k]_1 = k$, $[\tilde{k}]_1 = k$. Now introduce

$$[k]_q! = [1]_q! \cdot [2]_q! \cdots \cdot [k]_q!, \quad [0]_q! = 1.$$

and

$$[\tilde{k}]_q! = [\tilde{1}]_q! \cdot [\tilde{2}]_q! \cdots \cdot [\tilde{k}]_q!, \quad [\tilde{0}]_q! = 1.$$

By these factorial notations two q-generalizations of the exponential function e^x are defined as

$$\mathbf{exp}_q(x) = \sum_{k=0}^{\infty} \frac{x^k}{[k]_q!}, \quad \mathbf{Exp}_q(x) = \sum_{k=0}^{\infty} \frac{x^k}{[\tilde{k}]_q!}$$

Obviously, $\mathbf{Exp_q(x)} = \mathbf{exp_{1/q}(x)}$. If $0 < q < 1$, then the series representing $\mathbf{exp_q(x)}$ converges in the interval $x < (1 - q)^{-1}$. Its inverse is q-logarithm, $\mathbf{ln}_q(x)$. Obviously, these two functions coincide with standard exponential and logarithmic functions if $q = 1$, i.e., $\exp_1(x) = e^x$ and $\ln_1(x) = \ln(x)$. These functions found many applications in mathematics, physics, and other fields of modern science. For example, they played a central role in the development of hypergeometric functions, which are an important tool in physics, including statistical physics, quantum mechanics, etc. For the history and various generalizations of Euler's q-exponential functions we refer the reader to the following sources Ernst (2003, 2012), Diaz and Pariguan (2009), Cieśliński (2011, 2012), Lu (2009), Bohner and Guseinov (2010), Zhang (2014), Koelink and Van Assche (2009), and Blitvić (2012).

(3) The q-exponential and q-logarithmic functions

The q-exponential and q-logarithmic functions defined in Eqs. (2.11) and (2.13) were first introduced in 1994 (Tsallis, 1994) and presented in a seminar of the High Energy Group of the Michigan State University.

The fundamental relationship

$$e_q^{x \oplus y} = e_q^x \cdot e_q^y,$$

valid for the q-exponential function, was used to prove the Einstein's likelihood principle, first appeared in Einstein's paper (Einstein, 1910), for generalized q-entropies S_q; see details in Tsallis and Haubold (2015).

We note that the q-exponential and q-logarithmic functions introduced and studied in this chapter are totally different from the Euler's q-exponential and q-log functions. Therefore, they are not generalizations of Euler's q-exponential and q-logarithmic functions. As we will see in next chapters, the present q-exponential and q-log functions have found many exciting applications (Tsallis, 2009a; Tsallis and Haubold, 2015; Naudts, 2010), among many others. In Chapter 6, we will present some further applications of this relatively new theory for modeling processes in a low temperature environment (cold atoms), processes in fractal media, biophysics, neurophysics, astrophysics, etc.

(4) *q*-Trigonometric functions

q-Trigonometric functions are defined through the *q*-generalized Euler's formulas. In the literature there are various versions of *q*-trigonometric functions defined by Euler's *q*-exponential function and their generalizations (see, e.g., Ernst, 2012 and references therein). Borges (1998) studied some properties of the *q*-trigonometric functions introduced in Section 2.6. One can easily verify that the first and second derivatives of the *q*-sine function are

$$\frac{d\sin_q(x)}{dx} = -\sin_{2-1/q}(qx), \qquad (2.86)$$

and

$$\frac{d^2\sin_q(x)}{dx^2} = -q\sin_q\left((2q-1)x\right). \qquad (2.87)$$

Similarly, the first and second derivatives of the *q*-cosine function are

$$\frac{d\cos_q(x)}{dx} = \cos_{2-1/q}(qx), \qquad (2.88)$$

and

$$\frac{d^2\cos_q(x)}{dx^2} = -q\cos_q((2q-1)x). \qquad (2.89)$$

It follows from (2.86)–(2.89) that $\sin_q(x)$ and $\cos_q(x)$ satisfy the functional-differential equation

$$\frac{d^2f(x)}{dx^2} + qf((2q-1)x) = 0, \quad x \in \mathbb{R}. \qquad (2.90)$$

Functional-differential equations frequently emerge in the theory of *q*-calculus. For instance, in Chapter 3, we derive a functional-differential equation for the density functions of *q*-Gaussian random variables.

(5) *q*-Gaussian distribution for $\beta < 0$

As we have seen in Section 2.7, the *q*-Gaussian is normalizable for all values of $q < 3$ if $\beta > 0$. The *q*-Gaussian stays meaningful for $q > 3$ as well, if $\beta < 0$. As noted in Rodriguez and Tsallis (2014),

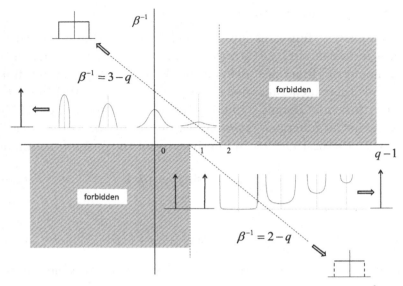

Fig. 2.4: Graphs of q-Gaussian density functions $G_q(\beta; x)$ on the $(\beta^{-1}, q-1)$-plane for different values of β and q. When $\beta > 0$ the q-Gaussians have compact support if $q < 1$ and are defined on $(-\infty, \infty)$ if $1 \leq q < 3$. For $\beta < 0$ the corresponding q-Gaussian densities defined for $q > 2$ and form U-shaped distributions over open intervals.

there is an entire region for q-Gaussians (see Fig. 2.4 taken from Rodriguez and Tsallis (2014)) with $q > 2$, where β must be negative, thus providing U-shaped distributions (which do appear in some specialized applications).

Expanding the region of meaningful cases of (q, β), and rewriting the q-Gaussian in the form $G_q(\beta; x) = C(\beta, q) \exp_q(-\beta x^2)$, we have that the normalizing constant $C(\beta, q)$ has the form (see (2.66) and (2.68))

$$C(\beta, q) = \begin{cases} \dfrac{(3-q)\sqrt{\beta(1-q)}\,\Gamma\left(\frac{3-q}{2(1-q)}\right)}{2\sqrt{\pi}\,\Gamma\left(\frac{1}{1-q}\right)}, & -\infty < q < 1, \beta > 0 \\[4pt] & \text{and } q > 2, \beta < 0, \\[10pt] \dfrac{\sqrt{\beta(q-1)}\,\Gamma\left(\frac{1}{q-1}\right)}{\sqrt{\pi}\,\Gamma\left(\frac{3-q}{2(q-1)}\right)}, & 1 < q < 3. \end{cases}$$

Moreover, the support of the q-Gaussian is $(-\infty, \infty)$ for $1 \leq q < 3$, $\beta > 0$, and $[-(\beta(1-q))^{-1}, (\beta(1-q))^{-1}]$ for $-\infty < q < 1, \beta > 0$, and for $q > 2, \beta < 0$ (excluding the interval endpoints in this case).

(6) Connections of the q-Gaussian distribution with Student's t-distribution and the r-distribution

As shown in Souza and Tsallis (2019), the q-Gaussian distribution for some values of q is connected with Student's t-distribution T and the r-distribution R. The density function of T with m degrees of freedom is given by Korn and Korn (1968)

$$F_T(t) = \frac{1}{\sqrt{m\pi}} \frac{\Gamma(\frac{1+m}{2})}{\Gamma(\frac{m}{2})} \frac{1}{(1 + \frac{t^2}{m})^{(1+m)/2}}, \quad t \in \mathbb{R}.$$

Setting

$$q = \frac{3+m}{1+m} \quad \text{and} \quad t = \sqrt{\frac{2\beta m}{1+m}} x,$$

in the q-Gaussian density function $G_q(\beta; x)$, one obtains $F_T(t)$, establishing the connection between the q-Gaussian and Student's t-distributions for $q > 1$.

For $q < 1$ a similar connection holds between the q-Gaussian distribution and R with $\ell - 2$ $(\ell > 4)$ degrees of freedom, the density function of which is given by Korn and Korn (1968)

$$f_R(r) = \frac{\Gamma(\frac{\ell-1}{2})}{\sqrt{\pi}\Gamma(\frac{\ell-2}{2})} (1 - r^2)^{\ell-4}/2, \quad -1 \leq r \leq 1, \ \ell > 4.$$

One can easily verify that substitutions $q = (\ell - 6)/(\ell - 4)$ and $r = \sqrt{\frac{2\beta}{\ell-4}} x$ reduce the q-Gaussian density $G_q(\beta; x)$ to $f_R(r)$.

(7) Comparison of α-stable and q-Gaussian distributions

The α-stable (see the definition in Section 1.5 of Chapter 1) and q-Gaussian distributions both have density functions with a power law decay at infinity. On the other hand, these two distributions

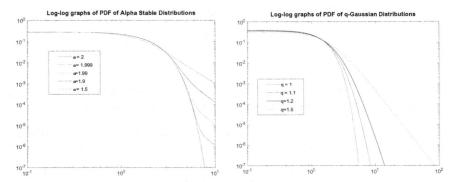

Fig. 2.5:   Log-log graphs of the probability density functions of symmetric α-stable distributions for $\alpha = 2, 1.5, 1, 0.75, 0.5$, and log-log graphs of the probability density functions of q-Gaussian distributions for $q = 1, 1.1, 1.2, 1.5$.

describe random variables with distinct natures. Indeed, one can see in Fig. 2.5 that log-log graphs of density functions of α-stable distributions have inflexion points for all $0 < \alpha < 2$. At the same time log-log graphs of q-Gaussian densities do not have inflexion points in a finite interval. For further details we refer the reader to paper Tsallis and Arenas (2014).

(8) General multivariate q-Gaussian density functions

In Section 2.8, we introduced two versions of the multivariate q-Gaussian density functions. q-Gaussians play an important role in the q-generalized central limit theorem discussed in Chapter 4. More general version of multivariate q-Gaussian functions are studied in Amari and Ohara (2011) in connection with information geometry. Vignat and Plastino (2005, 2006, 2007a) proved some nice properties of multivariate q-Gaussian functions.

For the q-Gaussian random vector $\mathcal{N}_q = (N_{1,q}, \ldots, N_{d,q})$ with independent components $N_{j,q}$, $j = 1, \ldots, d$, it is not hard to calculate the mean vector, covariance matrix, and higher-order moments. However, for the q-Gaussian density function $\mathbb{N}_q$, defined in Definition 2.13, with dependent components, the calculation of its covariance matrix and higher order moments require marginal densities. Let

$$G_{j,q}(\Sigma; x_j) = \int_{\mathbb{R}_j^{d-1}} G_q(\Sigma; x)dx_1 \ldots dx_{j-1}dx_{j+1} \ldots dx_d,$$

$$1 \leq j \leq d,$$

be the jth marginal density of $\mathbb{N}_q$. Here $\mathbb{R}_j^{d-1}$ is the $(d-1)$-dimensional Euclidean space of variables $(x_1, \ldots, x_{j-1}, x_{j+1}, \ldots, x_d)$. Then for the covariance matrix $Cov(\mathbb{N}_q)$ of $\mathbb{N}_q$ with entries $c_{j,k}$, $j, k = 1, \ldots, d$, we have

$$c_{j,k} = \int_{\mathbb{R}^d} x_k x_j G_q(\Sigma; x)dx = \begin{cases} 0, & \text{if } k \neq j, \\ \sigma_{q,j}^2, & \text{if } k = j, \end{cases}$$

where

$$\sigma_{q,j}^2 = \int_{-\infty}^{\infty} x_j^2 G_{j,q}(\Sigma; x_j)dx_j.$$

Similarly one can find higher-order moments of $\mathbb{N}_q$.

Chapter 3

q-Fourier Transform and Its Properties

3.1. Introduction

This chapter is devoted to the study of the q-generalized Fourier transform, called *q-Fourier transform*. The q-Fourier transform is an important mathematical tool. Recall that the classical Fourier transform is

$$F[f](\xi) = \int_{\mathbb{R}} f(x)e^{ix\xi}dx, \quad \xi \in \mathbb{R},$$

and defined for absolutely integrable functions $f(x)$, $x \in \mathbb{R}$, that is $\int_{\mathbb{R}} |f(x)|dx < \infty$. The class of functions satisfying this condition we denote by $L_1(\mathbb{R})$. With the norm

$$\|f\|_{L_1} = \int_{\mathbb{R}} |f(x)|dx.$$

$L_1(\mathbb{R})$ is a Banach space. For $f \in L_1(\mathbb{R})$, its Fourier transform $F[f](\xi)$ is a continuous function of $\xi \in \mathbb{R}$ and vanishes at infinity: $F[f](\xi) \to 0$ as $|\xi| \to \infty$ (see, e.g., Reed and Simon, 1972). Thus the mapping

$$F : L_1(\mathbb{R}) \to C_0(\mathbb{R}),$$

where $C_0(\mathbb{R})$ is the space of continuous functions vanishing at infinity, is continuous. Recall that $C_0(\mathbb{R})$ is a Banach space with the supremum-norm

$$\|f\|_{C_0} = \sup_{x \in \mathbb{R}} |f(x)|.$$

Definition 3.1. Let $q \in \mathbb{R}$. The q-Fourier transform F_q of $f \in L_1(\mathbb{R})$ is defined by

$$F_q[f](\xi) = \int\limits_{\mathrm{supp}\ [f]} [f(x)] \otimes_q \exp_q(ix\xi)dx, \quad \xi \in \mathbb{R}, \qquad (3.1)$$

where $\otimes_q$ is the q-product, $\mathrm{supp}[f]$ is the support of f, and $\exp_q(z)$ is the q-exponential function defined in Chapter 2.

If $q = 1$, then the q-Fourier transform coincides with $F[f](\xi)$. This immediately follows from the fact that the q-product and the q-exponential function coincide with the standard product and usual exponential function, respectively, if $q = 1$ (see Proposition 2.2 and (2.10)). The q-Fourier transform plays the same role in q-calculus as the Fourier transform does in modern analysis. However, unlike the classical Fourier transform, the q-Fourier transform for $q \neq 1$ is a nonlinear operator. This circumstance makes calculations connected with q-Fourier transform somewhat harder, but it is well adjusted to the q-operations introduced in Chapter 2. The q-Fourier transform of an absolutely integrable function, like the classical Fourier transform, is continuous and vanishes at infinity (see the proof in Section 3.2). However, the nonlinearity of the q-Fourier transform seriously affects the invertibility properties: for $q \neq 1$ the q-Fourier transform is not invertible. This issue will be discussed in Section 3.10 in detail.

One of the crucial properties of the q-Fourier transform is the fact that it transforms a q-Gaussian density function to another q-Gaussian up to a constant factor, but with a different q-value (see Theorem 3.4). This property is a key point in the proof of the q-Central Limit Theorem for $q > 1$ proved in Section 4.3. In Section 3.2, we study some basic properties of the q-Fourier transform. Section 3.4 discusses some mapping properties of the q-Fourier transform in the class of q-Gaussian densities. The q-Fourier transform in the case $q < 1$ behaves differently than the case $q \geq 1$. In Section 3.7, we discuss some issues arising in the case $q < 1$. In Section 3.8, we derive functional-differential equations for the q-Fourier transform of q-Gaussians. In Section 3.9, we give a comprehensive answer to the question — *Is the q-Fourier transform of a q-Gaussian a q'-Gaussian (with some q') up to a constant factor?* — for the whole range of $q' \in (-\infty, 3)$. Using the functional-differential equations approach we prove that the answer is affirmative if and only if $1 \leq q < 3$, excluding two particular cases of $q < 1$, namely, $q = \frac{1}{2}$ and $q = \frac{2}{3}$. Multidimensional versions of the q-Fourier transform are studied in Sections 3.12–3.15.

3.2. *q*-Fourier Transform for *q* > 1 and Its Properties

The q-Fourier transform for all $q \in \mathbb{R}$ is defined in Definition 3.1. We will see below that the properties of the q-Fourier transform for $q > 1$ and for $q < 1$ drastically differ. In this section, we assume that $q > 1$. The case $q < 1$ will be considered in Section 3.7. To avoid complications in calculations of q-Fourier transform, we only consider nonnegative functions $f \in L_1(\mathbb{R})$. The proposition below provides a different representation of the q-Fourier transform of $f \in L_1(\mathbb{R})$, which does not use the q-product.

Proposition 3.1. *Let* $f \in L_1(\mathbb{R})$ *and* $f(x) \geq 0$, $x \in \mathbb{R}$. *Then q-Fourier transform can be written in the form*

$$F_q[f](\xi) = \int_{\text{supp } [f]} f(x) \exp_q \left(ix\xi[f(x)]^{q-1} \right) dx. \qquad (3.2)$$

Proof. We have

$$e_q^{ix\xi} \otimes_q f(x) = [[\exp_q(ix\xi)]^{1-q} + [f(x)]^{1-q} - 1]^{\frac{1}{1-q}}$$

$$= \left[\exp\left[\frac{\text{Ln}(1 + (1-q)ix\xi)}{1-q} \right]^{1-q} + [f(x)]^{1-q} - 1 \right]^{\frac{1}{1-q}}$$

$$= [1 + (1-q)ix\xi + [f(x)]^{1-q} - 1]^{\frac{1}{1-q}}$$

$$= f(x)[1 + (1-q)ix\xi[f(x)]^{q-1}]^{\frac{1}{1-q}}$$

$$= f(x) \exp_q(ix\xi[f(x)]^{q-1}). \qquad (3.3)$$

Recall that here $\exp(z)$ is the complex exponential function for a complex number z, $Ln(z)$ is the principal value of the complex logarithm $\ln(z)$, cut alone the negative real axis $Re(z) < 0$, and $z^{1/(q-1)}$ is understood as

$$z^{1/(q-1)} = \exp\left(\frac{Ln(z)}{q-1}\right).$$

Equality (3.3) implies (3.2). Therefore, these two forms of the q-Fourier transform are equivalent.   □

Remark 3.1.

(1) It follows from (3.2) that the q-Fourier transform explicitly can be expressed as

$$F_q[f](\xi) = \int_{\text{supp } [f]} \frac{f(x)dx}{\left[1 + (1-q)ix\xi[f(x)]^{q-1}\right]^{\frac{1}{q-1}}}.$$

(2) Analogously, for a discrete function f_k, $k \in \mathcal{Z}$, in the space $\ell_1(\mathcal{Z})$ of absolutely convergent sequences, the discrete q-Fourier transform, or in other words the q-Fourier series, is defined by

$$F_q[f](x) = \sum_{k \in \mathcal{Z}} f_k \exp_q\left(ikx f_k^{q-1}\right) \quad x \in \mathbb{R}.$$

See Section 3.16 for more on q-Fourier series.

The following lemma will be used in establishing mapping properties of the q-Fourier transform. Introduce the function $g_q(a; u)$, $u \geq 0$, defined as

$$g_q(a; u) = u \exp_q(iau^{q-1}), \quad u \in \mathbb{R}_+, \tag{3.4}$$

where $a \in \mathbb{R}$.

Lemma 3.1. *Let $q > 1$ and b be a positive number. The function $g_q(a; u)$ defined in (3.4) possesses the following properties:*

(1) *it is continuous on the closed interval $[0, b]$;*
(2) *it is differentiable in the open interval $(0, b)$;*
(3) *for the derivative $g'_q(a; u)$ of $g_q(a; u)$ with respect to u the estimate*

$$|g'_q(a; u)| \leq C, \quad a \in \mathbb{R}, \ u \geq 0, \tag{3.5}$$

holds, where C is a constant independent of a and u;

(4) *for any u and v satisfying* $0 < u < v < b$ *there exists* $u_* \in (0, b)$ *such that the equality*

$$g_q(a; u) - g_q(a; v) = g_q'(a; u_*)(u - v) \tag{3.6}$$

holds.

Proof. Properties in Parts (1) and (2) easily follow from the continuity and differentiability of $\exp_q(z)$ and $\rho(u) = iau^{q-1}$, $u \geq 0$, since $g_q(a; u)$ is the composition of these two functions. To show Part (3), we notice that

$$g_q'(a; u) = \exp_q(iau^{q-1})[1 + ia(q-1)[u\exp_q(iau^{q-1})]^{q-1}].$$

Due to Proposition 2.10 the q-exponential function satisfies the estimate $|\exp_q(iy)| \leq 1$, for all $y \in \mathbb{R}$. Using this fact, it follows from the latter that

$$|g_q'(a; u)| \leq \left| 1 + ia(q-1) \left[u\exp_q(iau^{q-1}) \right]^{q-1} \right|$$

$$= \left| 1 + (q-1)\frac{iau^{q-1}}{1 + i(1-q)au^{q-1}} \right|$$

$$\leq 1 + \frac{(q-1)|a|u^{q-1}}{\left[1 + (q-1)^2 a^2 u^{2(q-1)} \right]^{1/2}} \leq 2,$$

proving Part (3). Part (4) is a simple implication of Parts (1) and (2), due to Lagrange's Average Theorem. □

Theorem 3.1. *Let* $q \geq 1$. *Then for the* q-*Fourier transform the following continuous mapping holds:*

$$F_q : L_1(\mathbb{R}) \to C_0(\mathbb{R}), \tag{3.7}$$

where $C_0(\mathbb{R})$ *is the space of continuous functions vanishing at infinity.*

Proof. The case $q = 1$ is well known (Reed and Simon, 1972). Let $q > 1$ and $f \in L_1(\mathbb{R})$. Using (3.2), we have the estimate

$$|F_q[f](\xi)| \le \int_{-\infty}^{\infty} |f(x)|| \exp_q(ix\xi[f(x)]^{q-1})|dx.$$

Further, due to Proposition 2.10, the latter implies

$$|F_q[f](\xi)| \le \int_{-\infty}^{\infty} |f(x)|dx = \|f\|_{L_1} < \infty, \quad \forall \xi \in \mathbb{R}.$$

To show continuity of $F_q[f](\xi)$ we estimate $|F_q[f](\xi + \varepsilon) - F_q[f](\xi)|$. We have

$$|F_q[f](\xi + \varepsilon) - F_q[f](\xi)|$$
$$\le \int_{\mathbb{R}} |f(x)| \left| \exp_q \left(ix(\xi + h)[f(x)]^{q-1} \right) - \exp_q \left(ix\xi[f(x)]^{q-1} \right) \right| dx.$$

Taking into account continuity of the q-exponential function and Lebesgue's dominated convergence theorem, we obtain for every fixed $\xi \in \mathbb{R}$

$$|F_q[f](\xi + \varepsilon) - F_q[f](\xi)| \to 0,$$

as $\varepsilon \to 0$. The fact that $F_q[f](\xi) \to 0$ as $|\xi| \to \infty$ follows from Proposition 2.10, Part(4), which implies $F_q[f](\xi) = O(|\xi|^{-\frac{1}{q-1}})$, $|\xi| \to \infty$.

Now we show the continuity of the mapping (3.7). Let $f_m \in L_1(\mathbb{R}), m = 1, 2, \ldots$, be a sequence such that $f_m \to f_0 \in L_1(\mathbb{R})$ as $m \to \infty$ in the norm of $L_1(\mathbb{R})$. We need to show that the sequence $F_q[f_m](\xi) \to F_q[f_0](\xi)$ uniformly. We have

$$|F_q[f_m](\xi) - F_q[f_0](\xi)|$$
$$\le \int_{\mathbb{R}} |f_m(x) \exp_q(ix\xi[f_m(x)]^{q-1}) - f_0(x) \exp_q(ix\xi[f_0(x)]^{q-1})|dx$$
$$= \int_{\mathbb{R}} |g_q(x\xi; f_m(x)) - g_q(x\xi; f_0(x))|dx,$$

where $g_q(a; u)$ is defined in (3.4). Further, making use of Lemma 3.1, Part (3), we have

$$|F_q[f_m](\xi) - F_q[f_0](\xi)| \leq \int_{\mathbb{R}} |g_q'(x\xi; u_*)||f_m(x) - f_0(x)|dx$$

$$\leq C \int_{\mathbb{R}} |f_m(x) - f_0(x)|dx.$$

Here we used the fact that $|g_q'(x\xi; u_*)| \leq C$, where constant C does not depend on x. Thus,

$$\sup_{\xi \in \mathbb{R}} |F_q[f_m](\xi) - F_q[f_0](\xi)| \to 0, \quad m \to \infty,$$

which completes the proof. □

Corollary 3.2. *The q-Fourier transform exists for any nonnegative* $f \in L_1(R)$ *and* $|F_q[f](\xi)| \leq \|f\|_{L_1}$. *Moreover, if* $f(x) = 0$ *almost everywhere in* $\mathbb{R}$, *then* $F_q[f](\xi) = 0$ *for all* $\xi \in \mathbb{R}$.

Corollary 3.3. *Assume* $f(x) \geq 0$, $x \in R$ *and* $F_q[f](\xi) = 0$ *for all* $\xi \in R$. *Then* $f(x) = 0$ *almost everywhere in* $\mathbb{R}$.

Proof. Proofs of these corollaries are simple implications of Theorem 3.1. □

Example 3.1.

(1) Consider the function $f(x) = A$, if $|x| \leq a$, and $f(x) = 0$, otherwise. Here $A > 0$, $a > 0$ are constants. Obviously $f \in L_1(\mathbb{R})$. Let $q = 2$. Then 2-Fourier transform of $f(x)$ is

$$F_2[f](\xi) = \frac{2}{\xi} \tan^{-1}(aA\xi), \quad \xi \in \mathbb{R}.$$

This function is continuous on $\mathbb{R}$ and one can easily verify that $F_2[f](0) = 2aA$ and $F_2[f](\xi) \to 0$, as $|\xi| \to \infty$, confirming Theorem 3.1 and Proposition 3.2, Part (1).

(2) Consider the sequence of functions $f_m(x) = \frac{1}{m}$, $m = 1, 2, \ldots$, if $|x| \leq a$, and $f_m(x) = 0$, otherwise. Obviously, $f_m \in L_1(\mathbb{R})$ for all $m = 1, 2, \ldots$, and $f_m(x) \to 0$, in $L_1(\mathbb{R})$. Let again $q = 2$. Then

2-Fourier transforms of functions f_m are

$$F_2[f_m](\xi) = \frac{2}{\xi} \tan^{-1}\left(\frac{a\xi}{m}\right), \quad \xi \in \mathbb{R}, \ m = 1, 2, \ldots,$$

which converges to 0 uniformly as $m \to \infty$.

As we will see in Section 3.7 these properties do not hold if $q < 1$.

Proposition 3.2. *For any constants $a > 0$, $b > 0$, the following relations hold:*

(1) $F_q[f](0) = \displaystyle\int_{supp\ [f]} f(x)dx$;

(2) $F_q[af(x)](\xi) = aF_q[f(x)](\frac{\xi}{a^{1-q}})$;

(3) $F_q[f(bx)](\xi) = \frac{1}{b}F_q[f(x)](\frac{\xi}{b})$.

Proof.

(1) Immediately follows from the definition of the q-Fourier transform.

(2) Using representation (3.2) for the q-Fourier transform, we have

$$F_q[af(x)](\xi) = \int_{supp\ [f]} af(x) \exp_q\left(ix\xi[af(x)]^{q-1}\right) dx$$

$$= a \int_{supp\ [f]} f(x) \exp_q\left(ixa^{q-1}\xi[f(x)]^{q-1}\right) dx$$

$$= aF_q[f(x)]\left(\frac{\xi}{a^{1-q}}\right).$$

(3) Again using representation (3.2), we have

$$F_q[f(bx)](\xi) = \int_{supp\ [f(bx)]} f(bx) \exp_q\left(ix\xi[f(bx)]^{q-1}\right) dx$$

$$= \frac{1}{b} \int_{supp\ [f]} f(x) \exp_q\left(i\frac{x}{b}\xi[f(x)]^{q-1}\right) dx$$

$$= \frac{1}{b}F_q[f(x)]\left(\frac{\xi}{b}\right). \qquad \square$$

Remark 3.2. We note that Proposition 3.2 holds for all $q \in \mathbb{R}$.

3.3. q-Fourier Transform of q-Gaussian Densities for $q > 1$

In this section, we find a relation between a q-Gaussian density and its q-Fourier transform in the case $q > 1$. This relation plays an important role in the proof of the q-Central Limit Theorem discussed in Chapter 4.

Recall that the q-Gaussian density is given by

$$G_q(\beta; x) = \frac{\sqrt{\beta}}{C_q} e_q^{-\beta x^2}, \quad x \in \mathbb{R}, \tag{3.8}$$

where β is a positive number and C_q is the normalization constant

$$C_q = \frac{\sqrt{\pi}\, \Gamma\left(\frac{3-q}{2(q-1)}\right)}{\sqrt{q-1}\, \Gamma\left(\frac{1}{q-1}\right)}, \quad 1 < q < 3.$$

Proposition 3.3. *The q-Fourier transform of the q-Gaussian density can be written in the form*

$$F_q[G_q(\beta; \cdot)](\xi) = \frac{\sqrt{\beta}}{C_q} \int_{\mathbb{R}} \exp_q \left(-\beta x^2 + \frac{i\beta^{(q-1)/2} x\xi}{C_q^{q-1}} \right) dx. \tag{3.9}$$

Proof. Using the property (b) in Proposition 3.2 with $a = \sqrt{\beta}/C_q$ and the property $\exp_q(z) \otimes \exp_q(\zeta) = \exp_q(z+\zeta)$ of the q-exponential function valid for all $z, \zeta \in \mathbb{C}$, we obtain (3.9). $\qquad\square$

Theorem 3.4. *Let $1 \le q < 3$. Then for the q-Fourier transform of the q-Gaussian density the following formula holds:*

$$F_q[G_q(\beta; x)](\xi) = e_{q_1}^{-\beta_* \xi^2}, \quad \xi \in \mathbb{R}, \tag{3.10}$$

where $q_1 = \frac{1+q}{3-q}$ and $\beta_* = \frac{3-q}{8\beta^{2-q} C_q^{2(q-1)}}$.

We first prove the following lemma:

Lemma 3.2. *Let $1 \le q < 3$. For the q-Fourier transform of a q-Gaussian, the following formula holds:*

$$F_q[G_q(\beta; x)](\xi) = \left(e_q^{-\frac{\xi^2}{4\beta^{2-q} C_q^{2(q-1)}}} \right)^{\frac{3-q}{2}}. \tag{3.11}$$

Proof. We denote $a = \frac{\sqrt{\beta}}{C_q}$ and use Proposition 3.3 and the property (b) in Proposition 3.2 to have

$$
F_q[ae_q^{-\beta x^2}](\xi) = a \int_{-\infty}^{\infty} e_q^{-\beta x^2 + ia^{q-1}x\xi} dx
$$

$$
= a \int_{-\infty}^{\infty} e_q^{-(\sqrt{\beta}x - \frac{ia^{q-1}\xi}{2\sqrt{\beta}})^2 - \frac{a^{2(q-1)}\xi^2}{4\beta}} dx
$$

$$
= a \int_{-\infty}^{\infty} e_q^{-(\sqrt{\beta}x - \frac{ia^{q-1}\xi}{2\sqrt{\beta}})^2} \otimes_q e_q^{-\frac{a^{2(q-1)}\xi^2}{4\beta}} dx.
$$

The substitution $y = \sqrt{\beta}x - \frac{ia^{q-1}\xi}{2\sqrt{\beta}}$ in the latter integral yields

$$
F_q[ae_q^{-\beta x^2}](\xi) = \frac{a}{\sqrt{\beta}} \int_{-\infty+i\eta}^{\infty+i\eta} e_q^{-y^2} \otimes_q e_q^{-\frac{a^{2(q-1)}\xi^2}{4\beta}} dy
$$

$$
= \lim_{R\to\infty} \frac{a}{\sqrt{\beta}} \int_{-R+i\eta}^{R+i\eta} e_q^{-y^2} \otimes_q e_q^{-\frac{a^{2(q-1)}\xi^2}{4\beta}} dy, \qquad (3.12)
$$

where $\eta = \frac{\xi a^{q-1}}{2\sqrt{\beta}}$. Consider the closed rectangular contour $\Gamma = \Gamma_1 \cup \Gamma_2 \cup \Gamma_3 \cup \Gamma_4$, where $\Gamma_1 = [-R, R]$, Γ_2 is the line segment joining points $z = -R + i\eta$, $z = R + i\eta$ in the complex plane, and Γ_3, Γ_4 are line segments joining points $z = R$, $z = R+i\eta$ and $z = -R$, $z = -R+i\eta$, respectively. As the positive orientation we take counterclockwise direction. Then in accordance with Cauchy's integral theorem (see, e.g., Conway, 1978), we have

$$
\int_{-R+i\eta}^{R+i\eta} e_q^{-y^2} \otimes_q e_q^{-\frac{a^{2(q-1)}\xi^2}{4\beta}} dy = \int_{\Gamma_2} e_q^{-y^2} \otimes_q e_q^{-\frac{a^{2(q-1)}\xi^2}{4\beta}} dy
$$

$$
= \int_{-R}^{R} e_q^{-y^2} \otimes_q e_q^{-\frac{a^{2(q-1)}\xi^2}{4\beta}} dy
$$

$$
+ \int_{\Gamma_3} e_q^{-z^2} \otimes_q e_q^{-\frac{a^{2(q-1)}\xi^2}{4\beta}} dz
$$

$$
+ \int_{\Gamma_4} e_q^{-z^2} \otimes_q e_q^{-\frac{a^{2(q-1)}\xi^2}{4\beta}} dz. \qquad (3.13)
$$

Last two integrals over Γ_3 and Γ_4 vanish as $R \to \infty$. Let us show this fact for the integral over Γ_3. Indeed, we have

$$\int_{\Gamma_3} e_q^{-z^2} \otimes_q e_q^{-\frac{a^{2(q-1)}\xi^2}{4\beta}} dz$$

$$= \int_0^\eta e_q^{-R^2+y^2+2Ryi} \otimes_q e_q^{-\frac{a^{2(q-1)}\xi^2}{4\beta}} idy$$

$$= \int_0^\eta e_q^{-R^2} \otimes_q e_q^{2Ryi} \otimes_q e_q^{y^2-\frac{a^{2(q-1)}\xi^2}{4\beta}} idy$$

$$= \int_0^\eta e_q^{-R^2} \otimes_q e_q^{y^2-\frac{a^{2(q-1)}\xi^2}{4\beta}} \exp_q \left(2Ryi \left[e_q^{y^2-\frac{a^{2(q-1)}\xi^2}{4\beta}} \right]^{q-1} \right) idy.$$

Here we used the equality (see Proposition 2.11)

$$\exp_q(z) \otimes_q \exp_q(\zeta) = \exp_q(\zeta) \exp_q(\zeta[\exp_q(z)]^{q-1}), \quad z, \zeta \in \mathbb{C}.$$
$$(3.14)$$

Now taking into account the fact $|\exp_q(iu)| \le 1$ (Proposition 2.10, Part (3)) valid for all $u \in \mathbb{R}$ if $q > 1$, we have

$$\left| \int_{\Gamma_3} e_q^{-z^2} \otimes_q e_q^{-\frac{a^{2(q-1)}\xi^2}{4\beta}} dz \right| \le \int_0^\eta e_q^{-R^2} \otimes_q e_q^{y^2-\frac{a^{2(q-1)}\xi^2}{4\beta}} dy.$$

Further, for fixed η (which means ξ is fixed) there is a constant $B > 0$, such that

$$e_q^{y^2-\frac{a^{2(q-1)}\xi^2}{4\beta}} \le B, \quad \forall y \in [0,\eta].$$

Therefore, using the property of the q-product given in Proposition 2.3, Part (5), we have

$$\left| \int_{\Gamma_3} e_q^{-z^2} \otimes_q e_q^{-\frac{a^{2(q-1)}\xi^2}{4\beta}} dz \right| \le |\eta| \, (e_q^{-R^2} \otimes B).$$

Since

$$e_q^{-R^2} \otimes B = B e_q^{-R^2 B^{q-1}}$$

$$= \frac{B}{[1 + (q-1)R^2 B^{q-1}]^{\frac{1}{q-1}}} \to 0, \quad R \to \infty.$$

the integral over Γ_3 on the right-hand side of (3.13) vanishes as $R \to \infty$. Similarly one can show that the integral over Γ_4 also vanishes when $R \to \infty$.

Hence, letting $R \to \infty$, it follows from (3.12) and (3.13) that

$$F_q[a e_q^{-\beta x^2}](\xi) = \frac{a}{\sqrt{\beta}} \int_{-\infty}^{\infty} e_q^{-y^2} \otimes_q e_q^{-\frac{a^{2(q-1)} \xi^2}{4\beta}} \, dy.$$

Using again equality (3.14), we have

$$F_q[G_q(\beta; \cdot)](\xi) = \frac{a e_q^{-\frac{a^{2(q-1)}}{4\beta} \xi^2}}{\sqrt{\beta}} \int_{-\infty}^{\infty} e_q^{-y^2 (e_q^{-\frac{a^{2(q-1)}}{4\beta} \xi^2})^{q-1}} \, dy.$$

Finally, using the fact that $\int_{\mathbb{R}} e_q^{-kx^2} dx = C_q/\sqrt{k}$, where C_q does not depend on k, we have

$$\int_{-\infty}^{\infty} e_q^{-y^2 (e_q^{-\frac{a^{2(q-1)}}{4\beta} \xi^2})^{q-1}} \, dy = C_q (e_q^{-\frac{a^{2(q-1)}}{4\beta} \xi^2})^{-(q-1)/2}.$$

Thus,

$$F_q[G_q(\beta; \cdot)](\xi) = \frac{a C_q}{\sqrt{\beta}} (e_q^{-\frac{a^{2(q-1)} \xi^2}{4\beta}})^{1-\frac{q-1}{2}}.$$

Replacing $a = \sqrt{\beta}/C_q$ and simplifying the latter we obtain (3.11). $\qquad\square$

Proof of Theorem. Recall that in Proposition 2.14 we proved the formula

$$(\exp_q(\kappa x))^p = \exp_{(p-1+q)/p}(p\kappa x),$$

valid for all $\kappa > 0$ and $p \in \mathbb{R}$. Let

$$\kappa = \frac{1}{4\beta^{2-q}C_q^{2(q-1)}}, \quad p = \frac{3-q}{2}.$$

Then it follows from Lemma 3.2 that

$$F_q[G_q(\beta;\cdot)](\xi) = e_{q_1}^{-\beta_*\xi^2},$$

where

$$q_1 = \frac{\frac{3-q}{2} - 1 + q}{\frac{3-q}{2}} = \frac{1+q}{3-q}, \tag{3.15}$$

and

$$\beta_* = \frac{3-q}{8\beta^{2-q}C_q^{2(q-1)}}, \tag{3.16}$$

which completes the proof. □

Remark 3.3.

(1) Theorem 3.4 says that for $q > 1$ the q-Fourier transform of a q-Gaussian density is a q_1-Gaussian up to a constant factor. Note that if $1 < q < 3$, then $q_1 > 1$. Moreover, if $1 < q < 2$, then $1 < q_1 < 3$, making q_1-Gaussian density normalizable.

(2) If $q = 1$, then setting $\beta = 1/(2\sigma^2)$, Theorem 3.4 recovers the well-known fact (see, e.g., Umarov, 2015)

$$\frac{1}{\sigma\sqrt{2\pi}} \int_{\mathbb{R}} e^{-\frac{x^2}{2\sigma^2}} e^{ix\xi} dx = e^{-\frac{\sigma^2\xi^2}{2}},$$

In light of relationship between q and q_1 it is useful to know properties of the function

$$u(s) = \frac{1+s}{3-s}, \quad 1 \le s < 3. \tag{3.17}$$

Denote the inverse function $u^{-1}(s)$ by $v(s)$. It can be easily verified that

$$v(s) = \frac{3s-1}{1+s}, \quad s \ge 1. \tag{3.18}$$

Proposition 3.4. *The functions $u(s)$ and $v(s)$ possess the following properties:*

(1) $u(\frac{1}{s}) = \frac{1}{v(s)}$ *and* $v(\frac{1}{s}) = \frac{1}{u(s)}$;

(2) $u(\frac{1}{u(s)}) = \frac{1}{s}$;

(3) $u(s)\, u(2 - s) = 1$;

(4) $u(2 - s) + v(s) = 2$.

Proof. Proofs are straightforward.   □

Let $q \geq 1$ be given. We set $q_1 = u(q)$ and $q_{-1} = v(q)$. Then, it follows from properties (1) and (2) of functions $u(s)$ and $v(s)$ in Proposition 3.4 that

$$u\left(\frac{1}{q_1}\right) = \frac{1}{q} \quad \text{and} \quad u\left(\frac{1}{q}\right) = \frac{1}{q_{-1}}. \tag{3.19}$$

Properties (3) and (4) yield $v(s) + 1/u(s) = 2$, which leads to the following important duality relation between q_1 and q_{-1}:

$$q_{-1} + \frac{1}{q_1} = 2. \tag{3.20}$$

Corollary 3.1. *For q-Gaussian densities the following q-Fourier transforms hold:*

$$F_q[G_q(\beta; x)](\xi) = e_{u(q)}^{-\beta_*(q)\xi^2}, \quad 1 \leq q < 3; \tag{3.21}$$

$$F_{v(q)}[G_{v(q)}(\beta; x)](\xi) = e_q^{-\beta_*\left(v(q)\right)\xi^2}, \quad 1 \leq q < 3, \tag{3.22}$$

where

$$\beta_*(q) = \frac{3 - q}{8\beta^{2-q}C_q^{2(q-1)}}.$$

Remark 3.4.

(1) Note that $\beta_*(q) > 0$ if $q < 3$. The parameter β is connected with the "width" of the corresponding density. Therefore, it is useful to know how β_* depends on β for various values of q. For $q = 1$

it is known that dependence of β_* on β is $\beta_* = 1/(4\beta)$, that is β_* decreases, when β increases. This tendency is kept for all $1 \le q < 2$. For $q = 2$ one has $\beta_* = \pi$, i.e., it does not depend on β. However, if $2 < q < 3$ (in this case q-Fourier transform is not normalizable; see Remark 3.3), then β_* becomes an increasing function of β.

(2) Note also that the q-generalization of the Heisenberg uncertainty property is connected with the remark above. Namely, up to $q = 2$, the q-Fourier transform of a fat q-Gaussian is a thin q_1-Gaussian and vice versa (see details in Tsallis, 2009a). We emphasize that when q increases above one, q_1 increases even more rapidly.

(3) The q-Fourier transform of the q-Gaussian density can be considered as the q-characteristic function of the q-Gaussian random variable N_q, that is $\mathbb{E}[\exp_q(i\xi N_q)]$. We will see in Section 3.5 that the q-characteristic function of N_q is in a close relation with q-moments of N_q.

3.4. Mapping Properties of the q-Fourier Transform in $\mathcal{G}_q$

In this section, we discuss mapping properties of the q-Fourier transform in the class $\mathcal{G}_q$ of q-Gaussian type functions (not necessarily density functions). We establish a representation formula for the inverse F_q-transform in this class as well.

Introduce the set of functions

$$\mathcal{G}_q = \{f : f(x) = ae_q^{-\beta x^2}, \, a > 0, \, \beta > 0\}. \tag{3.23}$$

and

$$\mathcal{G} = \cup_{-\infty < q < 3} \, \mathcal{G}_q. \tag{3.24}$$

Obviously, $\mathcal{G} \subset L_1(\mathbb{R})$ and the set of all q-Gaussian densities is a subset of $\mathcal{G}$.

Apart from the q-Gaussian operator F_q we introduce the operator

$$F_q^*[f](\xi) = \int_{-\infty}^{\infty} f(x)e_q^{-ix\xi[f(x)]^{q-1}}dx. \tag{3.25}$$

Operators F_q and F_q^* are connected through $F_q^*[f](\xi) = F_q[f](-\xi)$. This implies that the operator F_q^* is well defined on elements of $L_1(\mathbb{R})$, and, like for F_q,

$$\sup_{\xi \in \mathbb{R}} |F_q^*[f](\xi)| \leq \|f\|_{L_1}. \tag{3.26}$$

Moreover, if $f \in L_1(\mathbb{R})$, then $F_q^*[f](\xi)$ is continuous and $F_q^*[f](\xi) \to 0$ as $|\xi| \to \infty$. It follows immediately from (3.25) that

$$F_q^*[f](0) = \int_{R^1} f(x)dx. \tag{3.27}$$

One can define the q-Fourier transform in $L_2(\mathbb{R})$ as well, that is in the space of square integrable complex-valued functions with the inner product

$$(f,g) = \int_{\mathbb{R}^1} f(x)\bar{g}(x)dx,$$

where $\bar{g}$ is the complex conjugate of g, similar to the standard Fourier transform in $L_2(\mathbb{R})$. Namely,

$$F_q[f](\xi) = \text{l.i.m.}_{A \to \infty} \int_{|x|<A} f(x) \exp_q(ix\xi[f(x)]^{q-1})dx,$$

where the symbol l.i.m. means convergence in the sense of L_2. Evidently, in $L_2(\mathbb{R})$ the q-Fourier transform can be written through the inner product

$$F_q[f](\xi) = (f, \exp_q(-ix\xi[f(x)]^{q-1})). \tag{3.28}$$

Example 3.2. The last representation allows to extend F_q at least for those generalized functions in the sense of Schwartz distributions, for which (3.28) makes sense. For instance, using (3.28) combined with (3.27), one has

$$F_q[1](\xi) = (1, \exp_q(ix\xi)).$$

For the usual Fourier transform ($q = 1$) it is well known the formula (see, e.g., Umarov, 2015, p. 23)

$$F[1](\xi) = (1, \exp(ix\xi)) = 2\pi\delta(\xi), \tag{3.29}$$

or

$$\delta(\xi) = \frac{1}{2\pi} F[1](\xi),$$

where $\delta(\xi)$ is the Dirac delta-function. By analogy, we define $\delta_q(\xi)$ via the following expression:

$$\delta_q(\xi) = \frac{1}{2\pi} F_q[1](\xi) = \frac{1}{2\pi} \int\limits_{-\infty}^{\infty} \exp_q(-ix\xi)dx. \tag{3.30}$$

For $1 \leq q < 2$, as is shown in Jauregui and Tsallis (2011) and Plastino and Rocca (2011), the relation

$$\frac{2\pi}{2-q} f(0) = \int\limits_{\mathbb{R}^2} f(x) \exp_q(-ix\xi)d\xi dx$$

$$= \int\limits_{\mathbb{R}} f(x) \int\limits_{\mathbb{R}} \exp_q(-ix\xi)d\xi dx$$

holds for any smooth rapidly decreasing function f. The latter is the same as

$$\left(\frac{2\pi}{2-q}\delta(x), f(x) \right) = (2\pi\delta_q(x), f(x)),$$

meaning

$$\delta_q(x) = \frac{1}{2-q}\delta(x). \tag{3.31}$$

We call this function (in the sense of Schwartz distributions) q-Dirac delta function. It follows from (3.30) and (3.31) that

$$F_q[1](\xi) = 2\pi\delta_q(\xi) = \frac{2\pi}{2-q}\delta(\xi),\tag{3.32}$$

which concludes our example.

Reformulating Corollary 3.1, established in the previous section, we can assert that the following mappings are injective:

$$F_q : \mathcal{G}_q \to \mathcal{G}_{q_1}, \quad q_1 = u(q),\ 1 \le q < 3,\tag{3.33}$$

$$F_{q-1} : \mathcal{G}_{q-1} \to \mathcal{G}_q, \quad q_{-1} = v(q),\ 1 \le q < 3,\tag{3.34}$$

Similarly, we have

$$F_q^* : \mathcal{G}_q \to \mathcal{G}_{q_1}, \quad q_1 = u(q),\ 1 \le q < 3,\tag{3.35}$$

$$F_{q-1}^* : \mathcal{G}_{q-1} \to \mathcal{G}_q, \quad q_{-1} = v(q),\ 1 \le q < 3,\tag{3.36}$$

An important question arising here is whether mappings (3.33) and (3.34) are surjective as well. To answer this question we have to show that the inverse q-Fourier transform F_q^{-1} in the class $\mathcal{G}_{u(q)}$ does exist. Below we treat this question in the general setting.

For this purpose let us introduce the sequence $q_n,\ n = 0, 1, \ldots,$ as follows:

$$q_0 = u_0(q) = q,$$

$$q_n = u_n(q) = u(u_{n-1}(q)), \quad n = 1, 2, \ldots,\tag{3.37}$$

where

$$u(s) = \frac{1+s}{3-s}.$$

Thus,

$$q_n = \frac{1 + u_{n-1}(q)}{3 - u_{n-1}(q)} = \frac{1 + q_{n-1}}{3 - q_{n-1}}, \quad n = 1, 2, \ldots.\tag{3.38}$$

We can extend the sequence q_n for negative integers $n = -1, -2, \ldots$ as well, putting

$$q_{-n} = v_n(q) = v(v_{n-1}(q)), \quad n = 1, 2, \ldots.$$

with $v_0(q) = q_0$, and

$$v(s) = \frac{3s - 1}{1 + s}.$$

Thus,

$$q_{-n} = v_n(q) = \frac{3v_{n-1}(q) - 1}{1 + v_{n-1}(q)} = \frac{3q_{-(n-1)} - 1}{1 + q_{-(n-1)}}, \quad n = 1, 2, \ldots.$$

$$(3.39)$$

It is not hard to verify that the sequence q_n can explicitly be expressed as

$$q_n = \frac{2q + n(1 - q)}{2 + n(1 - q)} = 1 + \frac{2(q - 1)}{2 - n(q - 1)}, \quad n = 0, \pm 1, \pm 2, \ldots,$$

$$(3.40)$$

which, for $q \neq 1$, can be rewritten as

$$\frac{2}{1 - q_n} = \frac{2}{1 - q} + n. \qquad (3.41)$$

Remark 3.5.

(1) Note that if $q_0 = q = 1$, then $q_n = 1$ for all $n = 0, \pm 1, \pm 2, \ldots$.
(2) $\lim_{n \to \pm\infty} q_n = 1$ for all $q \neq 1$.
(3) For $q \in (1, 3)$ it follows from (3.40) that $q_n > 1$ for all $n < 2/(q - 1)$, and $q_n < 1$, otherwise. Moreover, obviously, $2/(q - 1) > 1$, if $q \in (1, 3)$, which implies $q_1 > 1$. Hence, it follows from (3.41) that the condition $q_n > 1$ guarantees $q_k > 1$, $k = n - 1, n, n + 1$, for three consequent members of the sequence (3.40).
(4) Let $q > 1$. As is seen from (3.40) that $q_n > 1$, if $n \leq \frac{2}{q-1} - 1$, and $q_n < 1$, if $n > \frac{2}{q-1} - 1$.

Notice now that if $q_{n_0} = 3$ for some $n = n_0$, then, as a consequence of (3.38), it follows that $q_{n_0+1}, q_{n_0+2}, \ldots$, are not defined. In this case we say that the sequence q_n is interrupted at $n = n_0$. In order to know for what values of q such an interruption occurs, consider the sequence q_{n_0-k}, $k = 0, 1, \ldots$. The members of this sequence are $v_k(q_{n_0})$, $k = 0, 1, \ldots$, and, as it follows from (3.39), are given by

$$\ldots, \frac{5}{4}, \frac{9}{7}, \frac{4}{3}, \frac{7}{5}, \frac{3}{2}, \frac{5}{3}, 2, 3.$$

In other words, the members of q_{n_0-k} consist of numbers of the form

$$(m+1)/m \quad \text{and} \quad (2m+1)/(2m-1), \quad m = 1, 2, \ldots.$$

For example, if $q_0 = q = 3/2$, then in 3 steps the interruption occurs, since in this case $q_1 = 5/3$, $q_2 = 2$, and $q_3 = 3$. In the general case, if $q_0 = q = (m+1)/m$, then in $(2m-1)$ steps we have $q_{2m-1} = 3$, and therefore the interruption occurs. Similarly, if $q_0 = q = (2m+1)/(2m-1)$, then in $2(m-1)$ steps we have $q_{2m-2} = 3$. This observation leads to the following proposition.

Proposition 3.5.

(1) *Let $q = q_0 = (m+1)/m$, where $m = 1, 2, \ldots$. Then $q_{2m-1} = 3$, thus the sequence $\{q_n\}$ is interrupted in $(2m-1)$ steps;*
(2) *Let $q = q_0 = (2m+1)/(2m-1)$, where $m = 1, 2, \ldots$. Then $q_{2m-2} = 3$, thus the sequence $\{q_n\}$ is interrupted in $(2m-2)$ steps;*
(3) *Let $q = q_0 \geq 1$ and $q_0 = q \neq (m+1)/m$, $(2m+1)/(2m-1)$, $m = 1, 2, \ldots$. Then the sequence $\{q_n\}$ is never interrupted.*
(4) *Let $q = q_0 > 1$ and N_* is the integer part of $\frac{2}{q-2} - 1$, that is*

$$N_* = \left\lfloor \frac{2}{q-1} - 1 \right\rfloor. \tag{3.42}$$

Then $q_n > 1$ for all $n \leq N_*$ with the limit $\lim_{n \to -\infty} q_n = 1 + 0$, and $q_n < 1$ for all $n > N_*$ with the limit $\lim_{n \to \infty} q_n = 1 - 0$.

Proposition 3.6. *For the members of the sequence q_n assuming it is not interrupted (if the interruption occurs see Proposition 3.5) the duality relations*

$$q_{n-1} + \frac{1}{q_{n+1}} = 2, \quad n = 0, \pm 1, \ldots, \tag{3.43}$$

hold.

Proof. Making use of Proposition 3.4, Parts (3) and (4), we obtain

$$v(q_n) + \frac{1}{u(q_n)} = 2.$$

The latter immediately implies (3.43). □

Further, we introduce a kth power of the q-Fourier transform as a composition of k successive q-Fourier transforms.

Definition 3.2. For $k = 1, 2, \dots$ and $n = 0, \pm 1, \dots$, we define the kth power of the q-Fourier transform as

$$F_{q_n}^k [f](\xi) = F_{q_{n+k-1}}[\dots F_{q_{n+1}}[F_{q_n}[f]]](\xi).$$

For the operator $F_{q_n}^k$ we also use the notation

$$F_{q_n}^k = F_{q_{n+k-1}} \circ \cdots \circ F_{q_n},$$

where the symbol "$\circ$" means the composition of corresponding q-Fourier transform operators. Additionally, for $k = 0$ we let $F_q^0 = I$, the identity operator.

Similarly, one can define kth power of F_q^* :

$$F_{q_n}^{*k} = F_{q_{n+k-1}}^* \circ \cdots \circ F_{q_n}^*, \quad F_q^{*,0} = I.$$

Now we can reformulate Theorem 3.4 in terms of the sequence $\{q_n\}$.

Proposition 3.7. *Let an element q_n of the sequence defined in (3.40) to satisfy the condition $1 \leq q_n < 3$. Then for the q_n-Fourier transform of the q_n-Gaussian density the following formula holds:*

$$F_{q_n}[G_{q_n}(\beta; x)](\xi) = e_{q_{n+1}}^{-\beta_* \xi^2}, \quad \xi \in \mathbb{R}, \qquad (3.44)$$

where $q_{n+1} = u(q_n)$ and $\beta_* = \frac{3-q_n}{8\beta^{2-q_n} C_{q_n}^{2(q_n-1)}}$.

The proposition below is a direct implication of the injective mapping (3.33).

Proposition 3.8.

(1) *Let $q_0 = q > 1$. Then the mappings*

$$F_{q_n} : \mathcal{G}_{q_n} \to \mathcal{G}_{q_{n+1}}, \quad \forall n < N_*,$$

and

$$F_{q_n}^* : \mathcal{G}_{q_n} \to \mathcal{G}_{q_{n+1}}, \quad \forall n < N_*,$$

are injective. Here N_* is defined in (3.42).
(2) *Let $q_0 = q > 1$. Then the mappings*

$$F_{q_n}^k : \mathcal{G}_{q_n} \to \mathcal{G}_{q_{n+k}}, \quad \forall n, \ k : n + k \leq N_*,$$

and

$$F_{q_n}^{* \ k} : \mathcal{G}_{q_n} \to \mathcal{G}_{q_{n+k}}, \quad \forall n, \ k : n + k \leq N_*,$$

are injective.

Now we prove that the mappings in the latter propositions are surjective as well.

Theorem 3.5 (Umarov and Tsallis, 2008). *Let $q_0 = q \geq 1$ and N_* be defined as in* (3.42). *Then the mappings*

$$F_{q_n} : \mathcal{G}_{q_n} \to \mathcal{G}_{q_{n+1}}, \quad n < N_*, \tag{3.45}$$

are invertible.

Proof. Denote by $\mathcal{T}$ the set

$$\mathcal{T} = \{(q, a, \beta) : q \geq 1, \ a > 0, \ \beta > 0\}.$$

Obviously, the sets $\mathcal{T}$ and $\mathcal{G}$ are in a one-to-one relationship, since every element $g(x) = a \exp_q(-\beta x^2)$ of $\mathcal{G}$ is uniquely determined by the triplet $(q, a, \beta) \in \mathcal{T}$. Hence, with the operator $F_{q_n} : \mathcal{G}_{q_n} \to \mathcal{G}_{q_{n+1}}$ we can associate the mapping $\varphi : \mathcal{T} \to \mathcal{T}$ defined as $\varphi(q_n, a, \beta) = (q_{n+1}, A, B)$, where $q_{n+1} = u(q_n)$, $A = \frac{aC_{q_n}}{\sqrt{\beta}}$, and

$B = \frac{a^{2(q_n-1)}(3-q_n)}{8\beta}$. Assuming that Q, A and B are given, consider the system of equations

$$\begin{cases} \dfrac{1+q_n}{3-q_n} = Q, \\[2mm] \dfrac{aC_{q_n}}{\sqrt{\beta}} = A, \\[2mm] \dfrac{a^{2(q_n-1)}(3-q_n)}{8\beta} = B, \end{cases}$$

for (q_n, a, β). The first equation is autonomous and has a unique solution $q_n = (3Q-1)/(Q+1)$. If the condition $n < N_*$ is fulfilled, then the other two equations have a unique solution as well, namely,

$$a = \left(\frac{A\sqrt{3-q_n}}{2C_{q_n}\sqrt{2B}}\right)^{\frac{1}{2-q_n}}, \quad \beta = \left(\frac{A^{2(q_n-1)}(3-q_n)}{8C_{q_n}^{2(q_n-1)}B}\right)^{\frac{1}{2-q_n}}. \tag{3.46}$$

It follows from (3.40) that Q and q_n are related as $q_n = v(Q)$. Hence, the inverse mapping $\varphi^{-1} : \mathcal{T} \to \mathcal{T}$ exists. This implies that the inverse operator $F_{q_n}^{-1} : \mathcal{G}_{q_{n+1}} \to \mathcal{G}_{q_n}$ exists and maps each element $Ae_Q^{-B\xi^2} \in \mathcal{G}_Q$ to the element $ae_{q_n}^{-\beta x^2} \in \mathcal{G}_{q_n}$ with $q_n = v(Q)$ and a and β defined in (3.46). $\qquad\square$

Now we find a representation formula for $F_{q_n}^{-1}$, the inverse operator of F_{q_n}. Denote by φ_0 the mapping $\varphi_0 : (a,\beta) \to (A,B)$, where $A = \frac{aC_{q_n}}{\sqrt{\beta}}$ and $B = \frac{a^{2(q_n-1)}(3-q_n)}{8\beta}$, as indicated above. We have seen, in particular, that φ_0 is invertible and $\varphi_0^{-1} : (A,B) \to (a,\beta)$ with a and β in (3.46). Assume that

$$(\bar{a}, \bar{\beta}) = \varphi_0^{-2}(A,B) = \varphi_0^{-1}(\varphi_0^{-1}(A,B)) = \varphi_0^{-1}(a,\beta).$$

Further, we introduce the operator $I_{(q_{n+1}, q_{n-1})} : \mathcal{G}_{q_{n+1}} \to \mathcal{G}_{q_{n-1}}$ defined by the formula

$$I_{(q_{n+1}, q_{n-1})}[Ae_{q_{n+1}}^{-B\xi^2}] = \bar{a}e_{q_{n-1}}^{-\bar{\beta}\xi^2}. \tag{3.47}$$

Consider the following composite operator:

$$H_{q_n} = F_{q_{n-1}}^* \circ I_{(q_{n+1}, q_{n-1})}. \tag{3.48}$$

By definition, it is clear that

$$I_{(q_{n+1}, q_{n-1})} : \mathcal{G}_{q_{n+1}} \to \mathcal{G}_{q_{n-1}}.$$

Moreover, since

$$F^*_{q_{n-1}} : \mathcal{G}_{q_{n-1}} \to \mathcal{G}_{q_n},$$

we have that H_{q_n} maps $\mathcal{G}_{q_{n+1}}$ to $\mathcal{G}_{q_n}$, that is

$$H_{q_n} : \mathcal{G}_{q_{n+1}} \to \mathcal{G}_{q_n}.$$

Let $\hat{f} \in \mathcal{G}_{q_{n+1}}$, that is $\hat{f}(\xi) = Ae_{q_{n+1}}^{-B\xi^2}$, where A and B are arbitrary positive numbers. Then, the explicit form of the operator H_{q_n} acting on the function $\hat{f}(\xi)$ is

$$H_{q_n}[\hat{f}(\xi)](x) = \int_{\mathbb{R}} \left(\bar{a} e_{q_{n-1}}^{-\bar{\beta}\xi^2} \right) \otimes_{q_{n-1}} e_{q_{n-1}}^{-ix\xi} d\xi$$

$$= \int_{\mathbb{R}} I_{(q_{n+1}, q_{n-1})}[\hat{f}(\xi)] \otimes_{q_{n-1}} e_{q_{n-1}}^{-ix\xi} d\xi. \qquad (3.49)$$

The following proposition states that, in fact, the operator H_{q_n} is the inverse to F_{q_n} in the space $\mathcal{G}_{q_{n+1}}$.

Proposition 3.9.

(1) *Let $f \in \mathcal{G}_{q_n}$ and $n < N_*$. Then $H_{q_n} \circ F_{q_n}[f] = f$;*
(2) *Let $f \in \mathcal{G}_{q_{n+1}}$ and $n < N_*$. Then $F_{q_n} \circ H_{q_n}[f] = f$.*

Proof.

(1) We need to show the validity of the equation

$$F^*_{q_{n-1}} \circ I_{(q_{n+1}, q_{n-1})} \circ F_{q_n} = I,$$

where I is the identity operator in $\mathcal{G}_{q_n}$. Due to the symmetric nature of functions in $\mathcal{G}_q$, this equation is equivalent to the equation $\varphi_0 \circ \varphi_0^{-2} \circ \varphi_0 = J$, where J is the identity operator in $\mathcal{T}$, with fixed q. The latter equation is correct by construction.

(2) Now the equation

$$F_{q_n} \circ F^*_{q_{n-1}} \circ I_{(q_{n+1}, q_{n-1})} = I'$$

where I' is the identity operator in $\mathcal{G}_{q_{n+1}}$, is equivalent to the identity $\varphi_0^2 \circ \varphi_0^{-2} = J$. □

Corollary 3.2. *The operator $H_{q_n} : \mathcal{G}_{q+1} \to \mathcal{G}_{q_n}$ is the inverse to the q_n-Fourier transform: $H_{q_n} = F^{-1}_{q_n}$.*

Thus, we have proved that, if $\hat{f}(\xi)$ is a function in $\mathcal{G}_{q_{n+1}}$, with q_n, $n < N_*$, defined in (3.40) for $q \in [1, 3)$, then the inverse q_n-Fourier transform is explicitly expressed by

$$F^{-1}_{q_n}[\hat{f}(\xi)](x) = \int_{-\infty}^{\infty} I_{(q_{k+1}, q_{n-1})}[\hat{f}(\xi)] \otimes_{q_{n-1}} e^{-ix\xi}_{q_{n-1}} d\xi, \qquad (3.50)$$

where the operator $I_{(q_{n+1}, q_{n-1})}$ is given in (3.47). This might constitute a useful step for finding a representation of $F^{-1}_q[\hat{f}(\xi)](x)$ for generic function $\hat{f}(\xi)$ in appropriate spaces of functions, which is a challenging open problem.

Corollary 3.3. *For $q = 1$ the representation (3.49) for the inverse operator F^{-1}_q, takes the form*

$$F^{-1}_1[\hat{f}(\xi)](x) = \frac{1}{2\pi} \int_{-\infty}^{\infty} \hat{f}(\xi) e^{-ix\xi} dx, \qquad (3.51)$$

i.e., $F^{-1}_1[\hat{f}(\xi)](x)$ coincides with the classical inverse Fourier transform.

Proof. If $q = 1$, then by definition one has $q_n = q_{n-1} = q_{n+1} = 1$. We find $\bar{a}$ and $\bar{\beta}$ taking $(A, B) = (1, 1)$. It follows from relationships (3.46) that $(a, \beta) = T^{-1}(1, 1) = (\frac{1}{2\sqrt{\pi}}, \frac{1}{4})$. Again using (3.46) we obtain $(\bar{a}, \bar{\beta}) = T^{-2}(1, 1) = T^{-1}(\frac{1}{2\sqrt{\pi}}, \frac{1}{4}) = (\frac{1}{2\pi}, 1)$. This means that

$$I_{(1,1)}\hat{f}(\xi) = \frac{1}{2\pi}\hat{f}(\xi), \quad \hat{f} \in \mathcal{G}_1. \qquad (3.52)$$

Hence, the formula (3.49) is reduced to (3.51), recovering the well known formula for the classic inverse Fourier transform for functions $\hat{f} \in \mathcal{G}_1$. □

Remark 3.6.

(1) Obviously, the operator $I_{(1,1)}$ in (3.52) is extendable to spaces of continuous or integrable functions, which in turn, allows to obtain well-known theorems on the inverse Fourier transform. For instance, (3.51) can be extended to $L_1(\mathbb{R})$ or $L_2(\mathbb{R})$, if F_1 is originally considered in $L_1(\mathbb{R})$ or $L_2(\mathbb{R})$, respectively.
(2) The extendability of the operator $I_{q,q'}$ to wider classes of functions is currently an open problem.

Corollary 3.6. *There exist the following inverse q-Fourier transforms*

$$F_q^{-1} : \mathcal{G}_{q_1} \to \mathcal{G}_q, \quad q_1 = u(q), \quad 1 \le q < 3,$$
$$F_{q-1}^{-1} : \mathcal{G}_q \to \mathcal{G}_{q-1}, q_{-1} = v(q), \quad 1 \le q < 3.$$

Corollary 3.4. *Let $q_0 = q \ge 1$ and N_* be defined as in (3.42). Then the operator $F_{q_n}^k$, where $n + k < N_*$, is invertible and its inverse is*

$$\left(F_{q_n}^k \right)^{-1} = F_{q_n}^{-k} = F_{q_n}^{-1} \circ \cdots \circ F_{q_{n+k-1}}^{-1}.$$

Moreover, the mapping

$$F_{q_n}^{-k} : \mathcal{G}_{q_{n+k}} \to \mathcal{G}_{q_n}$$

holds.

Summarizing, we have the following series of bijective mappings.

Proposition 3.10. *Let $q_0 = q \ge 1$ and N_* be defined as in (3.42). Then for any $n < N_*$ the following series of mappings are bijective:*

$$\cdots \overset{F_{q-2}}{\to} \mathcal{G}_{q-1} \overset{F_{q-1}}{\to} \mathcal{G}_{q_0} \overset{F_{q_0}}{\to} \mathcal{G}_{q_1} \overset{F_{q_1}}{\to} \mathcal{G}_{q_2} \overset{F_{q_2}}{\to} \cdots \overset{F_{q_n}}{\to} \mathcal{G}_{q_{n+1}}, \qquad (3.53)$$

$$\cdots \overset{F_{q-2}^{-1}}{\leftarrow} \mathcal{G}_{q-1} \overset{F_{q-1}^{-1}}{\leftarrow} \mathcal{G}_{q_0} \overset{F_{q_0}^{-1}}{\leftarrow} \mathcal{G}_{q_1} \overset{F_{q_1}^{-1}}{\leftarrow} \mathcal{G}_{q_2} \overset{F_{q_2}^{-1}}{\leftarrow} \cdots \overset{F_{q_n}^{-1}}{\leftarrow} \mathcal{G}_{q_{n+1}}. \qquad (3.54)$$

3.5. *q*-Characteristic Functions of Random Variables

Let X be a continuous random variable wth a range $\mathcal{X}$. Recall that the characteristic function of X is defined by

$$\varphi_X(\xi) = \mathbb{E}[\exp(iX\xi)] = \int_{\mathcal{X}} f_X(x)e^{ix\xi}dx, \quad \xi \in \mathbb{R},$$

where $f_X(x)$ is the probability density function of X. The characteristic function is a useful tool in the study of limiting processes. In particular, it can be used in the proof of the central limit theorem. If $\mathcal{X}$ is compact then the characteristic function of X can be expressed as a power series with coefficients connected with moments of X. If $\mathcal{X}$ is not compact, then a power series representation of the characteristic function is possible only if the density function $f_X(x)$ rapidly decays at infinity. The q-Gaussian random variables studied in Section 2.7 have density functions with compact support if $q < 1$. However, if $q > 1$ the density functions of q-Gaussians have no compact support and do not decay rapidly at infinity. Therefore, characteristic functions may not be effective in the study of limiting processes formed by q-Gaussian distributions.

Here below we introduce the notion of the q-characteristic function which acts more effectively in the study of limiting processes, connected with q-Gaussian distributions, to be compared with the traditional characteristic functions.

Definition 3.3. Let X be a random variable with a range $\mathcal{X}$ and a density function $f_X(x)$, $x \in \mathcal{X}$. The q-characteristic function of X is defined by

$$\varphi_{q,X}(\xi) = \mathbb{E}_q[\exp_q(iX\xi)] = \int_{\mathcal{X}} f(x) \otimes_q \exp_q(ix\xi)dx, \quad \xi \in \mathbb{R}.$$

It follows from this definition that

$$\varphi_{q,X}(\xi) = F_q[f_X](\xi) = \int_{\mathcal{X}} f_X(x)\exp_q\left(ix\xi[f_X(x)]^{q-1}\right)dx. \quad (3.55)$$

Further, using the power series representation (2.37) for the q-exponential function $\exp_q(z)$, we have

$$\varphi_{q,X}(\xi) = \sum_{n=0}^{\infty} \frac{i^n A_n \, \mu_{q,n}[X]}{n!} \, \xi^n, \qquad (3.56)$$

where $A_n, n = 0, 1, \ldots$, are defined in (2.38), and

$$\mu_{q,n}[X] = \int_{\mathcal{X}} x^n [f_X(x)]^{n(q-1)+1} dx, \qquad n = 0, 1, \ldots . \qquad (3.57)$$

Remark 3.7.

(1) It is worth to note that the integral (3.57) has better convergence properties than the integral

$$\mu_n[X] = \int_{\mathcal{X}} x^n [f_X(x)] dx,$$

expressing the usual moments, due to the power $n(q-1) + 1$ of $f(x)$, when $q > 1$. For example, if $f(x)$ decreases at infinity behaving like $|x|^{-1/(q-1)}$, then the integral (3.57) converges for all n, if $q < 2$, and diverges for all n, if $q \geq 2$; see Tsallis *et al.* (2009). Therefore, the notion of the q-characteristic function is an effective tool in our further analysis.

(2) We also note that the q-characteristic function $\varphi_{q,X}(\xi)$ of a random variable X defines uniquely the corresponding density function $f_X(x)$ of X. Indeed, if the q-characteristic functions $\varphi_{q,X}(\xi)$ and $\varphi_{q,Y}(\xi)$ of two random variables X and Y are convergent on a neighborhood of $\xi = 0$ and equal, then (3.56) implies that $\mu_{q,n}[X] = \mu_{q,n}[Y]$ for all $n = 0, 1, \ldots$. Moreover, since the set of functions $x^n, n = 0, 1, \ldots$, form a basis on any finite interval $(a, b) \subset \mathbb{R}$, one has

$$[f_X(x)]^{n(q-1)+1} = [f_Y(x)]^{n(q-1)+1}, \qquad x \in (a, b), \ n = 0, 1, \ldots .$$

This, in turn, implies that the densities are equal on any finite interval of $\mathbb{R}$. Thus, we conclude that $f_X(x) = f_Y(x)$ on the whole real axis $\mathbb{R}$, obtaining the uniqueness of the density function, or the corresponding random variable, through the q-characteristic function.

It is not hard to see that $\mu_{q,n}[X]$ is connected with the q-moment $\mathbb{E}_{n,q}[X^n]$ of order n of the random variable X. Indeed, let

$$\nu_{q,n}[X] = \int_{\mathcal{X}} [f_X(x)]^{n(q-1)+1} dx, \quad n = 0, 1, \dots.$$

Then the connection of $\mu_{q,n}[X]$ with the q-moment of order n of X is given by

$$\mu_{q,n}[X] = \nu_{q,n}[X]\, \mathbb{E}_{n,q}[X^n], \quad n = 0, 1, \dots.$$

Making use of (3.56) one can easily verify that

$$\frac{d^n \varphi_{q,X}(0)}{d\xi^n} = i^n A_n \mu_{q,n}[X], \quad n = 0, 1, \dots,$$

where $A_n, n = 0, 1, \dots$, are defined in (2.38). This leads to the following proposition.

Proposition 3.11. *Suppose the q-moment $\mathbb{E}_{n,q}[X^n]$ of order n of the random variable X exists. Then the following relations holds:*

$$\frac{d^n \varphi_{q,X}(0)}{d\xi^n} = i^n A_n \nu_{q,n}[X]\, \mathbb{E}_{n,q}[X^n], \quad n = 0, 1, \dots.$$

Proposition 3.12.

(1) *Let $\mu_{q,1}[X] < \infty$ for a random variable X. Then for its q-characteristic function the following asymptote near zero holds:*

$$\varphi_{q,X}(\xi) = 1 + i\mu_{q,1}\xi + o(|\xi|), \quad |\xi| \to 0.$$

(2) *Let $\mu_{q,1}[X] = 0$ and $\mu_{q,2}[X] < \infty$ for a random variable X. Then for its q-characteristic function the following asymptote near zero holds:*

$$\varphi_{q,X}(\xi) = 1 - \frac{q}{2}\,\mu_{q,2}[X]\,\xi^2 + o(|\xi|^2), \quad |\xi| \to 0.$$

Proof. The proof immediately follows from representation (3.56).

$\square$

For symmetric random variables $\mu_{q,n} = 0$ for odd n, since q moments of odd orders vanish, consequently, the power series in (3.56) takes the form

$$\varphi_{q,X}(\xi) = \sum_{n=0}^{\infty} \frac{(-1)^n \, A_{2n} \, \mu_{q,2n}[X]}{(2n)!} \, \xi^{2n}. \tag{3.58}$$

Proposition 3.13. *Let $\mu_{q,2}[X] < \infty$ for a symmetric random variable X. Then for its q-characteristic function the following asymptote near zero holds:*

$$\varphi_{q,X}(\xi) = 1 - \frac{q}{2} \, \mu_{q,2}[X] \, \xi^2 + o(|\xi|^3), \quad |\xi| \to 0,$$

Proof. The proof immediately follows from representation (3.58).

□

Example 3.3. Let $q > 1$ and $X = N_q$ be the q-Gaussian random variable with the density function $G_q(\beta; x)$ defined in (2.66). Taking into account the fact that $\mu_{q,2n+1} = 0, \; n = 0, 1, \ldots$, and

$$\mu_{q,2n}[N_q] = \frac{\Gamma(n + \frac{1}{2})\Gamma(n + \frac{1}{q-1} - \frac{1}{2})}{\sqrt{\pi}\beta^n (q-1)^n \Gamma(2n + \frac{1}{q-1} - \frac{1}{2})}, \quad n = 0, 1, \ldots,$$

we have the following power series representation for the q-characteristic function of the q-Gaussian distribution N_q :

$$\varphi_{q,N_q}(\xi) = \frac{1}{\sqrt{\pi}} \sum_{n=0}^{\infty} \frac{(-1)^n A_{2n} \, \Gamma(n + \frac{1}{2}) \, \Gamma(n + \frac{1}{q-1} - \frac{1}{2})}{\beta^n (q-1)^n (2n)! \, \Gamma(2n + \frac{1}{q-1} - \frac{1}{2})} \xi^{2n}. \tag{3.59}$$

For the q-characteristic function of the q-Gaussian distribution the following asymptotic relation near zero holds:

$$\varphi_{q,N_q}(\xi) = 1 - \frac{q}{q+1} \, \xi^2 + o(|\xi|^3), \quad |\xi| \to 0.$$

Similarly, one can find q-characteristic function of N_q and its asymptotic behavior near zero for $q < 1$. Also one can verify that $\varphi_{q,N_q}(\xi)$ recovers the characteristic function of the standard normal distribution if $q = 1$ and $\beta = 1/2$ (see Section 3.16 for the proof of this fact).

3.6. *q*-Fourier Transform of Symmetric Densities

In this section, we will study some useful properties of q-Fourier transforms of probability density functions of symmetric random variables with zero mean. Density functions of such variables are even functions. The first and obvious example of symmetric random variables is a q-Gaussian with the density function $G_q(\beta; x)$ defined in (2.66). In the subsequent chapters, we will see various other examples of symmetric random variables, including q-generalizations of Lévy stable distributions. Therefore, it is useful to know what properties possesses the q-Fourier transform of density functions symmetric about the origin.

In this section we continue assuming that $q \geq 1$.

Proposition 3.14. *Let $f(x)$ be a positive even function in $L_1(\mathbb{R})$. Then its q-Fourier transform can be written in the form*

$$F_q[f](\xi) = \int\limits_{-\infty}^{\infty} f(x) \cos_q(x\xi[f(x)]^{q-1}) dx$$

$$= 2 \int\limits_{0}^{\infty} f(x) \cos_q(x\xi[f(x)]^{q-1}) dx. \tag{3.60}$$

Proof. Notice that, because of the symmetry of f about the origin,

$$\int\limits_{-\infty}^{\infty} \exp_q(ix\xi) \otimes_q f(x) dx = \int\limits_{-\infty}^{\infty} \exp_q(-ix\xi) \otimes_q f(x) dx.$$

Taking this into account, we have

$$F_q[f](\xi) = \frac{1}{2} \int\limits_{-\infty}^{\infty} (\exp_q(ix\xi) \otimes_q f(x) + \exp_q(-ix\xi) \otimes_q f(x)) dx.$$

Now due to (3.25) and Definition (2.60) of the q-cosine, we obtain

$$F_q[f](\xi) = \int\limits_{-\infty}^{\infty} f(x) \frac{\exp_q(ix\xi[f(x)]^{q-1}) + \exp_q(-ix\xi[f(x)]^{q-1})}{2} dx,$$

which coincides with the first equality in (3.60). The second equality is a simple implication of the first one due to the even integrand.   □

For our further considerations we introduce the set of continuous absolute integrable functions with a specific asymptotic behavior at infinity, namely,

$$H_{q,\alpha}(\mathbb{R}) = \{f \in L_1(\mathbb{R}) \cap C(\mathbb{R}) : f(x) \sim C|x|^{-\frac{1+\alpha}{1+\alpha(q-1)}}, \quad |x| \to \infty\}, \tag{3.61}$$

where we assume that $0 < \alpha \leq 2$ and $C = C_f$ is a constant. Here the symbol "$h(x) \sim g(x)$, $|x| \to \infty$" means the asymptotic equivalence of functions $h(x)$ and $g(x)$, when $|x| \to \infty$, that is

$$\lim_{|x| \to \infty} \frac{h(x)}{g(x)} = 1.$$

For a given $f \in H_{q,\alpha}(\mathbb{R})$ the constant $C = C_f$ in (3.61) is defined uniquely by f. We use the notation

$$\phi(q, \alpha) = \frac{\alpha + 1}{1 + \alpha(q - 1)}. \tag{3.62}$$

It is readily seen that $\phi(q, \alpha) > 1$ for all $\alpha \in (0, 2]$ and $q \in [1, 2)$. Moreover,

$$\phi(q, \alpha)(2q - 1) < 3$$

for all $\alpha \in (0, 2)$ and $q \in [1, 2)$. The latter implies that if $\alpha \in (0, 2)$ and $q \in [1, 2)$, then for any $f \in H_{q,\alpha}(\mathbb{R})$

$$\sigma_{2q-1}^2[f] = \int_{\mathbb{R}} x^2 f_{2q-1}(x)dx = \infty,$$

where

$$f_{2q-1}(x) = \frac{[f(x)]^{2q-1}}{\int_{\mathbb{R}} [f(x)]^{2q-1}dx},$$

is the escort density. One can also notice that

$$\phi(q, \alpha) = 1 + \alpha^*(q, \alpha),$$

where

$$\alpha^* = \alpha^*(q, \alpha) = \frac{\alpha(2-q)}{1+\alpha(q-1)}.$$

Remark 3.8. In Chapter 5, we will see that the probability density function $g(x)$ of any α^*-stable Lévy distribution has the asymptotic behavior $g(x) \sim C/|x|^{1+\alpha^*}$, $|x| \to \infty$. Hence, for a fixed $q \in [1, 2)$ the set of functions $H_{q,\alpha}(\mathbb{R})$ is isomorphic to the set of densities of α^*-stable Lévy distributions.

The following proposition plays a key role in our further analysis. In this proposition the set Q_2 is the rectangle

$$Q_2 = \{(q, \alpha) : 1 \le q < 2, \ 0 < \alpha < 2\}.^{[1]}$$

Proposition 3.15. *Let $f(x)$, $x \in \mathbb{R}$, be a symmetric about the origin probability density function of a random variable X. Further, let either*

(i) *the $(2q-1)$-variance $\sigma_{2q-1}^2[f] < \infty$, (associated with $\alpha = 2$, and $1 \le q < 2$), or*
(ii) *$f(x) \in H_{q,\alpha}(\mathbb{R})$, where $(q, \alpha) \in Q_2$.*

Then, for the q-Fourier transform of $f(x)$, the following asymptotic relation holds:

$$F_q[f](\xi) = 1 - \mu_{q,\alpha}|\xi|^\alpha + o(|\xi|^\alpha), \xi \to 0, \tag{3.63}$$

where

$$\mu_{q,\alpha} = \begin{cases} \dfrac{q}{2}\sigma_{2q-1}^2[f] \ \nu_{2q-1}[f], & \text{if } \alpha = 2; \\ \dfrac{2^{2-\alpha}(1+\alpha(q-1))C_f}{2-q} \displaystyle\int_0^\infty \frac{-\Psi_q(y)}{y^{\alpha+1}} dy, & \text{if } (q, \alpha) \in Q_2. \end{cases}$$

$$\tag{3.64}$$

---

[1]Notice that $1 \le 2q - 1 < 3$, if $1 \le q < 2$.

with

$$\nu_{2q-1}[f] = \int\limits_{-\infty}^{\infty} [f(x)]^{2q-1}\, dx.$$

Proof.

(1) First, assume that $\alpha = 2$. By Proposition 3.14,

$$F_q[f](\xi) = \int\limits_{-\infty}^{\infty} \exp_q(ix\xi) \otimes_q f(x)dx$$

$$= \int\limits_{-\infty}^{\infty} f(x)\cos_q(x\xi[f(x)]^{q-1})dx \quad \xi \in \mathbb{R}. \qquad (3.65)$$

Further, since the q-Fourier transform of f coincides with the q-characteristic function of X, making use of the asymptotic relation in Proposition 3.13, we have

$$F_q[f](\xi) = \varphi_{q,X}(\xi) = 1 - \frac{q}{2}\,\mu_{q,2}[X]\xi^2 + o(|\xi|^3)$$

$$= 1 - \frac{q}{2}\,\sigma^2_{2q-1}[f]\,\nu_{2q-1}[f]\,\xi^2 + o(|\xi|^3), \quad \xi \to 0, \tag{3.66}$$

This implies Part (1) of the proposition due to the equality

$$\mu_{q,2}[X] = \nu_{q,2}[X]\,\mathbb{E}_{2,q}[X^2] = \nu_{2q-1}[f]\,\sigma^2_{2q-1}[f].$$

(2) Now, we assume $(q,\alpha) \in \mathcal{Q}_2$. Applying Proposition 3.14 and taking into account that $f(x)$ is a density function, we have

$$F_q[f](\xi) - 1 = \int\limits_{-\infty}^{\infty} f(x)[\cos_q(x\xi[f(x)]^{q-1}) - 1]dx.$$

Let N be a sufficiently large finite number. Taking into account that the integrand is even function, we have

$$F_q[f](\xi) - 1 = 2 \int_0^N f(x)\, h_q \left(\frac{x\xi[f(x)]^{q-1}}{2} \right) dx$$

$$+ 2 \int_N^\infty f(x)\, h_q \left(\frac{x\xi[f(x)]^{q-1}}{2} \right) dx, \qquad (3.67)$$

where

$$h_q(u) = \cos_q(u) - 1, \quad u \in \mathbb{R}.$$

In Propositions 2.23 and 2.24, we proved that the function $h_q(u)$ has the asymptotic behavior

$$h_q(u) = -2\,q\,u^2 + o(u^3), \quad u \to 0,$$

near zero and satisfies the estimate $-3 \le h_q(u) \le 0$ for all $u \in \mathbb{R}$. Therefore, for every finite x, one has

$$h_q \left(\frac{x\xi[f(x)]^{q-1}}{2} \right) = -\frac{q}{2}\, x^2\, [f(x)]^{2q-2}\, \xi^2 + o(|\xi|^3), \quad \xi \to 0.$$

Hence, in the first integral of the right-hand side of (3.67), we have

$$2 \int_0^N f(x) h_q \left(\frac{x\xi[f(x)]^{q-1}}{2} \right) dx$$

$$= -q\xi^2 \int_0^N x^2 f^{2q-1}(x)dx + o(|\xi|^3), \quad \xi \to 0, \qquad (3.68)$$

It follows from (3.68) that

$$2 \int_0^N f(x) h_q \left(\frac{x\xi[f(x)]^{q-1}}{2} \right) dx = o(|\xi|^\delta), \quad \xi \to 0, \qquad (3.69)$$

for any $0 < \delta < 2$.

By taking into account in the second integral the hypothesis of the proposition with respect to $f(x)$ when $|x| \to \infty$, we have

$$2 \int_N^\infty f(x) h_q \left(\frac{x\xi[f(x)]^{q-1}}{2} \right) dx$$

$$\sim 2C_f \int_N^\infty \frac{1}{x^{\frac{\alpha+1}{1+\alpha(1-q)}}} h_q \left(\frac{x^{1-\frac{(\alpha+1)(q-1)}{1+\alpha(q-1)}}\xi}{2C_f^{1-q}} \right) dx. \qquad (3.70)$$

By using now the substitution

$$x^{\frac{2-q}{1+\alpha(q-1)}} = \frac{2y}{C_f^{q-1}\xi}$$

in the integral in (3.70) we obtain

$$2 \int_N^\infty f(x) h_q \left(\frac{x\xi[f(x)]^{q-1}}{2} \right) dx = \mu_{q,\alpha}|\xi|^\alpha + o(|\xi|^\alpha), \quad \xi \to 0,$$

$$(3.71)$$

where

$$\mu_{q,\alpha} = -\frac{2^{2-\alpha}(1+\alpha(q-1))C_f}{2-q} \int_0^\infty \frac{h_q(y)}{y^{\alpha+1}} dy.$$

Thus, the obtained asymptotic relations (3.69) (with $\delta = \alpha$) and (3.71) complete the proof. □

Remark 3.9.

(1) For stable distributions, which will be studied in Chapter 5, $\mu_{q,\alpha}$ must be positive. We have seen in Proposition 2.24 that if $q \geq 1$, then $h_q(x) \leq 0$ (not being identically zero), which yields $\mu_{q,\alpha} > 0$.
(2) Note also that Part (1) of Proposition 3.15 is valid not only for symmetric density functions. In fact, as it follows from Proposition 3.12, Part (2), the conditions $\sigma_{2q-1}^2[f] < \infty$ and $\mathbb{E}_{1,q}[X] = 0$ are sufficient for the asymptote in (3.63) for $\alpha = 2$.

3.7. *q*-Fourier Transform for $q < 1$

In this section, we discuss some issues related to the q-Fourier transform in the case $q < 1$. Some statements provided below are true in the case when $q \geq 1$, as well. Therefore, in statements of this section we indicate the exact range of q.

In Section 3.2, we introduced the function $g_q(a; u)$ for $q > 1$ defined as

$$g_q(a; u) = u \exp_q(iau^{q-1}), \quad u \in \mathbb{R}_+, \tag{3.72}$$

where $a \in \mathbb{R}$. This function can be used in the case $q < 1$ as well, though it behaves differently than for the case $q > 1$. Namely, the following proposition holds.

Proposition 3.16. *Let* $q < 1$ *and* b *be a positive number. The function* $g_q(a; u)$ *defined in* (3.72) *possesses the following properties*:

(1) *it is continuous on the closed interval* $[0, b]$;
(2) *it is differentiable in the open interval* $(0, b)$;
(3) *there exists* $u_* \in (0, b)$ *such that for all* $0 < u < v < b$ *the equality*

$$g_q(a; u) - g_q(a; v) = g'_q(a; u_*)(u - v) \tag{3.73}$$

holds. Here $g'_q(a; u_*)$ *is the derivative of* $g_q(a; u)$ *with respect to the variable* u *evaluated at* $u = u_*$.

Proof. Let $q < 1$. We need to show that $g_q(a; u)$ has a finite (in the complex plane) limit L, when $u \to 0 +$. Using L'Hopital's rule, we have

$$L = \lim_{u \to 0+} g_q(a; u) = \lim_{u \to 0+} \frac{\exp_q\left(\frac{ia}{u^{1-q}}\right)}{\frac{1}{u}}$$

$$= ia(1 - q) \lim_{u \to 0+} \frac{[\exp_q\left(\frac{ia}{u^{1-q}}\right)]^q}{\frac{1}{u^q}}$$

$$= ia(1 - q) \left[\lim_{u \to 0+} \frac{\exp_q\left(\frac{ia}{u^{1-q}}\right)}{\frac{1}{u}}\right]^q$$

$$= ia(1 - q)[\lim_{u \to +} h_q(u)]^q = ia(1 - q)L^q.$$

It follows that[2]

$$L = [ia(1-q)]^{\frac{1}{1-q}},$$

and $|L| < \infty$, proving Part (1) for $q < 1$. Part (2) for $q < 1$ is similar to the case $q > 1$, and it follows from differentiability of $\exp_q(z)$ and $\rho(u) = iau^{q-1}$ for all $u > 0$. Part (3) is a simple implication of Parts (1) and (2), due to Lagrange's average theorem.   □

Remark 3.10. Unlike for the case $q > 1$, the function $g_q'(a; u)$ is not bounded on the interval $[0, \infty)$ if $q < 1$.

If $q < 1$, then the statement of Theorem 3.1 is not valid. Consider an example. Let $q = 0$ and $f(x) = 1$, if $x \in [0, 1]$, and $f(x) = 0$, otherwise. Then we have

$$F_0[f](\xi) = \int_0^1 1 \otimes_0 \exp_0(ix\xi)dx$$

$$= \int_0^1 \exp_0(ix\xi)dx$$

$$= \int_0^1 (1 + ix\xi)dx = 1 + \frac{1}{2}\xi i.$$

Obviously, this function does not belong to $C_0(\mathbb{R})$. However, it is continuous and bounded on every compact set of $\mathbb{R}$. It turns out that this conclusion is valid not only for this particular example, but it reflects the general case. In order to establish this fact, let us introduce the space $C_{\text{loc}}(\mathbb{R})$ consisting of continuous functions in $\mathbb{R}$. The convergence in $C_{\text{loc}}(\mathbb{R})$ is defined as follows: the sequence of functions $f_n(x) \in C_{\text{loc}}(\mathbb{R})$ is said to converge to a function $f(x) \in C_{\text{loc}}(\mathbb{R})$ if for any compact set $K \subset \mathbb{R}$ the uniform convergence $f_n(x) \to f(x)$ holds on K.

We also introduce the space $L_{\text{com}}(\mathbb{R})$ of absolutely integrable functions with compact support endowed with the convergence: the sequence of functions $f_n(x) \in L_{\text{com}}(\mathbb{R})$ is said to converge to a function $f(x) \in L_{\text{com}}(\mathbb{R})$, if

---

[2]Recall, $L = 0$, if $q \geq 1$.

(1) there exists a compact set $K \subset \mathbb{R}$ such that $\text{supp}[f_n] \subseteq K$ for all $n = 1, 2, \ldots$; and

(2) $f_n(x) \to f(x)$ in the norm of $L_1(\mathbb{R})$.

Proposition 3.17. *Let $q < 1$ and $f \in L_{\text{com}}(\mathbb{R})$. Then the q-Fourier transform $F_q[f](\xi)$ of $f(x)$ exists and belongs to the space $C_{\text{loc}}(\mathbb{R})$.*

Proof. Let $\text{supp}[f] = [a, b]$ and $f \in L_1[a, b]$. Then

$$F_q[f](\xi) = \int_a^b [[f(x)]^{1-q} + (1-q)ix\xi]^{\frac{1}{1-q}} \, dx.$$

Now the proof easily follows from Lebesgue's dominated convergence theorem. □

Remark 3.11. For $q < 1$ the mapping $F_q : L_{\text{com}}(\mathbb{R}) \to C_{\text{loc}}(\mathbb{R})$ is not continuous. Consider an example. Let $q = 0$ and the sequence $f_m(x), m = 1, 2, \ldots$, is defined by

$$f_m(x) = \begin{cases} \dfrac{1}{m}, & \text{if } x \in [0, 1], \\ 0, & \text{if } x \notin [0, 1]. \end{cases}$$

Obviously, $f_m \in L_{\text{com}}(\mathbb{R}), m = 1, 2, \ldots$, and $f_m \to 0$ in the sense of the convergence of $L_{\text{com}}(\mathbb{R})$. However, the sequence of 0-Fourier transforms of f_m are

$$F_0[f_m](\xi) = \frac{1}{m} + \frac{\xi}{2}i,$$

and the sequence $F_0[f_m](\xi)$ does not converge uniformly to 0 as $m \to \infty$ on any compact set of $\mathbb{R}$.

Proposition 3.18. *For all $q < 3$ the q-Fourier transform of $e_q^{-\beta x^2}$, $\beta > 0$, can be written in the form*

$$F_q[e_q^{-\beta x^2}](\xi) = \int_{-K_\beta}^{K_\beta} \exp_q(-\beta x^2 + ix\xi)dx, \qquad (3.74)$$

where

$$K_\beta = \begin{cases} [\beta(1-q)]^{-1/2}, & \text{if } q < 1, \\ \infty, & \text{if } q \geq 1. \end{cases}$$

Corollary 3.7. *Let $q < 3$. Then*

$$F_q[e_q^{-\beta x^2}](\xi) = 2 \int_0^{K_\beta} e_q^{-\beta x^2} \cos_q \left(\frac{x\xi}{[e_q^{-\beta x^2}]^{1-q}} \right) dx.$$

Proposition 3.19. *Let $q < 1$. Then*

$$F_q[G_q(\beta, x)](\xi) = e_{q_1}^{-\beta_*\xi^2} \left[1 - \frac{2}{C_q} \mathrm{Im} \int_0^{d_\xi} \exp_q(b_\xi + i\tau)d\tau \right],$$

$$\xi \in (-K_{\frac{1}{4\beta}}, K_{\frac{1}{4\beta}}),$$

where q_1 and β_* are as in Theorem 3.4 and

$$b_\xi + id_\xi = \frac{K_\beta\sqrt{\beta} - i\frac{\xi}{2\sqrt{\beta}}}{\left[e_q^{-\frac{\xi^2}{4\beta}} \right]^{\frac{1-q}{2}}}.$$

Proof. The proof of this statement can be obtained applying the Cauchy theorem to the integral of the function $\exp_q(-\beta z^2 + iz\xi)$ over the closed contour $C = C_0 \cup C_1 \cup C_- \cup C_+$, where $C_p = (-K_\beta + pi, K_\beta + ip)$, $p = 0, 1$, and $C_\pm = [\pm K_\beta, \pm K_\beta + i]$. $\quad\square$

Unifying Theorem 3.4 and Proposition 3.19, one has

$$F_q[G_q(\beta, x)](\xi) = e_{q_1}^{-\beta_*\xi^2} + I_{(-\infty,1)}(q) \, T_q(\xi),$$

where $I_{(a,b)}(\cdot)$ designates the indicator function of an interval (a, b), and

$$T_q(\xi) = -\frac{2}{C_q} e_{q_1}^{-\beta_*\xi^2} \mathrm{Im} \int_0^{d_\xi} e_q^{b_\xi + i\tau} d\tau.$$

Thus, for $q \geq 1$, operator F_q transforms a q-Gaussian to a q_1-Gaussian with the factor $C_{q_1}\beta^{-1/2}$. However, for $q < 1$ the additional tail $T_q(\xi)$ appears. The question whether the q-Fourier transform of the q-Gaussian density is again a q'-Gaussian with some q' in the case $q < 1$ will be studied in Section 3.9.

3.8. Functional-Differential Equation for the q-Fourier Transform of the q-Gaussian Density

In this section, we will derive a functional-differential equation for the q-Fourier transform of the q-Gaussian density $G_q(\beta; x)$. The obtained functional-differential equation will be used to study further important properties of $F_q[G_q(\beta; \cdot)](\xi)$.

Let $\widehat{G}_q(\beta, \xi)$ be the q-Fourier transform of a q-Gaussian $G_q(\beta, \xi)$, that is $\widehat{G}_q(\beta, \xi) = F_q[G_q(\beta, \cdot)](\xi)$, and $\widehat{G}_q(\xi) = \widehat{G}_q(1, \xi)$ for $\beta = 1$. Further, let $Y_q(\xi) = F_q[e_q^{-x^2}](\xi)$. In accordance with Proposition 3.18,

$$Y_q(\xi) = \int_{-K}^{K} \exp_q(-x^2 + ix\xi)dx, \quad \xi \in \mathbb{R}, \tag{3.75}$$

where

$$K = \begin{cases} \dfrac{1}{\sqrt{1-q}}, & \text{if } q < 1, \\ \infty, & \text{if } q \geq 1. \end{cases}$$

Lemma 3.8. *For any $q < 3$ and $\beta > 0$ the following relations hold:*

(1) $\widehat{G}_q(\beta, \xi) = \widehat{G}_q(\frac{\xi}{(\sqrt{\beta})^{2-q}})$;
(2) $\widehat{G}_q(\xi) = \frac{1}{C_q}Y_q(C_q^{1-q}\xi)$.

Proof.

(1) Using Proposition 3.2, Parts (2) and (3) with $a = b = \sqrt{\beta}$, we have

$$\widehat{G}_q(\beta, \xi) = F_q[G_q(\beta; x)](\xi) = F_q[\sqrt{\beta}G(1; \sqrt{\beta}x)](\xi)$$

$$= \frac{\sqrt{\beta}}{\sqrt{\beta}}F_q[G_q(1; x)]\left(\frac{\xi}{(\sqrt{\beta}\,\sqrt{\beta})^{1-q}}\right)$$

$$= \widehat{G}_q\left(\frac{\xi}{(\sqrt{\beta})^{2-q}}\right).$$

(2) Again using Proposition 3.2, Part (2) with $a = C_q^{-1}$, we obtain

$$\widehat{G}_q(\xi) = F_q[G_q(1;x)](\xi) = F_q\left[\frac{1}{C_q}e^{-x^2}\right](\xi) = \frac{1}{C_q}Y_q(C_q^{1-q}\xi).$$

$\square$

It is not hard to see that the two formulas in Lemma 3.8 yield

$$\widehat{G}_q(\beta, \xi) = \frac{1}{C_q}Y_q\left(\left(\frac{C_q}{\sqrt{\beta}}\right)^{1-q}\frac{\xi}{\sqrt{\beta}}\right). \tag{3.76}$$

Moreover, $\widehat{G}_q(\beta, 0) = 1$, which implies $\widehat{G}_q(0) = 1$, and therefore

$$Y_q(0) = C_q. \tag{3.77}$$

Thus, in order to know the properties of the q-Fourier transform of q-Gaussians it suffices to study the properties of $Y_q(\xi)$.

Theorem 3.9. *Let* $1 \leq q < 3$ *and* q_n, $n < N_* = \left\lfloor \frac{2}{q-1} - 1 \right\rfloor$, *be defined as in equation* (3.40). *Then* $Y_{q_n}(\xi)$ *satisfies the following functional-differential equation*

$$2\sqrt{q_n}\frac{dY_{q_n}(\xi)}{d\xi} + \xi Y_{q_{n-2}}(\sqrt{q_n}\xi) = 0, \quad \xi \in \mathbb{R}. \tag{3.78}$$

Proof. Let $1 \leq q < 3$. Then $K = \infty$. Differentiating

$$Y_q(\xi) = \int_{-\infty}^{\infty} \exp_q(-x^2 + ix\xi)dx$$

in the variable ξ under the integral sign (which is valid due to the theorem on differentiation under the integral sign), we have

$$\frac{dY_q(\xi)}{d\xi} = i\int_{-\infty}^{\infty} x\left[\exp_q(-x^2 + ix\xi)\right]^q dx$$

$$= i\lim_{K\to\infty} \int_{-K}^{K} x[\exp_q(-x^2 + ix\xi)]^q dx.$$

Further, integrating by parts,

$$\frac{dY_q(\xi)}{d\xi} = \frac{-i}{2} \lim_{K\to\infty} \int_{-K}^{K} d(\exp_q(-x^2 + ix\xi))$$

$$- \frac{\xi}{2} \lim_{K\to\infty} \int_{-K}^{K} (\exp_q(-x^2 + ix\xi))^q dx. \qquad (3.79)$$

For the first integral on the right-hand side of (3.79), we have the estimate

$$\left| \int_{-K}^{K} d(\exp_q(-x^2 + ix\xi)) \right|$$

$$\le |\exp_q(-K^2 + iK\xi)| + |\exp_q(-K^2 - iK\xi)|$$

$$= |e_q^{-K^2} \otimes_q \exp_q(iK\xi)| + |e_q^{-K^2} \otimes_q \exp_q(-iK\xi)|$$

$$\le 2|e_q^{-K^2}| \to 0, \quad K \to \infty. \qquad (3.80)$$

Here we used Corollary 2.5 and the fact that $e_q^{-x^2} \to 0$ as $|x| \to \infty$, if $q \ge 1$.

Further, using $(e_q^y)^q = e_{2-1/q}^{qy}$ (see equation (2.48)), which is valid for any $q < 3$, the second integral on the right-hand side of (3.79) can be represented in the form

$$\int_{-K}^{K} (\exp_q(-x^2 + ix\xi))^q dx = \frac{1}{\sqrt{q}} \int_{-K}^{K} \exp_{2-1/q}(-x^2 + ix\sqrt{q}\xi)dx.$$

Hence,

$$\lim_{K\to\infty} \int_{-K}^{K} (\exp_q(-x^2 + ix\xi))^q dx = \frac{1}{\sqrt{q}} Y_{2-1/q}\left(\sqrt{q}\xi\right). \qquad (3.81)$$

Thus, it follows from (3.79), (3.80), and (3.81) that for $q \ge 1$ the function $Y_q(\xi) = F_q[e_q^{-x^2}](\xi)$ satisfies the functional-differential equation

$$2\sqrt{q}\frac{dY_q(\xi)}{d\xi} + \xi Y_{2-1/q}(\sqrt{q}\xi) = 0, \quad \xi \in \mathbb{R}. \tag{3.82}$$

Finally, setting $q = q_n$, $n < N_*$, and taking into account the relation $2 - 1/q_n = q_{n-2}$ (Proposition 3.6), we obtain equation (3.78).   □

Theorem 3.10. *Let $0 < q < 1$ and $q \neq l/(l+1), l = 1, 2, \dots$. Then $Y_q(\xi)$ satisfies the following functional-differential equation*

$$2\sqrt{q}\frac{dY_q(\xi)}{d\xi} + \xi Y_{2-1/q}(\sqrt{q}\xi) = r_q \xi^{\frac{1}{1-q}}, \quad \xi \in \mathbb{R}, \tag{3.83}$$

where

$$r_q = 2\sqrt{q}(1-q)^{\frac{1}{2(1-q)}} \sin\frac{\pi}{2(1-q)}. \tag{3.84}$$

Proof. Assume that $0 < q < 1$ and $q \neq \frac{l}{l+1}, l = 1, 2, \dots$. In this case $K = 1/\sqrt{1-q}$. Differentiating

$$Y_q(\xi) = \int_{-K}^{K} \exp_q(-x^2 + ix\xi)dx$$

under the integral sign, we have

$$\frac{dY_q(\xi)}{d\xi} = i\int_{-K}^{K} x[\exp_q(-x^2 + ix\xi)]^q dx$$

$$= \int_{-K}^{K} d(\exp_q(-x^2 + ix\xi))$$

$$- \frac{\xi}{2}\int_{-K}^{K} (\exp_q(-x^2 + ix\xi))^q dx. \tag{3.85}$$

Unlike the case $q \geq 1$, now the first integral on the right-hand side of (3.85) does not vanish, and takes the form

$$\int_{-K}^{K} d(\exp_q(-x^2 + ix\xi)) = \exp_q(-K^2 + iK\xi) - \exp_q(-K^2 - iK\xi)$$

$$= 2i \operatorname{Im}(\exp_q(-K^2 + iK\xi)),$$

where $\operatorname{Im}(z)$ means the imaginary part of a complex number z. One can easily verify that $e_q^{-K^2} = 0$. Taking this into account, we have

$$\exp_q(-K^2 + iK\xi) = 0 \otimes_q \exp_q(iK\xi) = [i(1-q)K\xi]^{\frac{1}{1-q}}.$$

Further, substituting $K = 1/\sqrt{1-q}$, we obtain

$$\operatorname{Im}([i(1-q)K\xi]^{\frac{1}{1-q}}) = (1-q)^{\frac{1}{2(1-q)}} \sin\left(\frac{\pi}{2(1-q)}\right) \xi^{\frac{1}{1-q}}.$$

Note that the second integral on the right-hand side of equation (3.85) is the same as in the case of $1 < q < 3$. Consequently, $Y_q(\xi) = F_q[e_q^{-x^2}](\xi)$ satisfies the functional-differential equation

$$2\sqrt{q}\frac{dY_q(\xi)}{\partial \xi} + \xi Y_{2-1/q}(\sqrt{q}\xi) = r_q \xi^{\frac{1}{1-q}}, \tag{3.86}$$

where r_q is given in equation (3.84). $\qquad\square$

Remark 3.12.

(1) If $q = 0$ then it can be readily seen that

$$Y_0(\xi) = F_0[e_0^{-x^2}](\xi) = \int_{-1}^{1}(1 - x^2 + ix\xi)dx = 4/3,$$

i.e., a constant for all $\xi \in \mathbb{R}$. Obviously, such $Y_0(\xi)$ does not satisfy the functional-differential equation (3.83), and cannot be a q-Gaussian for any q.

(2) We will show later that the q-Fourier transform of any q-Gaussian with $q < 0$ can not be a function of the form $ae_{q'}^{-\beta\xi^2}$, for any $q' \in (-\infty, 3)$ (see Theorem 3.17).

Let us now consider the cases $q = \ell/(\ell+1)$, $\ell = 1, 2, ...,$ excluded from Theorem 3.10. For these values of q we have $K = \sqrt{\ell+1}$ and

$$Y_q(\xi) = F_q[e_q^{-x^2}](\xi) = \int_{-\sqrt{\ell+1}}^{\sqrt{\ell+1}} \left(1 - \frac{1}{\ell+1}x^2 + \frac{1}{\ell+1}ix\xi\right)^{\ell+1} dx.$$

The latter is a polynomial of order ℓ if ℓ is even, and of order $\ell+1$ if ℓ is odd.[3] In order to emphasize that $Y_q(\xi)$ is a polynomial, we use the notation $P_{\ell+1}(\xi) = Y_{\ell/(\ell+1)}(\xi)$ indicating the dependence on ℓ. Further, obviously $2 - \frac{1}{q} = \frac{\ell-1}{\ell}$. Therefore,

$$Y_{2-1/q}(\xi) = \int_{-\sqrt{\ell}}^{\sqrt{\ell}} \left(1 - \frac{1}{\ell}x^2 + \frac{1}{\ell}ix\xi\right)^{\ell} dx = P_{\ell}(\xi).$$

We notice that $P_{\ell}(\xi)$ is a polynomial of order ℓ if ℓ is even, and of order $\ell - 1$ if ℓ is odd. Moreover, $P_{\ell}(\xi)$ is an even function of ξ and $P_{\ell}(0) = C_{\frac{\ell-1}{\ell}} > 0$. Let ρ be the root of $P_{\ell}(\xi)$ closest to the origin. In our further analysis we will consider $P_{\ell}(\xi)$ only on the positivity interval $(-\rho, \rho)$.

Theorem 3.11. *Let* $q = \frac{2m-1}{2m}$, $m = 1, 2, ...$ *Then* $Y_q(\xi)$ *satisfies the functional-differential equation*

$$2\sqrt{q}\frac{dY_q(\xi)}{d\xi} + \xi Y_{2-1/q}(\sqrt{q}\xi) = 0, \quad \xi \in \mathbb{R}. \tag{3.87}$$

Proof. Assume $\ell+1 = 2m$, $m = 1, 2, \ldots$. In this case $Y_q(\xi) = P_{2m}(\xi)$ is a polynomial of order $2m$ and $Y_{2-1/q}(\xi) = P_{2m-1}(\xi)$ is a polynomial of order $2m - 2$. Moreover,

$$r_q = 2\sqrt{\frac{2m-1}{2m}}\left(\frac{1}{2m}\right)^m \sin m\pi = 0, \quad m = 1, 2, \ldots.$$

Thus, $Y_q(\xi)$ satisfies the equation (3.87).   □

Theorem 3.12. *Let* $q = \frac{2m}{2m+1}$, $m = 1, 2, \ldots$. *Then* $Y_q(\xi)$ *satisfies neither functional-differential equation* (3.83) *nor* (3.87).

---

[3]This polynomial does not contain odd order terms.

Proof. Let $\ell = 2m, m = 1, 2, \ldots$. Then $Y_q(\xi) = P_{2m+1}(\xi)$ is a polynomial of order $2m$, and so is $Y_{2-1/q}(\xi) = P_{2m}(\xi)$. Assume $Y_q(\xi)$ satisfies equation (3.87), which in this particular case takes the form

$$2\sqrt{q}\frac{dY_q(\xi)}{d\xi} = -\xi P_{2m}(\xi), \quad \xi \in \mathbb{R}. \tag{3.88}$$

Equation (3.88) is clearly inconsistent, since the left-hand side of this equation is a polynomial of order $2m - 1$, while the right hand side is a polynomial of order $2m + 1$. Analogously, $Y_q(\xi)$ cannot satisfy equation (3.83) either. Indeed, if $Y_q(\xi)$ would solve equation (3.83) then in this case the equation would read

$$2\sqrt{q}\frac{dY_q(\xi)}{d\xi} + \xi P_{2m}(\xi) = \frac{(-1)^m}{(2m-1)^{m-\frac{1}{2}}}\xi^{2m+1}. \tag{3.89}$$

But equation (3.89) is inconsistent as well, since the term of the highest order on the left is

$$\frac{2(-1)^m}{(2m+1)(2m)^{m-1/2}}\xi^{2m+1},$$

which is clearly distinct from the term of the highest order on the right. □

Remark 3.13. Equations (3.83) and (3.87) can easily be generalized for the q-Fourier transform of q-Gaussians with non-zero means. Namely, let $\mu \neq 0$ be a real number, and

$$Y_{\mu,q}(\xi) = \int_{\mu-K}^{\mu+K} e_q^{-(x-\mu)^2+ix\xi}dx.$$

Then the associated functional-differential equation for $Y_{\mu,q}$ with $q \in (0,3)$ takes the form

$$2\sqrt{q}\frac{dY_{\mu,q}(\xi)}{d\xi} + \xi Y_{\frac{\mu}{q},2-1/q}(\sqrt{q}\xi) - 2i\mu\sqrt{q}\,Y_{\mu,q}(\xi)$$

$$= I_{(0,1)}(q)r_q\xi^{\frac{1}{1-q}}, \quad \xi \in \mathbb{R}. \tag{3.90}$$

3.9. Is the q-Fourier Transform of a q-Gaussian Density a q'-Gaussian Again?

In this section, we discuss the above question, which is important from applications point of view. Namely, we prove for which values of q the q-Fourier transform of a q-Gaussian is a q'-Gaussian with some index q' in $(-\infty, 3)$. Recall the set of functions

$$\mathcal{G} = \bigcup_{q<3} \mathcal{G}_q, \quad \text{where } \mathcal{G}_q = \{f : f(x) = ae_q^{-\beta x^2}, \, a > 0, \, \beta > 0\}.$$

$$(3.91)$$

It follows from relationship (3.76) that, if the q-Fourier transform $F_q[G_q(\beta, x)](\xi)$ of a q-Gaussian is a q'-Gaussian with some $q' \in (-\infty, 3)$, then $Y_q(\xi)$ must belong to $\mathcal{G}$. Therefore we will study the existence of a solution of functional-differential equations (3.83) and (3.87) in the set $\mathcal{G}$.

First we provide two additional properties of the sequence q_n, defined in (3.40), which will be used in this section.

Proposition 3.20. *For all* $n < N_* = \left\lfloor \left| \frac{2}{q-1} - 1 \right| \right\rfloor$ *the following relations hold:*

(1) $(3 - q_n)q_{n+1} = (3 - q_{n-2})q_n$,
(2) $2C_{q_{n-2}} = \sqrt{q_n}\,(3 - q_n)\,C_{q_n}$, *where* C_q *is the normalizing constant of the q-Gaussian.*

Proof.

(1) It follows from the definition of q_n that $q_{n+1} = (1 + q_n)/(3 - q_n)$. This yields

$$(3 - q_n)q_{n+1} = 1 + q_n = \left(1 + \frac{1}{q_n}\right)q_n.$$

Further, using the equality $1/q_n = 2 - q_{n-2}$, we obtain Part (1).

(2) Obviously, if $q_0 = q = 1$, then the relationship (2) for all n reads $2\sqrt{\pi} = 2\sqrt{\pi}$. Let $q \neq 1$. Note that C_{q_n} is well defined for all q_n satisfying the condition $q_n < N_*$. Using the explicit form for C_q given in (2.68) and relationship $2 - q_{n-2} = 1/q_n$, one obtains:

in the case $1 < q < 3$,

$$\frac{2C_{q_{n-2}}}{C_{q_n}} = \frac{\sqrt{q_n}}{\frac{1}{2(q_n-1)}} \frac{\Gamma(\frac{1+q_n}{2(q_n-1)})}{\Gamma(\frac{3-q_n}{2(q_n-1)})} = \sqrt{q_n}(3-q_n);$$

and in the case $q < 1$,

$$\frac{2C_{q_{n-2}}}{C_{q_n}} = \frac{\sqrt{q_n}(3-q_n)}{\frac{1+q_n}{2(1-q_n)}} \frac{\Gamma(\frac{3-q_n}{2(1-q_n)})}{\Gamma(\frac{1+q_n}{2(1-q_n)})} = \sqrt{q_n}(3-q_n),$$

completing the proof of Part (2). ☐

Theorem 3.13. *Let $1 \le q < 3$ and q_n, $n < N_*$, be the sequence defined in (3.40). Then the functional-differential equation*

$$2\sqrt{q_n}\frac{dY_{q_n}(\xi)}{d\xi} + \xi Y_{q_{n-2}}(\sqrt{q_n}\xi) = 0, \quad \xi \in \mathbb{R}, \qquad (3.92)$$

has a unique solution $Y_{q_n}(\xi) \in \mathcal{G}$ satisfying the condition

$$Y_{q_n}(0) = C_{q_n}. \qquad (3.93)$$

This solution is specifically

$$Y_{q_n}(\xi) = C_{q_n}e_{q_{n+1}}^{-\frac{3-q_n}{8}\xi^2}. \qquad (3.94)$$

Proof. *Existence.* It follows immediately from representation (3.94) that $Y_{q_n}(0) = C_{q_n}$. Furthermore,

$$\frac{dY_{q_n}(\xi)}{d\xi} = -\frac{1}{4}(3-q_n)\,C_{q_n}\xi\,\left(e_{q_{n+1}}^{-\frac{3-q_n}{8}\xi^2}\right)^{q_{n+1}}, \qquad (3.95)$$

and

$$Y_{q_{n-2}}(\sqrt{q_n}\xi) = C_{q_{n-2}}e_{q_{n-1}}^{-q_n\frac{3-q_{n-2}}{8}\xi^2}. \qquad (3.96)$$

Due to equality $(e_q^y)^q = e_{2-1/q}^{qy}$ and Part (1) of Proposition 3.20, the equation in (3.95) can be rewritten as

$$\frac{dY_{q_n}(\xi)}{d\xi} = -\frac{1}{4}(3-q_n)\,C_{q_n}\xi\,e_{q_{n-1}}^{-q_n\frac{3-q_{n-2}}{8}\xi^2}. \qquad (3.97)$$

Substituting (3.96) and (3.97) into (3.92), we obtain

$$\left(-\sqrt{q_n}C_{q_n}\frac{3-q_n}{2}+C_{q_n-2}\right)e_{q_n-1}^{-\frac{q_n(3-q_n)}{8}\xi^2}=0. \tag{3.98}$$

Now taking into account Part (2) of Proposition 3.20 we conclude that $Y_{q_n}(\xi)$ in (3.94) satisfies (3.92).

Uniqueness. We recall that $|\cos_q(x)|\le 1$ for real x, if $q>1$ (see Proposition 2.10). This fact and Corollary 3.7 imply the following estimate

$$|Y_q(\xi)|=\left|\int_{-\infty}^{\infty}e_q^{-x^2+ix\xi}dx\right|\le\int_{-\infty}^{\infty}e_q^{-x^2}dx=C_q. \tag{3.99}$$

Assume that there are two solutions to problem (3.92) and (3.93), i.e., Y_{q_n} and $\tilde{Y}_{q_n}$. Then, it is not hard to verify that their difference $Z_{q_n}(\xi)=Y_{q_n}(\xi)-\tilde{Y}_{q_n}(\xi)$ also satisfies equation (3.92), and the condition $Z_{q_n}(0)=0$. Now estimate (3.99) yields $Z_{q_n}\equiv 0$, which, in turn, implies $Y_{q_n}\equiv\tilde{Y}_{q_n}$.   □

Corollary 3.5. *Let* $1\le q_n<3$. *Then*

$$F_{q_n}[G_{q_n}](\xi)=e_{q_n+1}^{-\frac{3-q_n}{8\beta^{2-q_n}C_{q_n}^{2(q_n-1)}}\xi^2}. \tag{3.100}$$

Remark 3.14.

(1) Representation (3.100) was obtained in Proposition 3.7 as a corollary of Theorem 3.4. The formula (3.10) in Theorem 3.4 corresponds to the particular case $n=0$ of (3.100) and obtained by the contour integration technique.

(2) If $q=1$ then the Cauchy problem (3.92) and (3.93) takes the linear form

$$2\frac{dY_1(\xi)}{d\xi}+\xi Y_1(\xi)=0,\quad Y_1(0)=\sqrt{\pi},$$

and its unique solution is $Y_1(\xi)=\sqrt{\pi}e^{-\xi^2/4}$. Moreover, the representation (3.100) reduces to

$$F\left[\frac{\sqrt{\beta}}{\sqrt{\pi}}e^{-\beta x^2}\right](\xi)=e^{-\frac{1}{4\beta}\xi^2}.$$

The latter with $\beta = 1/2$ is, in fact, the density function of the standard normal distribution.

Theorem 3.14. *Let $0 < q < 1$. Suppose that $q \neq m/(m+1)$, $m = 1, 2, \ldots$. Then the functional-differential equation*

$$2\sqrt{q}\frac{dY_q(\xi)}{d\xi} + \xi Y_{2-1/q}(\sqrt{q}\xi) = r_q\xi^{\frac{1}{1-q}}, \quad \xi \in \mathbb{R}, \tag{3.101}$$

has no solution in $\mathcal{G}$.

Proof. Let $0 < q < 1$ and $q \neq m/(m+1)$, $m = 1, 2, \ldots$. Recall that any function $f \in \mathcal{G}_q$ with $q < 1$ has a compact support. Further, we notice that a function with compact support cannot solve equation (3.101), because of the expression on the right-hand side. Thus, a solution of equation (3.101) cannot belong to $\mathcal{G}_q$, $q < 1$. Now assume that a solution $Y_q(\xi)$ of equation (3.101) is in $\mathcal{G}_{q'}$ and $Y_{2-1/q}(\xi) \in \mathcal{G}_{q''}$, with $1 < q' < 3$ or $1 < q'' < 3$, respectively.[4] In accordance with definition (3.91),

$$Y_q(\xi) = ae_{q'}^{-b\xi^2} \quad \text{and} \quad Y_{2-1/q}(\xi) = Ae_{q''}^{-B\xi^2},$$

with some a, b, A, B, real positive numbers, which may depend on q. Then, the growth rate of the function $\frac{dY_q(\xi)}{dx}$ is $2q'/(1-q')$, and the growth rate of the function $\xi Y_{2-1/q}(\xi)$ is $2/(1-q'')+1$. Therefore, for equation (3.101) to be consistent with the right-hand side, one needs to require at least one of the following two consistency equations:

$$\frac{2q'}{1-q'} = \frac{1}{1-q} \quad \text{or} \quad \frac{2}{1-q''} + 1 = \frac{1}{1-q}.$$

The first equation has no solution, since its left side is negative for all $q' > 1$, while its right side is positive due to $q < 1$. Solving the second equation for $q'' - 1$, one obtains

$$q'' - 1 = -\frac{2(1-q)}{q}.$$

This equation has no solution if $0 < q < 1$, since in this case the right side is negative, but the left side is positive for any $q'' > 1$. $\square$

---

[4]The reader can easily verify that $q' \neq 1$ and $q'' \neq 1$.

Next, we consider the cases $q = \frac{1}{2}, \frac{2}{3}, \ldots, \frac{m}{m+1}, \ldots$, excluded from Theorem 3.14. Direct computations show that in two specific cases, namely, if $q = 1/2$ and $q = 2/3$, then $Y_q(\xi) \in \mathcal{G}_0$ considered on the positivity intervals. Indeed,

$$Y_{\frac{1}{2}}(\xi) = F_{\frac{1}{2}}[e_{\frac{1}{2}}^{-x^2}](\xi) = \frac{16\sqrt{2}}{15}\left(1 - \frac{5}{16}\xi^2\right), \qquad (3.102)$$

which is non-negative for $\xi \in [-4/\sqrt{5}, 4/\sqrt{5}]$. Therefore on this interval we can associate it with an element of $\mathcal{G}_0$, writing $Y_{1/2}(\xi) = \frac{16\sqrt{2}}{15}e_0^{-(5/16)\xi^2} \in \mathcal{G}_0$. Similarly,

$$Y_{\frac{2}{3}}(\xi) = F_{\frac{2}{3}}[e_{\frac{2}{3}}^{-x^2}](\xi) = \frac{32\sqrt{3}}{35}\left(1 - \frac{7}{24}\xi^2\right) \in \mathcal{G}_0, \qquad (3.103)$$

on the positivity interval $\left(-2\sqrt{\frac{6}{7}}, 2\sqrt{\frac{6}{7}}\right)$.

However, $Y_q(\xi)$ does not belong to $\mathcal{G}$ for any other value of $q = 3/4, 4/5, \ldots$. In order to show this fact, first we derive an explicit form for $P_{m+1}(\xi) = Y_{m/(m+1)}(\xi)$. Recall that $P_{m+1}(\xi)$ is a polynomial of order $m + 1$ if $m + 1$ is even. Otherwise it is a polynomial of order m.

Theorem 3.15. *Let* $q = m/(m + 1), m = 1, 2, \ldots$. *Then* $Y_q(\xi) = P_{m+1}(\xi)$ *has representation*

$$P_{m+1}(\xi) = \sum_{k=0}^{\lfloor \frac{m+1}{2} \rfloor} \frac{(-1)^k}{(m+1)^{k-\frac{1}{2}}} \binom{m+1}{2k} B\left(k + \frac{1}{2}, m - 2k + 2\right) \xi^{2k}, \qquad (3.104)$$

where $\lfloor u \rfloor$ *is the integer part of* u, *and* $B(\cdot, \cdot)$ *is the Euler's beta-function.*

Proof. Recall that if $q = \frac{m}{m+1}$, then $Y_q(\xi)$ has the form

$$Y_q(\xi) = P_{m+1}(\xi) = \int_{-\sqrt{m+1}}^{\sqrt{m+1}} \left(1 - \frac{1}{m+1}x^2 + \frac{1}{m+1}ix\xi\right)^{m+1} dx.$$

Using the binomial formula, we have

$$P_{m+1}(\xi) = \sum_{k=0}^{m+1} \binom{m+1}{k} D_k(m) \frac{(i\xi)^k}{(m+1)^k},$$

where

$$D_k(m) = \int_{-\sqrt{m+1}}^{\sqrt{m+1}} \left(1 - \frac{1}{m+1}x^2\right)^{m-k+1} x^k dx.$$

It is not hard to verify that $D_k(m) = 0$ if k is odd and

$$D_{2k}(m) = (m+1)^{k+1/2} B(k+1/2, m-2k+2)$$

for $k = 0, \ldots, \lfloor \frac{m+1}{2} \rfloor$, which implies representation (3.104). $\square$

Theorem 3.16. *Let* $q = m/(m+1)$, $m = 3, 4, \ldots$ *. Then* $Y_q(\xi) \notin \mathcal{G}$.

Proof. It follows from representation (3.104) that polynomial $Y_q(\xi) = P_{m+1}(\xi)$, with the first three (non-zero) terms indicated, reads

$$Y_q(\xi) = D_0(m)\left[1 - (m+1)^2 \frac{B(\frac{3}{2}, m)}{B(\frac{1}{2}, m+2)}\xi^2\right.$$

$$\left. + \frac{m(m+1)^3}{2} \frac{B(\frac{5}{2}, m-2)}{B(\frac{1}{2}, m+2)}\xi^4 + \cdots \right]$$

$$= D_0(m)\left[1 - \frac{2m+3}{8(m+1)}\xi^2 + \frac{(2m+3)(2m+1)}{8(m+1)^2}\xi^4 + \cdots \right],$$

$$(3.105)$$

where

$$D_0(m) = C_{\frac{m}{m+1}} = \sqrt{m+1} B\left(\frac{1}{2}, m+2\right)$$

$$= \frac{\sqrt{m+1}(m+1)! 2^{m+2}}{(2m+3)!!}.$$

Now assume that $Y_q(\xi) \in \mathcal{G}_{q'}$ for some $q' < 3$. Then $1/(1-q') = (m+1)/2$, or $q' = (m-1)/(m+1)$. Therefore,

$$Y_q(\xi) = D_0(m) \left[1 - \beta(m)\xi^2\right]^{\left[\frac{m+1}{2}\right]},$$

where $\beta(m) > 0$ and $|\xi| \le 1/\sqrt{\beta(m)}$. Applying the binomial formula and indicating the first three terms, one has

$$Y_q(\xi) = D_0(m) \left[1 - \frac{(m+1)\beta(m)}{2}\xi^2 + \frac{(m^2-1)[\beta(m)]^2}{8}\xi^4 + \cdots\right].$$

$$(3.106)$$

Comparing the second and third terms in (3.105) and (3.106), one obtains contradictory relations

$$\beta(m) = \frac{2m+3}{4(m+1)^2}$$

and

$$[\beta(m)]^2 = \frac{(3m+3)(2m+1)}{(m-1)(m+1)^3} \ne \frac{(2m+3)^2}{16(m+1)^4} = [\beta(m)]^2,$$

$$m = 3, 4, \ldots,$$

which completes the proof.   □

Remark 3.15. The formula (3.104) for $q = 1/2$ and $q = 2/3$ gives

$$Y_{\frac{1}{2}}(\xi) = \frac{16\sqrt{2}}{15}(1 - \frac{5}{16}\xi^2) = \frac{16\sqrt{2}}{15}e_0^{-(5/16)\xi^2}, \quad \xi \in \left[-\frac{4\sqrt{5}}{5}, \frac{4\sqrt{5}}{5}\right],$$

and

$$Y_{\frac{2}{3}}(\xi) = \frac{32\sqrt{3}}{35}\left(1 - \frac{7}{24}\xi^2\right) = \frac{32\sqrt{3}}{35}e_0^{-\frac{7}{24}\xi^2}, \quad \xi \in \left[-2\sqrt{\frac{7}{6}}, 2\sqrt{\frac{7}{6}}\right].$$

which coincide with (3.102) and (3.103), respectively. Both functions belong to $\mathcal{G}_0$.

Theorem 3.17. *Let $q \leq 0$. Then $Y_q(\xi) \notin \mathcal{G}$.*

Proof. For $q = 0$ see Remark 3.12. Let $q < 0$. Repeating calculations used in proofs of Theorems 3.9 and 3.10, it is not hard to verify that the derivative of $Y_q(\xi)$ can be represented in the form

$$\frac{dY_q(\xi)}{d\xi} = -\frac{\xi}{2} \int_{-K}^{K} (e_q^{-x^2+ix\xi})^q dx + R_q \xi^{\frac{1}{1-q}}, \qquad (3.107)$$

where

$$R_q = (1-q)^{\frac{1}{2(1-q)}} \sin \frac{\pi}{2(1-q)}. \qquad (3.108)$$

We notice that the condition $q < 0$ implies two key instances that are important for the continuation of the proof, namely $0 < \frac{1}{1-q} < 1$, and $R_q \neq 0$. Now assume that $Y_q \in \mathcal{G}_{q'}$ with some $q' \in (-\infty, 3)$. In other words, there are positive numbers a and b, such that $Y_q(\xi) = a e_{q'}^{-b\xi^2}$. Taking the first derivative of the latter, equating it to the right-hand side of (3.107), and dividing both sides by $\xi^{\frac{1}{1-q}}$ ($\xi \neq 0$), we obtain

$$\xi^{-\frac{q}{1-q}} \left[\frac{1}{2} \int_{-K}^{K} (\exp_q(-x^2 + ix\xi))^q dx - 2ab(e_{q'}^{-b\xi^2})^{q'} \right] = R_q, \qquad (3.109)$$

which must be valid for all $\xi \in \mathbb{R}$. However, the left-hand side becomes arbitrarily small for ξ small, since $-\frac{q}{1-q} > 0$ and expression in brackets has finite limit as $\xi \to 0$, while the right-hand side is a non-zero constant. This contradiction completes the proof. $\square$

Now let us summarize the obtained results. The general picture for the q-Fourier transform of q-Gaussians is the following:

1. The case $1 \leq q < 3$:

   (1a) the q-Fourier transform acts as $F_q : \mathcal{G}_q \to \mathcal{G}_{q'}$;
   (1b) the relation between q and q' is as follows:

$$q' = \frac{1+q}{3-q}. \qquad (3.110)$$

2. The case $q = \frac{1}{2}$ or $q = \frac{2}{3}$:

   (2a) the q-Fourier transform acts as $F_q : \mathcal{G}_q \to \mathcal{G}_0$;

   (2b) the relationship (3.110) fails.

3. The case $q < 1$, but $q \neq \frac{1}{2}, \frac{2}{3}$:

   (3a) for any $f \in \mathcal{G}_q$ there is no q' such that the q-Fourier transform of a f would belong to $\mathcal{G}_{q'}$.

3.10. q-Fourier Transform is not Invertible. Hilhorst's Invariance Principle

We have seen in Section 3.4 that the q-Fourier transform is invertible in the set $\mathcal{G}_{q_1}$. However, the q-Fourier transform is not invertible in the entire space of density functions, or in $L_1(\mathbb{R})$. This is an effect of nonlinearity of the operator F_q. Hilhorst (2009, 2010) constructed examples using the invariance principle and showing non-invertibility of the q-Fourier transform. Therefore, in this section, we briefly discuss Hilhorst's invariance principle, which provides a general approach for construction of families of density functions with the same q-Fourier transform. We start with the following example.

Example 3.4. Consider an example of two functions with the same q-Fourier transform. Let the first function $f_1(x)$ be the density function $f_1(x) = G_q(x) = C_q^{-1} e_q^{-x^2}$ of the q-Gaussian distribution with $1 < q < 2$. Then, we know from Theorem 3.4 that its q-Fourier transform is

$$F_q[G_q(\cdot)](\xi) = e_{q_1}^{-\beta_* \xi^2}, \quad \xi \in \mathbb{R},$$

where $\beta_* = \frac{3-q}{8C_q^{2(q-1)}}$. As the second function consider

$$f_2(x) = \begin{cases} \dfrac{\left(x^{\frac{q-2}{q-1}} - 1\right)^{\frac{1}{q-2}}}{C_q x^{\frac{1}{q-1}} \left[1 + (q-1)(x^{\frac{q-2}{q-1}} - 1)^{2\frac{q-1}{q-2}}\right]^{\frac{1}{q-1}}}, & \text{if } 0 < x \leq 1, \\[2em] 0, & \text{if } x > 1, \\[0.5em] f_2(-x), & \text{if } x < 0. \end{cases}$$

This function is symmetric about the origin and has a compact support supp $[f_2] = [-1, 1]$. We show that the q-Fourier transform of f_2 coincides with the q-Fourier transform of f_1. Indeed,

$$F_q[f_2](\xi) = 2 \int_0^1 f_2(x) \cos_q(x\xi[f_2(x)]^{q-1}) dx$$

$$= 2 \int_0^1 \frac{(x^{\frac{q-2}{q-1}} - 1)^{\frac{1}{q-2}}}{C_q x^{\frac{1}{q-1}} \left[1 + (q-1)(x^{\frac{q-2}{q-1}} - 1)^{2\frac{q-1}{q-2}} \right]^{\frac{1}{q-1}}}$$

$$\times \cos_q \left(\frac{x\xi(x^{\frac{q-2}{q-1}} - 1)^{\frac{q-1}{q-2}}}{C_q^{q-1} x \left[1 + (q-1)(x^{\frac{q-2}{q-1}} - 1)^{2\frac{q-1}{q-2}} \right]} \right) dx$$

Using the change of variable

$$y = (x^{\frac{q-2}{q-1}} - 1)^{\frac{q-1}{q-2}}$$

we reduce the integral to

$$F_q[f_2](\xi) = 2 \int_0^\infty \frac{1}{C_q[1 + (q-1)y^2]^{\frac{1}{q-1}}} \cos_q \left(\frac{\xi y}{C_q^{q-1}[1 + (q-1)y^2]} \right) dy$$

$$= \int_{-\infty}^\infty G_q(y) \otimes_q \exp_q(iy\xi) dy$$

$$= F_q[f_1](\xi).$$

Thus, both f_1 and f_2 have the same q-Fourier transform. In fact, there is a whole family of density functions with the same q-Fourier transform as the q-Fourier transform of f_1. Hilhorst's invariance principle can be used to produce such a family of density functions.

The main idea of Hilhorst's invariance principle, published in his paper (Hilhorst, 2010), is as follows. Let $f(x)$, $x \in (-\infty, \infty)$, be a continuous symmetric density function, such that $\lambda(x) = x[f(x)]^{q-1}$, restricted to the semiaxis $[0, \infty)$, has a unique (local) maximum m at a point x_m. In other words, $\lambda(x)$ has two strictly monotonic pieces:

(1) $\lambda_+(x)$, increasing on the interval $0 \leq x < x_m$, and
(2) $\lambda_-(x)$, decreasing on the interval $x_m < x < \infty$.

These two functions as monotone functions have inverses (see Fig. 3.1). Let $x_\pm(\lambda)$, $0 \leq \lambda \leq m$, denote the inverses of $\lambda_\pm(x)$,

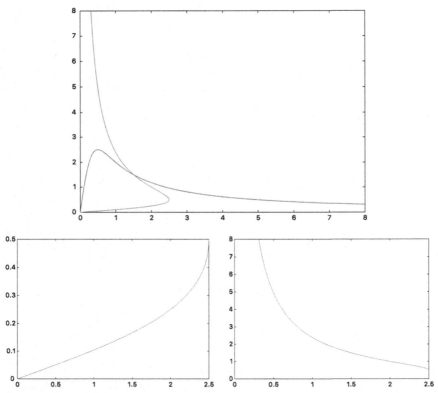

Fig. 3.1: Schematic representation of functions $\lambda_\pm(x)$ and $x_\pm(\lambda)$. The increasing piece of the blue curve on the figure on the top show $\lambda_+(x)$, and its decreasing piece shows $\lambda_-(x)$. The red curve on the top figure is the reflection of the blue curve about the line $\lambda = x$. The figures on the bottom show the functions $x_\pm(\lambda)$. The left figure is $x_+(\lambda)$, which is the inverse of $\lambda_+(x)$, and the right figure is $x_-(\lambda)$, which is the inverse of $\lambda_-(x)$.

respectively. The q-Fourier transform of f, we can write in the form (Proposition 3.14)

$$F_q[f](\xi) = 2 \int_0^\infty f(x) \cos_q \left(\xi x [f(x)]^{q-1} \right) dx$$

$$= 2 \int_0^{x_m} f(x) \cos_q \left(\xi x [f(x)]^{q-1} \right) dx$$

$$+ 2 \int_{x_m}^\infty f(x) \cos_q (\xi x [f(x)]^{q-1}) dx. \qquad (3.111)$$

Using the change of variable $x = x_+(\lambda)$ on the interval $(0, x_m)$ and the change of variable $x = x_-(\lambda)$ on the interval (x_m, ∞), we have

$$F_q[f](\xi) = 2 \int_0^m \left[\frac{\lambda}{x_+(\lambda)} \right]^{\frac{1}{q-1}} \cos_q(\xi\lambda) dx_+(\lambda)$$

$$- 2 \int_0^m \left[\frac{\lambda}{x_-(\lambda)} \right]^{\frac{1}{q-1}} \cos_q(\xi\lambda) dx_-(\lambda)$$

$$= 2 \int_0^m \cos_q(\xi\lambda) \left(\left[\frac{\lambda}{x_+(\lambda)} \right]^{\frac{1}{q-1}} \frac{dx_+(\lambda)}{d\lambda} \right.$$

$$\left. - \left[\frac{\lambda}{x_-(\lambda)} \right]^{\frac{1}{q-1}} \frac{dx_-(\lambda)}{d\lambda} \right) d\lambda$$

$$= 2 \int_0^m F(\lambda) \cos_q(\xi\lambda) d\lambda, \qquad (3.112)$$

where[5]

$$F(\lambda) = [\frac{\lambda}{x_+(\lambda)}]^{\frac{1}{q-1}} \frac{dx_+(\lambda)}{d\lambda} - [\frac{\lambda}{x_-(\lambda)}]^{\frac{1}{q-1}} \frac{dx_-(\lambda)}{d\lambda}$$

$$= \frac{q-1}{q-2} \lambda^{\frac{1}{q-1}} \frac{d}{d\lambda} [x_+^{\frac{q-2}{q-1}}(\lambda) - x_-^{\frac{q-2}{q-1}}(\lambda)], \quad \lambda \in [0, m]. \quad (3.113)$$

The key point of the invariance principle is that if we replace $x_+^{\frac{q-2}{q-1}}(\lambda)$ by $x_+^{\frac{q-2}{q-1}}(\lambda) + H(\lambda)$ and $x_-^{\frac{q-2}{q-1}}(\lambda)$ by $x_-^{\frac{q-2}{q-1}}(\lambda) + H(\lambda)$, then $F(\lambda)$ in (3.113) does not change. Here $H(\lambda)$ is a suitable function defined on the interval $(0, m)$, such that the functions $x_\pm(\lambda) + H(\lambda)$ are invertible. Now, let us denote

$$F_H(\lambda) = \frac{q-1}{q-2} \lambda^{\frac{1}{q-1}} \frac{d}{d\lambda} [X_-^{\frac{q-2}{q-1}}(\lambda) - X_+^{\frac{q-2}{q-1}}(\lambda)], \ \lambda \in [0, m], \quad (3.114)$$

where

$$X_\pm^{\frac{q-2}{q-1}}(\lambda) = x_\pm^{\frac{q-2}{q-1}}(\lambda) + H(\lambda), \quad (3.115)$$

Further, denote by $\Lambda(x)$ the function defined by the two pieces of inverses of $X_\pm(\xi)$, namely

$$\Lambda_H(x) = \begin{cases} X_+^{-1}(x), & \text{if } 0 \le x \le x_{m,H}, \\ X_-^{-1}(x), & \text{if } x > x_{m,H}, \end{cases}$$

where $x_{m,H} = [x_m^{\frac{q-2}{q-1}} + H(m)]^{\frac{q-1}{q-2}}$. The function $\Lambda_H(x)$ is continuous, since $X_+^{-1}(x_{m,H}) = X_-^{-1}(x_{m,H})$. Then, since $F_H(\lambda) = F(\lambda)$ for all $\lambda \in [0, m]$, the q-Fourier transform of the function

$$f_H(x) = \left(\frac{\Lambda(x)}{x}\right)^{\frac{1}{q-1}} \quad (3.116)$$

coincides with the q-Fourier transform of the function f. The function f_H coincides with f if and only if $H(\lambda)$ is identically zero.

---

[5]Notice that there is an inadvertence in equation (3.9) in Hilhorst (2010).

Now assume that $f(x)$ is a q-Gaussian density (with $\beta = 1$),

$$f(x) = G_q(x) = \frac{1}{C_q \left[1 + (q-1)x^2\right]^{\frac{1}{q-1}}}, \quad 1 < q < 2, \quad (3.117)$$

where C_q is the normalization constant. Obviously, $G_q(x)$ is symmetric, and the function $\lambda_q(x) = x[G_q(x)]^{q-1}$ considered on the semiaxis $[0, \infty)$ has a unique maximum $m = \frac{1}{2\sqrt{q-1}C_q^{q-1}}$, attained at the point

$$x_m = (q-1)^{-\frac{1}{2}}. \quad (3.118)$$

Moreover, the functions $x_{\pm}(\lambda)$ in this case take the forms

$$x_{\pm}(\lambda) = \frac{C_q^{1-q} \mp \left[C_q^{2(1-q)} - 4(q-1)\lambda^2\right]^{\frac{1}{2}}}{2\lambda(q-1)}, \quad 0 < \lambda \le m. \quad (3.119)$$

We denote the density $f_H(x)$ and the function $\Lambda_H(x)$ corresponding to the q-Gaussian by $G_{q,H}(x)$ and $\Lambda_{q,H}(x)$, respectively. In particular, selecting $H(\lambda) = A \ge 0$ constant, we have

$$\Lambda_{q,A}(x) = \frac{\left(x^{\frac{q-2}{q-1}} - A\right)^{\frac{q-1}{q-2}}}{C_q^{q-1}\left[1 + (q-1)(x^{\frac{q-2}{q-1}} - A)^{2\frac{q-1}{q-2}}\right]},$$

if $x^{\frac{q-2}{q-1}} \ge A$, and $\Lambda_{q,A} = 0$, otherwise. The corresponding family of density functions $G_{q,A}(x)$ is

$$G_{q,A}(x) = \begin{cases} \dfrac{\left(x^{\frac{q-2}{q-1}} - A\right)^{\frac{1}{q-2}}}{C_q x^{\frac{1}{q-1}}\left[1 + (q-1)(x^{\frac{q-2}{q-1}} - A)^{2\frac{q-1}{q-2}}\right]^{\frac{1}{q-1}}}, & 0 < x \le (1/A)^{\frac{q-1}{2-q}}, \\ 0, & (1/A)^{\frac{q-1}{2-q}} < x < \infty, \\ G_{q,A}(-x), & x < 0. \end{cases}$$

$$(3.120)$$

By construction, the q-Fourier transform of $G_{q,A}(x)$ for all A is one and the same as the q-Fourier transform of the q-Gaussian density $G_q(x)$. In Example 3.4 the function $f_1(x) = G_q(x)$ corresponds to $A = 0$, and the function $f_2(x)$ to $A = 1$.

Remark 3.16.

(1) Of course the family $G_{q,A}(x)$ does not exhaust all possible density functions having the same q-Fourier transform $F_q[G_q](\xi)$. Choosing $H(\xi) = H_0(\xi) + A$, where $H_0(\xi)$ is a suitable function and A is an arbitrary constant, one finds another family, as well.

(2) In the description of Hilhorst's invariance principle we assumed that the function $\lambda(x) = x[f(x)]^{q-1}$ has only one local maximum. If it has more local maximum and minimum points, then (3.111) takes the form

$$F_q[f](\xi) = 2 \sum_{k=1}^{M} \int_{x_{k-1}}^{x_k} f(x) \cos_q(\xi x[f(x)]^{q-1}) dx, \qquad (3.121)$$

where $x_1, x_2, \ldots, x_{M-1}$, are local extremum points of $\lambda(x)$, and $x_0 = 0$, $x_M = \infty$. In this case, $\lambda(x)$ is invertible in each interval $[x_{k-1}, x_k]$, $k = 1, \ldots, M$. These inverses provide appropriate changes of variable in each interval. We will establish some other facts on Hilhorst's invariance principle in Chapter 4 in light of the discussion of uniqueness of the limiting process in the q-generalized central limit theorem (see Section 4.3).

3.11. Representation Formulas for "The Inverse q-Fourier Transform"

In this section, we look for representation formulas for "the inverse q-Fourier transform". The expression "inverse q-Fourier transform" is taken into quotes, since we know from the previous section that the q-Fourier transform is not invertible in general. However, the operator F_q considered in special subspaces of $L_1(\mathbb{R})$ may be inverted. We have seen one example in Section 3.4 for the inverse q-Fourier transform in the set of q-Gaussian density functions.

The invertibility of the operator F_q is connected with the question: how large is the class of functions $B_f \subset C(\mathbb{R}) \cap L_1(\mathbb{R})$ providing the same q-Fourier transform as that of f? Hilhorst's invariance principle sheds light on this question. The size of the class of functions with the same q-Fourier transform depends on the value of q and on a representative of this class. In Section 3.10, we assumed that the graph of the function $\lambda(x) = x[f(x)]^{q-1}$ has two monotonic pieces. Inverses of functions representing these two pieces determine the class B_f. If $\lambda(x)$ has more extremum points then the class B_f enlarges (see Remark 3.16). The class B_f coincides with the set $\{f\}$, containing only the function f, if $\lambda(x)$ has no extremum points, that is, it is increasing on the interval $(0, \infty)$. Consider some examples. If $q = 1$, then $\lambda(x) = x$, and hence, in this case for any function $f \in L_1(\mathbb{R})$, one has $B_f = \{f\}$. Therefore, the classic Fourier transform is, as well known, invertible as an operator

$$F_1 : L_1(\mathbb{R}) \to C_0(\mathbb{R}).$$

Now, let $1 < q < 2$. Then one can easily verify that for any function of the form

$$\varphi_\alpha(x) = \begin{cases} 1, & \text{if } 0 \le x \le 1, \\ \dfrac{1}{x^{1+\alpha}}, & \text{if } 1 < x < \infty, \\ \varphi_\alpha(-x), & \text{if } x < 0, \end{cases}$$

where $1 < \alpha < \frac{2-q}{q-1}$, the function $\lambda(x) = x[\varphi_\alpha(x)]^{q-1}$ is strictly increasing. Therefore, the q-Fourier transform $F_q[\varphi_\alpha](\xi)$ is defined by the unique φ_α.

In fact, the q-Fourier transform for $q > 1$ is invertible as an operator

$$F_q : X_q \to C_0(\mathbb{R}),$$

where X_q is a subset of $L_1(\mathbb{R})$, defined as

$$X_q = \{f \in L_1(\mathbb{R}) : \frac{d}{dx}\left(x[f(x)]^{q-1}\right) > 0\}.$$

Below we prove this fact. Denote by Y_q the subset of $C_0(\mathbb{R})$ defined as

$$Y_q = F_q[X_q] \equiv \{F(\xi) : F(\xi) = F_q[f](\xi), \ f \in X_q\}.$$

Theorem 3.18. *Let* $1 \leq q < 2$. *The operator* $F_q : X_q \to Y_q$ *is invertible. Moreover, if* $F_q[f](\xi) = \hat{f}(\xi)$, *then the relation*

$$(\lambda^{-1}(x))' f\left(\lambda^{-1}(x)\right) = \frac{2-q}{2\pi} \int\limits_{-\infty}^{\infty} \hat{f}(\xi) e_q^{ix\xi} d\xi, \tag{3.122}$$

where $\lambda^{-1}(x)$ *is the inverse of* $\lambda(x) = x[f(x)]^{q-1}$, *holds.*

Proof. Let $f \in X_q$, that is $\lambda(x) = x[f(x)]^{q-1}$ is strictly increasing on $(0, \infty)$. Then the function $y = \lambda(x)$ has the inverse $x = \lambda^{-1}(y) = g(y)$, which is also strictly increasing on $(0, \infty)$. Therefore, in this case, equality (3.112) takes the form

$$F_q[f](\xi) = 2 \int\limits_{0}^{\infty} F(\lambda) \cos_q(i\lambda\xi) d\lambda$$

$$= \int\limits_{\mathbb{R}} F(\lambda) \exp_q(i\lambda\xi) d\lambda, \quad \xi \in \mathbb{R}, \tag{3.123}$$

where

$$F(\lambda) = \left[\frac{\lambda}{g(\lambda)}\right]^{\frac{1}{q-1}} \frac{dg(\lambda)}{d\lambda} = f(g(\lambda))g'(\lambda), \quad \lambda \in [0, \infty), \tag{3.124}$$

is uniquely defined by f. Thus, the operator F_q is one-to-one as an operator $F_q : X_q \to Y_q$.

To prove relation (3.122) we multiply both sides of equation (3.123) by $\exp_q(-ix\xi)$ and integrate over $\mathbb{R}$ (in the sense of distributions). Then we have

$$\int\limits_{\mathbb{R}} F_q[f](\xi) \exp_q(-ix\xi) d\xi = \int\limits_{\mathbb{R}} \int\limits_{\mathbb{R}} F(\lambda) \exp_q(i(\lambda - x)\xi) d\lambda d\xi$$

$$= \int\limits_{\mathbb{R}} F(\lambda) \int\limits_{\mathbb{R}} \exp_q(i(\lambda - x)\xi) d\xi d\lambda$$

$$= 2\pi \int\limits_{\mathbb{R}} F(\lambda)\delta_q(\lambda - x) d\lambda, \tag{3.125}$$

where

$$\delta_q(z) = \frac{1}{2\pi} \int_{\mathbb{R}} \exp_q(iz\xi) d\xi,$$

in the sense of distributions. By using the relationship (see Example 3.2)

$$\delta_q(z) = \frac{1}{2-q} \delta(z),$$

between the q-Dirac delta function $\delta_q(x)$ and the Dirac delta function $\delta(x)$, one obtains from (3.125) that

$$\int_{\mathbb{R}} F_q[f](\xi) \exp_q(-ix\xi) d\xi = \frac{2\pi}{2-q} \int_{\mathbb{R}} F(\lambda)\delta(\lambda - x) d\lambda = \frac{2\pi}{2-q} F(x).$$

Now taking into account the relation (3.124), we obtain (3.122). □

Remark 3.17.

(1) We note that the formula (3.122) can be interpreted as a representation of the inverse q-Fourier transform acting from Y_q to X_q. However, in order to find $f(x)$ one needs to solve the nonlinear algebraic equation

$$f(g(x))g'(x) = F(x)$$

for $f(x)$. Recall that $g(x)$ is the inverse of the function $\lambda(x) = x[f(x)]^{q-1}$, and hence, depends on $f(x)$. If $q = 1$, then $g(x) = x$, implying $f(x) = F(x)$ and recovering the classical inverse Fourier transform formula

$$f(x) = \frac{1}{2\pi} \int_{\mathbb{R}} \hat{f}(\xi) e^{-ix\xi} d\xi. \tag{3.126}$$

(2) Note also that the subspace X_q of $L_1(\mathbb{R})$ is rather narrow. As we have seen in Section 3.10, even q-Gaussian densities do not belong to X_q.

Is it possible to find f analytically similarly to the classical inverse Fourier transform if one possesses more information about $F_q[f](\xi)$?

It turns out that this question can be answered affirmatively in some cases. Below we provide two results (without proofs) in this context.

The first result due to Plastino and Rocca (2012). Assume we know the q-Fourier transforms $F_q[f](\xi)$ of a positive function $f \in L_1(\mathbb{R})$ for all $1 \le q < 2$. Then one can determine the function f in the integral form. Namely, the following theorem holds.

Theorem 3.19 (Plastino and Rocca, 2012). *Let* $0 \le f \in L_1(\mathbb{R})$ *and* $F(q,\xi) = (H(q-1) - H(q-2))F_q[f](\xi)$ *be the function of two variables* $(q,\xi) \in [1,2) \times \mathbb{R}$. *Then, the formula*

$$f(x) = \frac{1}{2\pi} \int_{\mathbb{R}} \left[\lim_{\varepsilon \to 0^+} \int_1^2 F(q,\xi)\delta(q-1-\varepsilon)dq \right] e^{ix\xi}d\xi \qquad (3.127)$$

holds.

The second result is obtained in the paper Jauregui and Tsallis (2011). Suppose we know the q-Fourier transform of $\tau_y f(x) = f(x+y)$ for all $y \in \mathbb{R}$. Then, again we are able to reestablish the function $f(x)$ through the q-Fourier transform of $\tau_y f$ in the integral form.

Theorem 3.20 (Jauregui and Tsallis, 2011). *Let* $0 \le f \in L_1(\mathbb{R})$ *and* $\mathbb{F}(y,\xi) = F_q[\tau_y f](\xi)$ *be the function of two variables* $(y,\xi) \in \mathbb{R}^2$. *Then, the formula*

$$f(y) = \left[\frac{2-q}{2\pi} \int_{\mathbb{R}} \mathbb{F}(y,\xi)d\xi \right]^{\frac{1}{2-q}} \qquad (3.128)$$

holds.

Remark 3.18. It is not hard to verify that both representations (3.127) and (3.128) reduce to (3.126) if $q = 1$.

3.12. Multidimensional q-Fourier Transforms

The q-Fourier transform can be defined for functions $f(x_1, \ldots, x_d)$, dependent on d independent variables $x_1, \ldots, x_d$. Below we consider two different versions of multidimensional generalizations of F_q, called *direct* and *sequential* q-Fourier transforms. We will consider only the case $q > 1$ to avoid complexities arising in the case $q < 1$ (see Section 3.7)

Definition 3.4. Let $q \geq 1$. By *direct* d-dimensional q-Fourier transform of a function $f(x) = f(x_1, \ldots, x_d)$ we understand the integral

$$\mathcal{F}_q[f](\xi) = \int_{\text{supp } [f]} f(x) \otimes_q e_q^{i(x,\xi)} dx$$

$$= \int_{\text{supp } [f]} f(x) e_q^{i(x,\xi)[f(x)]^{q-1}} dx, \qquad (3.129)$$

where $\xi = (\xi_1, \ldots, \xi_d) \in \mathbb{R}^d$, supp $[f]$ is the support of the function f, and

$$(x, \xi) = x_1 \xi_1 + \cdots + x_d \xi_d,$$

provided the integral exists.

As in the one-dimensional case, if $q = 1$, then (3.129) recovers the classic Fourier transform

$$\mathcal{F}[f](\xi) = \int_{\mathbb{R}^d} f(x) e^{i(x,\xi)} dx.$$

Moreover, in this case $\mathcal{F} = F_1 \circ \cdots \circ F_d$, where $F_j, \ j = 1, \ldots, d$, are the one-dimensional Fourier transforms with respect to the variables $x_j, j = 1, \ldots, d$, and the symbol "$\circ$" means the composition of operators $F_j, \ j = 1, \ldots, n$. This property fails if $q \neq 1$. That is, generically

$$\mathcal{F}_q \neq F_{1,q} \circ \cdots \circ F_{d,q},$$

where $F_{j,q}$ is the q-Fourier transform with respect to $x_j, \ j = 1, \ldots, d$. This is a reason why two different versions of q-Fourier transform arise in multidimensional case.

Assume $q \geq 1$ and define the sequence

$$q_n^d = \frac{2q - dn(q-1)}{2 - dn(q-1)}, \quad n = 0, \pm 1, \ldots. \tag{3.130}$$

It follows from the definition of q_n^d that $q_0^d = q$, and

$$\frac{2}{1 - q_n^d} = \frac{2}{1 - q} + dn, \quad n = 0, \pm 1, \ldots.$$

In the one-dimensional case the sequence q_n^d coincides with the sequence q_n defined in (3.40). Therefore, we write q_n if $d = 1$, that is $q_n^1 = q_n$.

Definition 3.5. Let Q_k be an ordered set of d numbers $Q_k = \{q_k, q_{k+1}, \ldots, q_{k+d-1}\}$, such that $q_{k+d-1} \geq 1$. The composition

$$\mathcal{F}_{Q_k}[f](\xi) = F_{d, q_{k+d-1}} \circ \cdots \circ F_{1, q_k}[f(x_1, \ldots, x_d)](\xi_1, \ldots, \xi_d) \tag{3.131}$$

is called a Q_k-*sequential Fourier transform* of a given function $f(x)$, provided the right-hand side of (3.131) exists.

The direct and sequential q-Fourier transforms coincide only if $q = 1$. In this case $Q_k = \{1, \ldots, 1\}$. Both versions of the q-Fourier transform are well-defined on functions $f(x) \in L_1(\mathbb{R}^d)$, $f(x) \geq 0$, and transfer them to functions in $C_0(\mathbb{R}^d)$, which are continuous and vanish at infinity. That is the following mappings are continuous:

$$\mathcal{F}_q : L_1(\mathbb{R}^d) \to C_0(\mathbb{R}^d),$$

$$\mathcal{F}_{Q_k} : L_1(\mathbb{R}^d) \to C_0(\mathbb{R}^d).$$

Proposition 3.21. *For all $0 \neq a \in \mathbb{R}$, $b > 0$, and invertible $(d \times d)$-matrix H the following properties hold:*

(1) $\mathcal{F}_q[f(ax)](\xi) = \frac{1}{a^d}\mathcal{F}_q[f(x)](\frac{\xi}{a})$;

(2) $\mathcal{F}_q[bf(x)](\xi) = b\mathcal{F}_q[f(x)](b^{q-1}\xi)$;

(3) $\mathcal{F}_q[f(Hx)](\xi) = \frac{1}{|H|}\mathcal{F}_q[f(x)]((H^{-1})^T\xi)$, *where $|H|$ is the determinant of H and $(H^{-1})^T$ is the transpose of the inverse matrix H^{-1}.*

The multidimensional generalizations of Theorem 3.4 for d-dimensional direct and sequential q-Fourier transforms are important in the study of multivariate q-central limit theorem. To prove the corresponding statements for both versions of q-Fourier transforms we need some auxiliary lemmas.

We recall that the d-dimensional q-Gaussian density function with a covariance matrix Σ is defined by (see (2.80))

$$G_q(\Sigma; x) = \frac{\sqrt{|\Sigma|}}{C_{d,q}} e_q^{-(\Sigma x, x)}, \quad x \in \mathbb{R}^d, \tag{3.132}$$

where the normalizing constant $C_{d,q}$ for $1 < q < 1 + d/2$ is (see (2.81))

$$C_{d,q} = \frac{\pi^{d/2} \Gamma(\frac{1}{q-1} - \frac{d}{2})}{(q-1)^{d/2} \Gamma(\frac{1}{q-1})}. \tag{3.133}$$

Lemma 3.3. *Let $1 < q < 1 + 2/d$. The normalizing constant $C_{d,q}$ of the d-dimensional q-Gaussian density $\mathbb{N}_q$ is connected with normalizing constants $C_{q_0}, \ldots, C_{q_{d-1}}$ ($q_0 = q$) of one-dimensional q_n-Gaussian densities through*

$$C_{d,q} = \frac{1}{(q-1)^{d/2}} \prod_{n=0}^{d-1} C_{q_n} \sqrt{q_n - 1}. \tag{3.134}$$

Proof. For $q = q_0$ satisfying the condition $1 \le q < 1 + 2/d$ we have $1 \le q_n < 3$ for all $n = 1, \ldots, d-1$. In accordance with representation (2.68) of C_q for $1 < q < 3$,

$$C_{q_n} = \frac{\sqrt{\pi} \Gamma\left(\frac{3-q_n}{2(q_n-1)}\right)}{\sqrt{q_n - 1} \Gamma\left(\frac{1}{q_n-1}\right)}, \quad n = 0, \ldots, d-1.$$

Substituting $q_n = 1 + \frac{2(q-1)}{2-n(q-1)}$ (see (3.40)), we obtain

$$C_{q_n} = \frac{\sqrt{\pi} \sqrt{2 - n(q-1)}}{\sqrt{2(q-1)}} \frac{\Gamma\left(\frac{1}{q-1} - \frac{n+1}{2}\right)}{\Gamma\left(\frac{1}{q-1} - \frac{n}{2}\right)}, \quad n = 0, \ldots, d-1.$$

$$\tag{3.135}$$

Making use of (3.135), we have

$$C_{q_0} \cdots \cdots C_{q_{d-1}} = \prod_{n=0}^{d-1} C_{q_n}$$

$$= \frac{(\pi)^{d/2}}{2^{d/2}(q-1)^{d/2}} \frac{\Gamma\left(\frac{1}{q-1} - \frac{d}{2}\right)}{\Gamma\left(\frac{1}{q-1}\right)} \prod_{n=0}^{d-1} \sqrt{2 - n(q-1)}$$

$$= \frac{1}{2^{d/2}} C_{d,q} \prod_{n=0}^{d-1} \sqrt{2 - n(q-1)}.$$

Now, taking into account equalities

$$2 - n(q-1) = \frac{2(q-1)}{q_n - 1}, \quad n = 0, \ldots, d-1, \tag{3.136}$$

we obtain (3.134). □

3.13. Direct q-Fourier Transform of Multivariate q-Gaussian Density Functions

Lemma 3.4. *Let $q \geq 1$. For any real $\beta > 0$ and $A \in \mathbb{C}$ the equality*

$$\int_{\mathbb{R}} A \otimes_q \exp_q(-\beta x^2 + \beta i x \xi) dx = \frac{C_q}{\sqrt{\beta}} \left[A \otimes_q e_q(-\beta \xi^2/4) \right]^{\frac{3-q}{2}} \tag{3.137}$$

holds. Here C_q is the normalizing constant of the one-dimensional q-Gaussian density (see (2.68)).

Proof. Let the conditions of the lemma be fulfilled. Then we have

$$\int_{\mathbb{R}} A \otimes_q \exp_q(-\beta x^2 + \beta i x \xi) dx = A \int_{\mathbb{R}} \exp_q(-\beta x^2 + i \beta x \xi) A^{q-1} dx$$

$$= A \int_{\mathbb{R}} \exp_q\left(-\beta(x - \frac{i\xi}{2})^2 A^{q-1} - \beta A^{q-1} \frac{\xi^2}{4}\right) dx.$$

Using the change of variable

$$x = \frac{y}{\sqrt{\beta} A^{\frac{q-1}{2}}} + \frac{i\xi}{2},$$

and Cauchy's integral theorem as we did in the proof of Lemma 3.2, we obtain

$$\int\limits_{\mathbb{R}} A \otimes_q \exp_q(-\beta x^2 + \beta i x \xi) dx = \frac{A}{\sqrt{\beta} A^{\frac{q-1}{2}}} \int\limits_{\mathbb{R}} e_q^{-y^2} \otimes_q e_q^{-\beta A^{q-1} \frac{\xi^2}{4}} dy$$

$$= \frac{A^{\frac{3-q}{2}}}{\sqrt{\beta}} e_q^{-\beta A^{q-1} \frac{\xi^2}{4}} \int\limits_{\mathbb{R}} e_q^{-y^2 \left[e_q^{-\beta A^{q-1} \frac{\xi^2}{4}} \right]^{q-1}} dy.$$

Further, making use of the change of variable

$$y = \frac{s}{\left[e_q^{-\beta A^{q-1} \frac{\xi^2}{4}} \right]^{(q-1)/2}},$$

the latter integral can be reduced to

$$\int\limits_{\mathbb{R}} A \otimes_q \exp_q(-\beta x^2 + \beta i x \xi) dx = \frac{A^{\frac{3-q}{2}}}{\sqrt{\beta}} [e_q^{-\beta A^{q-1} \frac{\xi^2}{4}}]^{1 - \frac{q-1}{2}} \int\limits_{\mathbb{R}} e_q^{-s^2} ds.$$

Finally, taking into account the fact that $\int_{\mathbb{R}} e_q^{-s^2} ds = C_q$, we have

$$\int\limits_{\mathbb{R}} A \otimes_q \exp_q(-\beta x^2 + \beta i x \xi) dx = \frac{C_q}{\sqrt{\beta}} [A e_q^{-\beta A^{q-1} \frac{\xi^2}{4}}]^{\frac{3-q}{2}}$$

$$= \frac{C_q}{\sqrt{\beta}} [A \otimes_q e_q(-\beta \xi^2/4)]^{\frac{3-q}{2}},$$

completing the proof. $\qquad\qquad\qquad\qquad\qquad\qquad\qquad\qquad\qquad\square$

Lemma 3.5. *Let* $1 \leq q \leq 1 + \frac{2}{d}$ *and* $\xi \in \mathbb{R}^d$, $d \geq 1$. *Then the following equality holds:*

$$\int\limits_{\mathbb{R}^d} \exp_q(-|x|^2 + i(x, \xi)) dx = C_{q,d} \, e_{q_{1,d}}^{-b|\xi|^2}, \qquad (3.138)$$

where $b = \frac{2 - d(q-1)}{8}$ *and* $q_{1,d} = \frac{2q - d(q-1)}{2 - d(q-1)}$.

Proof. The integral on the left-hand side of (3.138) can be written in the form

$$\int_{\mathbb{R}^d} \exp_q(-|x|^2 + i(x, \xi))dx \tag{3.139}$$

$$= \int_{\mathbb{R}^{d-1}} \int_{-\infty}^{\infty} \exp_q(-x_2 + i(x_1, \xi_1)) \otimes_q \exp_q(-|x'|^2 + i(x', \xi'))dx_1 dx', \tag{3.140}$$

where $x' = (x_2, \ldots, x_d) \in \mathbb{R}^{d-1}$ and $\xi' = (\xi_2, \ldots, \xi_d) \in \mathbb{R}^{d-1}$. Setting $\beta = 1$ and $A = \exp_q(-|x'|^2 + i(x', \xi'))$ in Lemma 3.4, we obtain

$$\int_{\mathbb{R}^d} \exp_q(-|x|^2 + i(x, \xi))dx$$

$$= C_q \int_{\mathbb{R}^{d-1}} [\exp_q(-|x'|^2 + i(x', \xi')) \otimes_q e_q^{-\frac{\xi_1^2}{4}}]^{\frac{3-q}{2}} dx'$$

$$= C_q \int_{\mathbb{R}^{d-1}} \left[\exp_q \left(-|x'|^2 + i(x', \xi') - \frac{\xi_1^2}{4} \right) \right]^{\frac{3-q}{2}} dx' \tag{3.141}$$

Due to Proposition 2.14 (with $\beta = 1$ and $a = (3-q)/2$), the integrand on the right-hand side of (3.141) can be expressed as

$$\left[\exp_q \left(-|x'|^2 + i(x', \xi') - \frac{\xi_1^2}{4} \right) \right]^{\frac{3-q}{2}}$$

$$= \exp_{q_1} \left(\frac{3-q}{2} \left(-|x'|^2 + i(x', \xi') - \frac{\xi_1^2}{4} \right) \right)$$

$$= \exp_{q_1} \left(\frac{3-q}{2} \left(-x_2^2 + i x_2 \xi_2 \right) \right)$$

$$\otimes_q \exp_q \left(\frac{3-q}{2} \left(-|x''|^2 + i(x'', \xi'') - \frac{\xi_1^2}{4} \right) \right),$$

where $x'' = (x_3, \ldots, x_d) \in \mathbb{R}^{d-2}$, $\xi'' = (\xi_3, \ldots, \xi_d) \in \mathbb{R}^{d-2}$, and

$$q_1 = u(q) = (1+q)/(3-q).$$

Taking this fact into account one can reduce the integral in (3.141) to the form

$$\int_{\mathbb{R}^d} \exp_q(-|x|^2 + i(x, \xi)) dx$$

$$= C_q \int_{\mathbb{R}^{d-2}} \int_{-\infty}^{\infty} \exp_{q_1}\left(\frac{3-q}{2}\left(-x_2^2 + ix_2\xi_2\right)\right)$$

$$\otimes_{q_1} \exp_{q_1}\left(\frac{3-q}{2}\left(-|x''|^2 + i(x'', \xi'') - \frac{\xi_1^2}{4}\right)\right) dx_2 dx''$$

$$(3.142)$$

Again using Lemma 3.4 in (3.142) with $\beta = \frac{3-q}{2}$ and

$$A = \exp_{q_1}\left(\frac{3-q}{2}\left(-|x''|^2 + i(x'', \xi'') - \frac{\xi_1^2}{4}\right)\right),$$

similarly as above, we obtain

$$\int_{\mathbb{R}^d} \exp_q(-|x|^2 + i(x, \xi)) dx$$

$$= \sqrt{\frac{2}{3-q_0}} C_{q_0} C_{q_1} \int_{\mathbb{R}^{d-2}} \exp_{q_2}\left(\frac{(3-q_0)(3-q_1)}{2^2}\left(-\frac{\xi_1^2 + \xi_2^2}{4}\right)\right)$$

$$\otimes_{q_2} \exp_{q_2}\left(\frac{(3-q_0)(3-q_1)}{2^2}(-|x''|^2 + i(x'', \xi''))\right) dx''$$

$$(3.143)$$

Here we notice that

$$\frac{(3-q_0)(3-q_1)}{2^2} = \frac{4-2q}{2}.$$

Doing this process d times, we ultimately arrive at the equality

$$\int_{\mathbb{R}^d} \exp_q(-|x|^2 + i(x,\xi))dx$$

$$= \prod_{n=0}^{d-1} \sqrt{\frac{2d}{2-n(q-1)}} C_{q_n} e_{q_{1,d}}^{-\frac{d+2-dq}{2} \cdot \frac{\xi_1^2+\cdots+\xi_d^2}{4}}, \qquad (3.144)$$

where $q_{1,d} = u_d(q) = q_d$ (see (3.40)). Here we used the equality

$$\frac{3-q_0}{2} \frac{3-q_1}{2} \cdots \frac{3-q_{d-1}}{2} = \frac{d+2-dq}{2}.$$

Finally, taking into account equality (3.136) and Lemma 3.3, we obtain (3.138). □

Lemma 3.6. *Let* $1 \le q \le 1 + \frac{2}{d}$ *and* $\xi \in \mathbb{R}^d$, $d \ge 1$. *Then the following equality holds:*

$$\mathcal{F}_q \left[\frac{e_q^{-|x|^2}}{C_{d,q}} \right](\xi) = e_{q_{1,d}}^{-B^*|\xi|^2},$$

where $B^* = \frac{2-d(q-1)}{8 C_{d,q}^{2(q-1)}}$.

Lemma 3.7. *Let* $1 \le q \le 1 + \frac{2}{d}$ *and* $\xi \in \mathbb{R}^d$, $d \ge 1$. *Then the following equality holds:*

$$\mathcal{F}_q \left[\frac{\beta^{d/2} e_q^{-\beta|x|^2}}{C_{d,q}} \right](\xi) = e_{q_{1,d}}^{-B^{**}|\xi|^2},$$

where $B^{**} = B^*/\beta^{2-q}$.

Proof. The proofs of these lemmas immediately follow from Lemma 3.5 and Proposition 3.21. □

Now we are ready to prove the following d-dimensional generalization of Theorem 3.4 in the case of the direct q-Fourier transform $\mathcal{F}_q$.

Theorem 3.21. *Let* $1 \leq q < 1 + \frac{2}{d}$ *and* Σ *be a symmetric positively defined* $d \times d$ *matrix with a determinant* $|\Sigma| > 0$. *Then for the direct q-Fourier transform of the multidimensional q-Gaussian density* $G_q(\Sigma; x)$ *defined in* (3.132), *the following formula holds:*

$$\mathcal{F}_q \left[G_q(\Sigma; x) \right](\xi) = e_{q_{1,d}}^{-\alpha(\Sigma^{-1}\xi, \xi)}, \tag{3.145}$$

where

$$\alpha = \frac{\left(2 - d(q-1) \right)|\Sigma|^{q-1}}{8[C_{d,q}]^{2(q-1)}}, \tag{3.146}$$

and $q_{1,d}$ *is defined in* (3.130).

Proof. Applying Proposition 3.21, Part (2), we have

$$\mathcal{F}_q[G_q(\Sigma; x)](\xi) = \mathcal{F}_q \left[\frac{\sqrt{|\Sigma|}}{C_{d,q}} e_q^{-(\Sigma x, x)} \right](\xi)$$

$$= \frac{\sqrt{|\Sigma|}}{C_{d,q}} \mathcal{F}_q \left[e_q^{-(\Sigma x, x)} \right] \left(\left[\frac{\sqrt{|\Sigma|}}{C_{d,q}} \right]^{q-1} \xi \right) \tag{3.147}$$

Therefore, it suffices to evaluate $\mathcal{F}_q \left[e_q^{-(\Sigma x, x)} \right](\xi)$. By assumption, Σ is a symmetric positive matrix. It is well known (Lay *et al.*, 2014) that for such a matrix there exists an orthogonal matrix U, such that $D = U^{-1}\Sigma U$ is a diagonal matrix. Assume $\delta_1, \ldots, \delta_d$ are the diagonal entries of D. The positive definiteness of Σ implies that $\delta_j > 0$, $j = 1, \ldots, d$, and the determinant $|\Sigma| = \delta_1 \cdots \delta_d$.

Thus, changing the variable $x = Uy$, we have

$$\mathcal{F}_q[e_q^{-(\Sigma x, x)}](\xi) = \int_{\mathbb{R}^d} e_q^{-(\Sigma Uy, Uy) + i(Uy, \xi)} dUy$$

$$= \int_{\mathbb{R}^d} e_q^{-(Dy, y) + i(Uy, \xi)} dUy.$$

Since D is a diagonal matrix, obviously $(Dy, y) = \delta_1 y_1^2 + \cdots + \delta_d y_d^2$. Therefore, setting $y_j = \delta_j^{-1/2} z_j, j = 1, \ldots, d$, we have

$$\mathcal{F}_q[e_q^{-(\Sigma x, x)}](\xi) = \frac{1}{\sqrt{|\Sigma|}} \int_{\mathbb{R}^d} e_q^{-|z|^2 + i(z, D^{-1/2} U^{-1} \xi)} \, dU z$$

$$= \frac{1}{\sqrt{|\Sigma|}} \int_{\mathbb{R}^d} e_q^{-|Uz|^2 + i(Uz, UD^{-1/2} U^{-1} \xi)} \, dU z,$$

where $D^{-1/2}$ is the diagonal matrix with diagonal entries $\delta_j^{-1/2}, j = 1, \ldots, d$. Here we also used the fact that $|Uz|^2 = z^2$ for any orthogonal matrix U. Now making the change $w = Uz$ in the latter integral, we reduce it to

$$\mathcal{F}_q[e_q^{-(\Sigma x, x)}](\xi) = \frac{1}{\sqrt{|\Sigma|}} \int_{\mathbb{R}^d} e_q^{-|w|^2 + i(w, UD^{-1/2} U^{-1} \xi)} \, dw.$$

Due to Lemma 3.4 the latter implies

$$\mathcal{F}_q[e_q^{-(\Sigma x, x)}](\xi) = \frac{C_{d,q}}{\sqrt{|\Sigma|}} e_{q1,d}^{-b \frac{\left| UD^{-1/2} U^{-1} \xi \right|^2}{4}} = \frac{C_{d,q}}{\sqrt{|\Sigma|}} e_{q1,d}^{-b \frac{(\Sigma^{-1} \xi, \xi)}{4}},$$

where $b = \frac{2 - d(q-1)}{2}$. Finally, using the relation (3.147), we have

$$\mathcal{F}_q \left[\frac{\sqrt{|\Sigma|}}{C_{d,q}} e_q^{-(\Sigma x, x)} \right](\xi) = e_{q1,d}^{-\frac{b|\Sigma|^{q-1}}{4 C_{d,q}^{2(q-1)}} (\Sigma^{-1} \xi, \xi)},$$

which proves the theorem. □

3.14. Sequential q-Fourier Transform of Multivariate q-Gaussian Density Functions

Consider for $1 < q < 1 + 2/d$ the density function

$$G_q(B; x) = \frac{1}{dC_q} e_q^{-(\beta_1 x_1^2 + \cdots + \beta_d x_d^2)},$$

of the d-dimensional q-Gaussian, where B is the diagonal matrix B with positive entries $\beta_1, \ldots, \beta_d$, and $_dC_q$ is the normalizing constant

$$_dC_q = \int_{\mathbb{R}^d} e_q^{-(\beta_1 x_1^2 + \cdots + \beta_d x_d^2)} dx.$$

It follows from Lemma 3.3 that for $_dC_q$ the representation

$$_dC_q = \sqrt{|B|} C_{d,q} = \frac{1}{(q-1)^{d/2}} \prod_{n=0}^{d-1} C_{q^n} \sqrt{\beta_{n+1}(q^n - 1)}$$

holds. We also use the notation

$$_{d-m}C_{q^m} = \begin{cases} \dfrac{1}{(q-1)^{(d-m)/2}} \displaystyle\prod_{n=m}^{d-1} C_{q^n} \sqrt{\beta_{n+1}(q^n - 1)}, \\ \qquad\qquad\qquad\qquad\qquad \text{if } m = 1, \ldots, d-1, \\ 1, \qquad\qquad\qquad\qquad\quad \text{if } m = d. \end{cases} \tag{3.148}$$

The following lemma plays a central role in the derivation of the formula for the sequential q-Fourier transform of the density function $G_q(B; x)$.

Lemma 3.22. *For any positive real constants β, C_0, and A, the following relation holds:*

$$F_q[C_0 e_q^{-\beta x^2 - A}](\xi) = C_1 e_{q_1}^{-\alpha_1 \frac{\xi^2}{4} - A_1}, \tag{3.149}$$

where

$$C_1 = \frac{C_0 C_q}{\sqrt{\beta}}, \quad \alpha_1 = \frac{(3-q)C_0^{2(q-1)}}{2\beta}, \quad and \quad A_1 = \frac{(3-q)A}{2}. \tag{3.150}$$

Here F_q is the one-dimensional q-Fourier transform, and C_q is the constant defined in (2.67).

Proof. Using Proposition 3.2 one can easily verify that

$$F_q[C_0 e_q^{-\beta x^2 - A}](\xi) = \frac{C_0}{\sqrt{\beta}} F_q[e_q^{-x^2 - A}]\left(\frac{C_0^{q-1}\xi}{\sqrt{\beta}}\right). \tag{3.151}$$

Therefore, it suffices to evaluate $F_q[e_q^{-x^2-A}](\xi)$. We have

$$F_q[e_q^{-x^2-A}](\xi) = \int_{\mathbb{R}} \exp_q(-x^2 - ix\xi - A)dx$$

$$= \int_{\mathbb{R}} \exp_q\left(-\left(x - \frac{i\xi}{2}\right)^2 - \frac{\xi^2}{4} - A\right)dx$$

$$= \int_{\mathbb{R}} \exp_q\left(-\left(x - \frac{i\xi}{2}\right)^2\right) \otimes_q e_q^{(-\frac{\xi^2}{4}-A)}dx$$

Further, using the change of variable $x = y + i\xi/2$ and applying Cauchy's integral theorem, we reduce the latter integral to

$$F_q[e_q^{-x^2-A}](\xi) = \int_{\mathbb{R}} e_q^{-y^2} \otimes_q e_q^{(-\frac{\xi^2}{4}-A)}dy$$

$$= e_q^{-\frac{\xi^2}{4}-A} \int_{\mathbb{R}} e_q^{-y^2}[e_q^{-\frac{\xi^2}{4}-A}]^{q-1}dy$$

$$= C_q[e_q^{-\frac{\xi^2}{4}-A}]^{\frac{3-q}{2}} = C_q e_{q_1}^{-\frac{3-q}{2}(\frac{\xi^2}{4}+A)}.$$

Now using the relationship (3.151), we obtain (3.149) and (3.150).

□

Theorem 3.23. *Let* $\beta_j > 0, j = 1,\ldots, d,$ *and* $Q = Q_0 = \{q_0,\ldots,q_{d-1}\}$ *with* $q_0 = q \in (1, 1+2/d)$. *Then for the d-dimensional sequential Q-Fourier transform the following relation holds:*

$$\mathcal{F}_Q[_d C_q e_q^{-(\beta_1 x_1^2+\cdots+\beta_d x_d^2)}](\xi) = e_{q_d}^{-(\beta_1^* \xi_1^2+\cdots+\beta_d^* \xi_d^2)}, \tag{3.152}$$

where

$$\beta_j^* = \frac{[\ _{d-j+1}C_{q_{j-1}}]^{2(q_{j-1}-1)} \displaystyle\prod_{n=j-1}^{d-1} \frac{3-q_n}{2}}{4\beta_j \displaystyle\prod_{n=0}^{j-2} \frac{3-q_n}{2}}, \quad j=1,\ldots,d. \quad (3.153)$$

Remark 3.19.

(1) In equation (3.153) it is assumed that $\prod_{n=0}^{-1} = 1$.
(2) In the one-dimensional case ($d = 1$) equation (3.153) recovers β^* in Theorem 3.4. Indeed, if $d = 1$, then $j = 1$, $\beta_1 = \beta$, $_1C_{q_0} = \sqrt{\beta}C_q$, and therefore,

$$\beta_1^* = \frac{(\sqrt{\beta}C_q)^{2(q-1)}}{8\beta} = \beta*.$$

Proof. The proof can be achieved by d successive applications of Lemma 3.22. Indeed, by definition

$$\mathcal{F}_Q[\ _dC_q e_{q_k}^{-(\beta_1 x_1^2 + \cdots + \beta_d x_d^2)}](\xi)$$

$$= F_{d,q_{d-1}} \circ \cdots \circ F_{2,q_1} \circ F_{1,q_0}[\ _dC_q e_{q_k}^{-(\beta_1 x_1^2 + \cdots + \beta_d x_d^2)}](\xi_1, \xi').$$

Applying Lemma 3.22 to the expression

$$F_{1,q}[\ _dC_q e_q^{-(\beta_1 x_1^2 + \cdots + \beta_d x_d^2)}](\xi_1, x'),$$

with $\beta = \beta_1$ and $A = \beta_2 x_2^2 + \cdots + \beta_d x_d^2$, we obtain

$$F_{1,q}[\ _dC_q e_q^{-(\beta_1 x_1^2 + \cdots + \beta_d x_d^2)}](\xi_1, x') = {}_{d-1}C_{q_1} e_{q_1}^{-\alpha_1 \frac{\xi_1^2}{4} - A_1^*},$$

where

$$\alpha_1 = \frac{3-q}{2} \frac{{}_dC_q^{2(q-1)}}{\beta_1}, \quad \text{and} \quad A_1^* = \frac{(3-q)(\beta_2 x_2^2 + \cdots + \beta_d x_d^2)}{2}.$$

Further, by applying again Lemma 3.22 to the expression

$$F_{2,q_1}[\ _{d-1}C_{q_1} e_{q_1}^{-\frac{(3-q)\beta_2}{2} x_2^2 - A_1}](\xi_1, \xi_2, x''),$$

with $\beta = \frac{(3-q)\beta_2}{2}$ and

$$A = A_1 = \frac{3-q}{2}(\beta_3 x_3^2 + \cdots + \beta_d x_d^2) - \alpha_1 \frac{\xi_1^2}{4},$$

we have

$$F_{2,q_1}\left[\,_{d-1}C_{q_1}e_{q_1}^{-\frac{(3-q)\beta_2}{2}x_2^2 - A_1}\right](\xi_1, \xi_2, x'') =_{d-2}C_{q_2}e_{q_2}^{-\alpha_2\frac{\xi_2^2}{4} - A_2^*},$$

where

$$\alpha_2 = \frac{3-q_1}{2}\frac{_{d-1}C_1^{2(q_1-1)}}{\frac{3-q}{2}\beta_2},$$

$$A_2^* = \frac{3-q_1}{2}A_1 = \frac{3-q_1}{2}\left[\frac{3-q}{2}(\beta_3 x_3^2 + \cdots + \beta_d x_d^2) - \alpha_1 \frac{\xi_1^2}{4}\right].$$

By continuing this process m times we obtain

$$F_{m,q_{m-1}}\left[\,_{d-m+1}C_{q_{m-1}}e_{q_{m-1}}^{-a_{m-1}(\beta_m x_m^2 + A_m) - \sum_{j=1}^{m-1}\gamma_{m-1,j}\xi_j^2}\right](\xi^{(m-1)}, x^{(m)})$$
$$=_{d-m}C_{q_m}e_{q_m}^{-a_m(\beta_{m+1}x_{m+1}^2 + A_{m+1}) - \sum_{j=1}^{m}\gamma_{m,j}\xi_j^2}, \qquad (3.154)$$

where

$$a_m = \prod_{n=0}^{m-1}\frac{3-q_n}{2},$$

$$A_m = \beta_{m+1}x_{m+1}^2 + \cdots + \beta_d x_d^2,$$

$$\gamma_{m,j} = \frac{[\,_{d-j+1}C_{q_{j-1}}]^{2(q_{j-1}-1)}}{4\beta_j}\frac{\prod_{n=j-1}^{m-1}\frac{3-q_n}{2}}{\prod_{n=0}^{j-2}\frac{3-q_n}{2}},$$

$$= \frac{[\,_{d-j+1}C_{q_{j-1}}]^{2(q_{j-1}-1)}}{4\beta_j}\frac{a_m^2}{a_d}\qquad j = 1, \ldots, m.$$

Thus, by performing the above process d times, we arrive at the result in equations (3.152) and (3.153) with $\beta_j^* = \gamma_{d,j}$. □

Remark 3.20. It is easy to verify that Theorems 3.21 and 3.23 recover Theorem 3.4 in the one-dimensional case ($d = 1$).

For an arbitrary collection $Q_k = \{q_k, \ldots, q_{k+d}\}$ Theorem 3.23 takes the following formulation.

Theorem 3.24. *Let $\beta_j > 0$, $j = 1, \ldots, d$, and $Q_k = \{q_k, \ldots, q_{d+k-1}\}$ with $q_k \in (1, 1 + 2/d)$. Then for the d-dimensional sequential Q_k-Fourier transform the following relation holds:*

$$\mathcal{F}_{Q_k}[\,_dC_{q_k}e_{q_k}^{-(\beta_1 x_1^2 + \cdots + \beta_d x_d^2)}](\xi) = e_{q_{d+k}}^{-(\beta_1^* \xi_1^2 + \cdots + \beta_d^* \xi_d^2)}, \qquad (3.155)$$

where

$$\beta_j^* = \frac{[\,_{d-j+1}C_{q_{k+j-1}}]^{2(q_{k+j-1}-1)}}{4\beta_j} \frac{\displaystyle\prod_{n=k+j-1}^{d+k-1} \frac{3 - q_n}{2}}{\displaystyle\prod_{n=k}^{k+j-2} \frac{3 - q_n}{2}}, \qquad j = 1, \ldots, d.$$

$$(3.156)$$

3.15. Mapping Properties of Multidimensional q-Fourier Transforms

Let $1 \leq q < 1 + 2/d$. Then, both versions of multidimensional q-Fourier transform are defined on functions of $L_1(\mathbb{R}^d)$ and belong to the space $C_0(\mathbb{R}^d)$. Moreover, mappings

$$\mathcal{F}_q : L_1(\mathbb{R}^d) \to C_0(\mathbb{R}^d),$$

$$\mathcal{F}_Q : L_1(\mathbb{R}^d) \to C_0(\mathbb{R}^d), \quad (Q = \{q, q_1, \ldots, q_{d-1}\})$$

are continuous.

Below we discuss mapping properties of $\mathcal{F}_q$ and $\mathcal{F}_Q$ in subsets of $L_1(\mathbb{R}^d)$. Namely, we introduce the set $\mathcal{G}_{d,q} \subset L_1(\mathbb{R}^d)$, defined as

$$\mathcal{G}_{d,q} = \{f = Ae_q^{(\Sigma x, x)}, \ A > 0, \ \Sigma \text{ is symmetric positive definite}\}.$$

Further, consider the sequence $q_n^d = u_{dn}(q), n = 1, 2, \ldots,$ with a given $q_0^d = q$, $q < 1 + \frac{2}{d}$. The function $u_n(\cdot)$ is defined in (3.37).

One can extend the sequence q_n for negative integers $n = -1, -2, \ldots$ as well, putting $q_{-n} = u_{nd}^{-1}(q), n = 1, 2, \ldots$. It is not hard to verify that

$$q_n^d = \frac{2q + dn(1-q)}{2 + dn(1-q)} = 1 + \frac{2(q-1)}{2 - dn(q-1)}, \quad n = 0, \pm 1, \pm 2, \ldots \tag{3.157}$$

or, equivalently,

$$\frac{1}{1 - q_n^d} = \frac{1}{1-q} + \frac{dn}{2}, \quad n = 0, \pm 1, \pm 2, \ldots \tag{3.158}$$

Note $q_n^d \equiv 1$ for all $n = 0, \pm 1, \pm 2, \ldots$, if $q = 1$ and $\lim_{n \to \pm\infty} q_n^d = 1$ for all $q \neq 1$.

Remark 3.21. The sequence (3.158) has appeared in a quite different context in Tsallis *et al.* (2005a); see Footnote of p. 15378. In the one-dimensional case, i.e., for $d = 1$, the relationship (3.157) is obtained in Umarov *et al.* (2008). For $n = 1$, in the d-dimensional case, the relationship

$$q_1^d = \frac{2q - d(q-1)}{2 - d(q-1)} = 1 + \frac{2(q-1)}{2 - d(q-1)}, \tag{3.159}$$

between components of elliptically invariant Gaussians was recorded in Vignat and Plastino (2007b).

Lemma 3.25. *The members of the sequence q_n^d satisfy the following duality relations*

$$dq_{n-1}^d + \frac{1}{q_{n+1}^d} = 1 + d, \quad n = 0, \pm 1, \pm 2, \ldots \tag{3.160}$$

Proof. By elementary calculations we have

$$1 + d - dq_{n-1}^d = 1 + \frac{2d(1-q)}{2 + d(n-1)(1-q)}$$

$$= \frac{1}{\frac{2q+d(n+1)(1-q)}{2+d(n+1)(1-q)}} = \frac{1}{q_{n+1}^d}. \qquad \square$$

Remark 3.22. In the one-dimensional case ($d = 1$) the duality relations (3.160) recover the relationship established in Proposition 3.6:

$$q_{n-1} + \frac{1}{q_{n+1}} = 2, \quad n = 0, \pm 1, \dots .$$

It follows from Theorems 3.21 and 3.24 that the multidimensional direct and sequential q-Fourier transforms perform the following mappings.

Proposition 3.22. *Let* $1 < q_k < 1 + 2/d$. *Then the following mappings hold:*

$$\mathcal{F}_{q_k} : \mathcal{G}_{d,q_k} \to \mathcal{G}_{d,q_{k+1}^d},$$

$$\mathcal{F}_{Q_k} : \mathcal{G}_{d,q_k} \to \mathcal{G}_{d,q_{k+d}}.$$

Proposition 3.23. *Let* $1 < q_k < 1 + 2/d$. *Then there exist the inverse operators*

$$\mathcal{F}_{q_k}^{-1} : \mathcal{G}_{d,q_{k+1}^d} \to \mathcal{G}_{d,q_k},$$

$$\mathcal{F}_{Q_k}^{-1} : \mathcal{G}_{d,q_{k+d}} \to \mathcal{G}_{d,q_k}.$$

Moreover, for these inverse operators, the following representations hold:

$$\mathcal{F}_{q_k}^{-1} = \mathcal{F}_{q_{k-1}^d}^* \circ I_{q_{k+1}^d, q_{k-1}^d}, \tag{3.161}$$

$$\mathcal{F}_{Q_k}^{-1} = (F_{1,q_{k-1}}^* \circ I_{q_{k+1}, q_{k-1}}) \circ \cdots \circ (F_{d,q_{d-2}}^* \circ I_{q_{k+d}, q_{k+d-2}}), \tag{3.162}$$

where operator $I_{p,q}$ *is defined in* (3.47) *and* $F_{j,q}^*$ *acting on the j-th variable in* (3.25), *respectively, and*

$$\mathcal{F}_q^*[f](\xi) = \int_{\mathbb{R}^d} f(x) \exp(-i(x,\xi)[f(x)]^{q-1}) dx.$$

We omit the proof of this statement, since it is similar to the proof of Theorem 3.5. We note that if $q = 1$, then both representations (3.161) and (3.162) recover the classical formula for the d-dimensional inverse

Fourier transform

$$\mathcal{F}_1[\hat{f}](x) = \frac{1}{(2\pi)^d} \int_{\mathbb{R}^d} \hat{f}(\xi)e^{-i(x,\xi)}d\xi.$$

We also note that both versions of the multidimensional q-Fourier transforms are not invertible in the entire space $L_1(\mathbb{R}^d)$.

3.16. Additional notes

(1) Historical notes

The Fourier transform is the continuous version of the Fourier series introduced by Jean–Baptiste Joseph Fourier[6] in the beginning of the 18th century. Fourier published his work related to a solution of heat distribution problem using trigonometric series. His complete investigation was published in 1822. In the one-dimensional case trigonometric series appeared even earlier in works by Leonhard Euler, Daniel Bernoulli.[7] Fourier studied his famous "Fourier series" in the three-dimensional case (for three-dimensional torus). Nowadays, both Fourier series and Fourier transform are invaluable mathematical tools, not only in mathematics, but also in theoretical and applied sciences and engineering. In modern mathematics the Fourier series and the Fourier transform gave rise to the harmonic analysis, spectral theory of linear (and nonlinear) operators, theory of pseudo-differential operators, theory of approximation including discrete ones, and many more.

In the literature there are q-generalizations of the Fourier series and Fourier transforms based on the q-generalizations of exponential and trigonometric functions in the Euler's spirit (see Section 2.9 of Chapter 2). Note that all these generalizations are linear operators acting in suitable spaces. The q-Fourier transform theory developed in the current chapter are based on the q-exponential functions and q-algebra developed in Chapter 2. The q-Fourier transform was first introduced by Umarov *et al.* (2008) in the one-dimensional case, and

---

[6] Joseph Fourier, 1768–1830.
[7] Daniel Bernoulli, 1700–1782.

in Umarov and Tsallis (2007) in the multi-dimensional case. The striking difference of this q-Fourier transform from others is the fact that it is nonlinear and noninvertible. The non-invertibility of the q-Fourier transform was first detected by Hilhorst (2009, 2010). The invertibility question was discussed in papers by Jauregui and Tsallis (2011), Jauregui *et al.* (2011), and Plastino and Rocca (2012, 2013).

(2) q-Trigonometric series. q-Cosine and q-sine series

In this chapter we developed the theory of only the q-Fourier transform, which is well defined for positive absolutely integrable functions f. In particular, f can be a density function of some continuous random variable. The q-Fourier transform, obviously, can be defined in the more general case of functions containing discrete and/or singularity points. Suppose f is a μ-measurable function, where μ is a Borel measure defined on $\mathbb{R}$. Then, the q-Fourier transform of f is defined by

$$
\begin{aligned}
F_q[f](\xi) &= \int_{\mathbb{R}} f(x) \otimes_q \exp(ix\xi) d\mu(x) \\
&= \int_{\mathbb{R}} f(x) \exp(ix\xi[f(x)]^{q-1}) d\mu(x).
\end{aligned}
$$

In particular, if f is a probability mass function with masses f_n of a discrete random variable X, then one can define the corresponding q-Fourier transform (q-characteristic function or q-Fourier series) through the infinite series

$$
\sum_{n=-\infty}^{\infty} f_n \exp_q(in f_n^{q-1} x). \tag{3.163}
$$

Similarly, one can define q-cosine and q-sine series by

$$
\sum_{n=0}^{\infty} f_n \cos_q(n f_n^{q-1} x) \quad \text{and} \quad \sum_{n=0}^{\infty} f_n \sin_q(n f_n^{q-1} x). \tag{3.164}
$$

To the authors best knowledge, such q-trigonometric series are not well studied yet. An important question arising in this context is what is the relation between the sequence f_n and the function $F(x)$ representing q-trigonometric series in (3.163) or (3.164). Another important question is the convergence of q-trigonometric series in appropriate spaces of functions. We note that similar questions have been studied for q-trigonometric series defined through the q-trigonometric functions in Euler's q-generalized approach (see, e.g., Feinsilver, 1989; Ol'shanetskii and Rogov, 1999; Fitouhi and Bouzffour, 2012; Ernst, 2012).

(3) Issues connected with the noninvertibility of the q-Fourier transform

The q-Fourier transform can be used for solution of problems of complex nonlinear nature. An example of such an application is the proof of the central limit theorem for strongly correlated sums of identically distributed random variables, discussed in Chapter 4. Another example is a solution of nonlinear Fokker–Planck equation using the q-Fourier transform, which will be discussed in Chapter 6. However, as noted above (see Section 3.10), the q-Fourier transform, in general, is not invertible. Hilhorst (2010), by using the invariance principle, constructed examples which show that the q-Fourier transform can be the same for an entire class of density functions. This obstacle creates some issues in applications of the q-Fourier transform. Namely, if one uses the q-Fourier transform for a solution of a mathematical problem, then the existence of solution and its representation in the q-Fourier domain can be obtained. However, the q-Fourier transform does not provide the uniqueness of this solution. Therefore, some other approach should be used for establishing the uniqueness. In next chapter, we will discuss the uniqueness of a limiting process in the q-generalized central limit theorem, existence of which will be established using the q-Fourier transform.

In some occasions, for uniqueness, one can use "representation formulas" (3.127) and (3.128). Going into further details we note that Plastino and Rocca in a series of papers (Plastino and Rocca, 2012, 2013) defined the q-Fourier transform in the space of ultra-distributions, which is a subspace of the space of hyper-functions.

The space of hyper-functions was introduced by Sato (1959, 1960), in 1959; see also Graf (2010). By definition, a hyper-function, defined on an interval $I \subset \mathbb{R}$, is an equivalence class of differences $g(x) = G_+(x + i0) - G_-(x - i0)$, where $G_+(z)$ and $G_-(z)$ are analytic functions on upper and lower complex neighborhoods of the interval I, respectively. The functions $G_+(z)$, $G_-(z)$ are called defining functions of the hyper-function $g(x)$. Any Schwartz distribution, or more generally, any ultra-distribution is also a hyper-function. Moreover, there are hyper-functions which are not ultra-distributions. Hence, the space of hyper-functions is wider than the space of Schwartz distributions, and even the space of ultra-distributions.

Plastino and Rocca defined the q-Fourier transform of f as a hyper-function

$$F(q, \xi) = G_+(q, \xi) - G_-(q, \xi), \qquad (3.165)$$

of variables $(q, \xi) \in (1, 2) \times \mathbb{R}$, with defining functions

$$G_+(q, \xi) = H^*(q) \left[H(Im(\xi)) \int_0^\infty f(x) \exp_q(ix\xi[f(x)]^{q-1}) dx \right],$$

and

$$G_-(q, \xi) = H^*(q) \left[H(-Im(\xi)) \int_{-\infty}^0 f(x) \exp_q(ix\xi[f(x)]^{q-1}) dx \right],$$

where $H^*(q) = (H(q-1) - H(q-2))$, $Im(\xi)$ is the imaginary part of ξ, $H(\cdot)$ is the Heaviside function, and $1 \le q < 2$, $\xi \in \mathbb{R}$. The meaning of this q-Fourier transform is different from the q-Fourier transform $F_q[f](\xi)$ discussed above. Here q is not fixed. Unlike $F_q[f](\xi)$, where q is fixed, the q-Fourier transform $F(q, \xi)$ defined in (3.165) must be valid for all $q \in [1, 2)$. This stronger requirement makes the q-Fourier transform invertible. Moreover, the original function $f(x)$ can be reconstructed through the known q-Fourier transforms for all $q \in [1, 2)$ by the formula in (3.127).

Let us finally mention that, by using further information about some q-moments, it is possible to remove the degeneracy indicated by Hilhorst in his examples (see details in Jauregui *et al.*, 2011).

(4) Other q-generalized Fourier transforms

The definition of the q-Fourier transform is based on the q-exponential and q-product introduced in the previous chapter. In modern literature there are various generalizations of this kind (Kaniadakis and Scarfone, 2002; Kaniadakis *et al.*, 2005; Scarfone, 2017; Kalogeropoulos, 2018). For example, the τ_q-Fourier transform introduced and studied in Kalogeropoulos (2018) and the κ-Fourier transform introduced in Scarfone (2017). We note that these two generalizations are linear, isomorphic to the classical Fourier transform, and therefore, invertible. Due to the linear nature of these transformations, they can not be directly applied for modeling strongly correlated random variables.

(5) Proof of the fact that the q-characteristic function $\varphi_{q,N_q}(\xi)$ of the q-Gaussian equals $e^{-\frac{1}{2}\xi^2}$ if $q = 1$, recovering the characteristic function of the standard normal distribution

Indeed, if $q = 1$, then for A_{2n} in (3.59), we have $A_{2n} = 1$ for all $n = 0, 1, \ldots$. Using the known formula [Abramowitz and Stegun (1964)]

$$\Gamma\left(n + \frac{1}{2}\right) = \frac{(2n)!}{4^n n!}\sqrt{\pi}, \quad n = 0, 1, \ldots,$$

we have

$$\frac{\Gamma(n + \frac{1}{2})}{\beta^n (2n)!} = \frac{1}{2^n\, n!}. \tag{3.166}$$

Next, we show that

$$\lim_{q \to 1+0} \frac{\Gamma(n + \frac{1}{q-1} - \frac{1}{2})}{(q - 1)^n \Gamma(2n + \frac{1}{q-1} - \frac{1}{2})} = 1. \tag{3.167}$$

To show this we use the change $z = \frac{1}{q-1}$. Obviously, $z \to \infty$ if $q \to 1 + 0$. Then, the expression on the left-hand side of (3.167) takes the form

$$\lim_{z \to \infty} \frac{z^n \Gamma(z + n - \frac{1}{2})}{\Gamma(z + 2n - \frac{1}{2})}.$$

Now, making use of the limit (Abramowitz and Stegun, 1964)

$$\lim_{z \to \infty} \frac{\Gamma(z + \alpha)}{\Gamma(z) z^\alpha} = 1,$$

we get

$$\lim_{z \to \infty} \frac{z^n \Gamma(z + n - \frac{1}{2})}{\Gamma(z + 2n - \frac{1}{2})}$$

$$= \lim_{z \to \infty} z^n \frac{\Gamma(z + n - \frac{1}{2})}{\Gamma(z) z^{n - \frac{1}{2}}} \frac{\Gamma(z) z^{2n - \frac{1}{2}}}{\Gamma(z + 2n - \frac{1}{2})} \frac{z^{n - \frac{1}{2}}}{z^{2n - \frac{1}{2}}} = 1,$$

proving (3.167). Finally, taking into account (3.166) and (3.167), it follows from (3.59) that

$$\varphi_{1,N_1}(\xi) = \sum_{n=0}^{\infty} \frac{(-1)^n}{n!} \xi^{2n} = e^{-\frac{1}{2}\xi^2},$$

obtaining the characteristic function of the standard normal distribution.

Chapter 4

q-Central Limit Theorems

4.1. Introduction

This chapter is devoted to the q-generalizations of the central limit theorem (CLT); see Section 1.2. CLT plays an important role in modern science and engineering, both theoretical and applied parts, dealing with modeling of random processes. CLT essentially teaches that the appropriately scaling limit of sequences of sums of any independent and identically distributed random variables with a finite variance is the standard normal random variable. In Section 4.3, we prove a q-generalization of CLT. The proof uses the technique based on the q-Fourier transform studied in detail in Chapter 3. We have seen that the q-Fourier transform does not have a unique preimage. Therefore, the q-Fourier transform technique does not provide uniqueness of the limiting distribution in the q-CLT. In Section 4.4, we establish the uniqueness theorem. Section 4.5 discusses multidimensional versions of the q-CLT.

The fundamental distinction of q-CLT from other central limit theorems is that it is valid for strongly dependent (correlated) random variables. An attempt to describe strong dependencies is taken in Section 4.2. For such dependencies we use the generic notion "q-independence", even though they represent strong dependence. The reason why it is called q-independence is that it becomes the standard statistical independence if $q = 1$.

There are other forms of describing strong dependencies via exchangeability or mixing independent variables with the help of a positive distribution (see Section 4.6), which are asymptotically $(n \to \infty)$ equivalent to the q-independence.

In the literature, there are other versions of the q-CLT, which do not use the q-Fourier transform. In Section 4.6, we present two approaches which do not use the q-Fourier transform: one is based on the exchangeability of random variables and published in Jiang *et al.* (2010), and the other one is based on mixtures of independent and identically distributed random variables with the help of χ^2-distribution and published in Vignat and Plastino (2007b). In Sections 4.7–4.9, we consider some implications of the q-CLT relevant to non-extensive statistical mechanics. Finally, in Section 4.10, we briefly discuss the construction of some versions of q-Gaussian stochastic processes.

4.2. *q*-Independence and *q*-Weak Convergence of Dependent Random Variables

In this section, we introduce the notion of *q-independence* of random variables and two types of convergence, which will be used in our further analysis. We call these convergences the q-convergence and the weak q-convergence. We establish their equivalence and relevance to the standard weak convergence.

Definition 4.1. Two random variables X and Y are said to be (q', q, q'')-*independent* if

$$F_{q'}[X + Y](\xi) = F_q[X](\xi) \otimes_{q''} F_q[Y](\xi). \tag{4.1}$$

In terms of densities, equation (4.1) can be rewritten as follows. Let f_X and f_Y be densities of X and Y, respectively, and let f_{X+Y} be the density of $X + Y$. Then

$$\int_{-\infty}^{\infty} e_{q'}^{ix\xi} \otimes_{q'} f_{X+Y}(x)dx = F_q[f_X](\xi) \otimes_{q''} F_q[f_Y](\xi). \tag{4.2}$$

If all three parameters q', q and q'' coincide, i.e., $q = q' = q''$, then we simply call it q-independent. For $q = 1$ the condition (4.1) recovers the well-known relation

$$F[f_X * f_Y] = F[f_X] \cdot F[f_Y]$$

between the convolution (noted $*$) of two densities and the multiplication of their (classical) characteristic functions, and holds for independent X and Y. If $q \neq 1$, then (q', q, q'')-independence describes a version of *strong dependence* between X and Y. As we will see later, the word "strong" indeed describes this dependence correctly.

Now assume that X_j, $j = 1, 2, \ldots$, be a sequence of identically distributed random variables. Denote

$$Y_n = X_1 + \cdots + X_n, \quad n = 1, 2, \ldots.$$

Definition 4.2. The infinite sequence of random variables $\{X_j\}_{j=1}^{\infty}$, is said to be (q', q, q'')-independent (or (q', q, q'')-*i.i.d.*) if the relations

$$F_{q'}[Y_n](\xi) = F_q[X_1](\xi) \otimes_{q''} \cdots \otimes_{q''} F_q[X_n](\xi) \qquad (4.3)$$

hold for all $n = 2, 3, \ldots$.

Remark 4.1. For $q = q' = q'' = 1$ the condition (4.3) turns into the condition for the sequence X_j, $j = 1, 2, \ldots$, usually referred to as i.i.d. If $q = q' = q''$ then we call the sequence X_j simply a q-i.i.d.

Example 4.1. Consider example of a (q', q, q)-i.i.d. sequence of random variables, where $q \in (1, 3)$ and $q' = (3q - 1)/(q + 1)$. Assume X_j, $j = 1, \ldots$, is the sequence of identically distributed random variables with the associated q'-Gaussian density

$$G_{q'}(\beta, x) = \frac{\sqrt{\beta}}{C_{q'}} e_{q'}^{-\beta x^2},$$

where $C_{q'}$ is the normalizing constant (see Section 2.7). Further, assume the sums $X_1 + \cdots + X_n$, $n = 2, 3, \ldots$, are distributed according to the density $G_{q'}(\alpha, x)$, where $\alpha = n^{-\frac{1}{2-q'}} \beta$. Then the sequence X_j satisfies (4.3) for all $n = 2, 3, \ldots$, with $q = q''$, thus being a (q', q, q)-independent identically distributed sequence of random variables.

Now we introduce two types of convergence which will be used in q-CLTs. By definition, a sequence of random variables V_n with the corresponding density functions f_{V_n} is said to be *q-convergent* to a random variable V_0 with the corresponding density function f_{V_0}, if

$$\lim_{n \to \infty} F_q[f_{V_n}](\xi) = F_q[f_{V_0}](\xi),$$

locally uniformly in ξ.

Evidently, this definition is equivalent to the weak convergence (denoted by "$\Rightarrow$") of random variables, if $q = 1$. For $q \neq 1$, denote by W_q the set of continuous functions ϕ satisfying the condition

$$|\phi(x)| \leq C(1 + |x|)^{-\frac{q}{q-1}}, \quad x \in \mathbb{R}.$$

A sequence of random variables V_n is called *weakly q-convergent* to a random variable V_0, if

$$\int_{\mathbb{R}} f_{V_n}(x)dm_q \to \int_{\mathbb{R}} f_{V_0}(x)dm_q$$

for arbitrary measure m_q defined as $dm_q(x) = \phi_q(x)dx$, where $\phi_q \in W_q$. We denote the weak q-convergence by the symbol $\overset{q}{\Rightarrow}$.

Proposition 4.1. *Let $q > 1$. Then $V_n \Rightarrow V_0$ yields $V_n \overset{q}{\Rightarrow} V_0$.*

The proof of this statement immediately follows from the obvious fact that W_q is a subset of the set of bounded continuous functions. Recall that a sequence of probability measures μ_n, $n = 1, 2, \ldots$, is called *tight* if, for an arbitrary $\epsilon > 0$, there is a compact K_ϵ and an integer n_ϵ^* such that $\mu_n(\mathbb{R} \setminus K_\epsilon) < \epsilon$ for all $n \geq n_\epsilon^*$.

Proposition 4.2. *Let $1 < q < 2$. Assume a sequence of random variables V_n, defined on a probability space with a probability measure P, and associated densities f_{V_n}, is q-convergent to a random variable V_0 with an associated density f_{V_0}. Then the sequence of associated probability measures $\mu_n = P(V_n^{-1})$, $n = 1, 2, \ldots$, is tight.*

Proof. Assume that $1 < q < 2$ and V_n is a q-convergent sequence of random variables with associated densities f_{V_n} and associated probability measures μ_n. Then for $R > 0$ we have

$$\frac{1}{R} \int\limits_{-R}^{R} \left(1 - F_q[f_{V_n}](\xi)\right) d\xi$$

$$= \frac{1}{R} \int\limits_{-R}^{R} \left(1 - \int\limits_{\mathbb{R}} f_{V_n}(x) e_q^{ix\xi[f_{V_n}(x)]^{q-1}} dx\right) d\xi$$

$$= \int\limits_{\mathbb{R}} \left(\frac{1}{R} \int\limits_{-R}^{R} \left(1 - e_q^{ix\xi[f_{V_n}(x)]^{q-1}}\right) d\xi\right) d\mu_n(x). \qquad (4.4)$$

It is not hard to verify that

$$\frac{1}{R} \int\limits_{-R}^{R} e_q^{ix\xi t} d\xi = \frac{2 \sin_{\frac{1}{2-q}}(Rx(2-q)t)}{Rx(2-q)t}. \qquad (4.5)$$

It follows from (4.4) and (4.5) that

$$\frac{1}{R} \int\limits_{-R}^{R} (1 - F_q[f_{V_n}](\xi)) d\xi$$

$$= 2 \int\limits_{-\infty}^{\infty} \left(1 - \frac{\sin_{\frac{1}{2-q}}(x(2-q)R[f_{V_n}(x)]^{q-1})}{Rx(2-q)[f_{V_n}(x)]^{q-1}}\right) d\mu_n(x). \qquad (4.6)$$

Since $1 < q < 2$ by assumption, $\frac{1}{2-q} > 1$. As we have seen (see Proposition 2.22), for any $q' > 1$ the properties $\sin_{q'}(x) \leq 1$ and $(\sin_{q'}(x))/x \to 1$, $x \to 0$, hold. Moreover, $(\sin_{q'}(x))/x \leq 1$, for all $x \in \mathbb{R}$. Let us now suppose,

$$\lim_{|x| \to \infty} |x|[f_n(x)]^{q-1} = L_n, \quad n \geq 1.$$

Divide the set $\{n \geq n_0\}$ into two subsets

$$A = \{n_j \geq n_0 : L_{n_j} > 1, \ j = 1, 2, \dots\}$$

and

$$B = \{n_k \geq n_0 : L_{n_k} \leq 1, \ k = 1, 2, \dots \}.$$

If $n \in A$, since $\sin \frac{1}{2-q} \leq 1$, there is a number $a > 0$ such that

$$\frac{1}{R} \int_{-R}^{R} (1 - F_q[f_{V_n}](\xi)) d\xi$$

$$\geq 2 \int_{|x| \geq a} \left(1 - \frac{1}{R|x|(2-q)[f_{V_n}]^{q-1}} \right) d\mu_n(x)$$

$$\geq C\mu_n \left(|x| \geq a \right), \quad C > 0 \ \forall n \in A,$$

for R small enough. Now taking into account the q-convergence of V_n to V_0 and, if necessary, taking R smaller, for any $\varepsilon > 0$, we obtain

$$\mu_n \left(|x| \geq a \right) \leq \frac{1}{CR} \int_{-R}^{R} (1 - F_q[f_{V_0}](\xi)) d\xi < \varepsilon, \quad \forall n \in A.$$

If $n \in B$, then there exist constants $b > 0$, $\delta > 0$, such that

$$f_{V_n}(x) \leq \frac{L_n + \delta}{|x|^{\frac{1}{q-1}}} \leq \frac{1 + \delta}{|x|^{\frac{1}{q-1}}}, \quad |x| \geq b, \ \forall n \in B.$$

Hence, we have

$$\mu_n(|x| > b) = \int_{|x| > b} f_{V_n}(x) dx$$

$$\leq (1 + \delta) \int_{|x| > b} \frac{dx}{|x|^{\frac{1}{q-1}}}, \quad n \in B.$$

Since, $1/(q-1) > 1$, for any $\varepsilon > 0$ we can choose a number $b_\varepsilon \geq b$ such that $\mu_n(|x| > b_\varepsilon) < \varepsilon$, $n \in B$. As far as $A \cup B = \{n : n \geq n_0\}$, the proof of the statement is complete. $\qquad\square$

To prove the fact that the q-weak convergence implies the q-convergence we need an additional property of the function $g_q(a; u)$ introduced in (3.4). Recall that

$$g_q(a; t) = t \exp_q(iat^{q-1})$$

$$= t(1 + i(1-q)at^{q-1})^{-\frac{1}{q-1}}, \quad t \in \mathbb{R}_+, \qquad (4.7)$$

where $1 < q < 2$ and a is a fixed real number. As it was established in Lemma 3.1, this function is continuous on $[0, b]$, $b > 0$ and differentiable in the interval $(0, b)$.

Consider the following Cauchy problem for the Bernoulli equation

$$y' - \frac{1}{t}y = \frac{ia(q-1)}{t}y^q, \quad y(0) = 0, \qquad (4.8)$$

It is not hard to verify that $y(t) = g_q(a; t)$ is a solution to problem (4.8).

Proposition 4.3. *For $g'_q(a; t)$ the estimate*

$$|g'_q(a; t)| \le C(1 + |a|)^{-\frac{q}{q-1}}, \quad t \in (0, 1], \ a \in \mathbb{R}, \qquad (4.9)$$

holds, where constant C does not depend on t.

Proof. It follows from (4.7) and (4.8) that

$$|y'(t)| \le t^{-1}|y + ia(q-1)y^q|$$

$$= |e_q^{iat^{q-1}} + ia(q-1)t^{q-1}(e_q^{iat^{q-1}})^q|$$

$$= |1 + ia(1-q)t^{q-1}|^{-\frac{q}{q-1}}$$

$$\le C(1 + |a|)^{-\frac{q}{q-1}}, \quad t \in (0, 1]. \qquad \square$$

Now we are in a position to formulate the following two theorems on the relationship between q-convergence and weak q-convergence.

Theorem 4.1. *Let $1 < q < 2$ and a sequence of random variables V_n be weakly q-convergent to a random variable V_0. Then V_n is q-convergent to V_0.*

Proof. Assume V_n, with associated densities f_{V_n}, is weakly q-convergent to a V_0, with an associated density f_{V_0}. The difference $\mathcal{F}_q[f_{V_n}](\xi) - \mathcal{F}_q[f_{V_0}](\xi)$ can be written in the form

$$\mathcal{F}_q[f_{V_n}](\xi) - \mathcal{F}_q[f_{V_0}](\xi)$$
$$= \int_{\mathbb{R}} [g_q(ix\xi; f_{V_n}(x)) - g_q(ix\xi; f_{V_0}(x))]dx, \qquad (4.10)$$

where $g_q(a;t)$ is defined in (4.7), with $a = ix\xi$. It follows from Lemma 3.1 and (4.9) that

$$|\mathcal{F}_q[f_{V_n}](\xi) - \mathcal{F}_q[f_{V_0}](\xi)|$$
$$\leq C \int_{\mathbb{R}} |(1 + |x|)^{-\frac{q}{q-1}} (f_{V_n}(x) - f_{V_0}(x))|dx, \quad \xi \in \mathbb{R},$$

where $C > 0$ does not depend on n. This estimate yields the convergence

$$\mathcal{F}_q[f_{V_n}](\xi) \to \mathcal{F}_q[f_{V_0}](\xi)$$

for all $\xi \in \mathbb{R}$, since V_n q-weakly converges to V_0 as $n \to \infty$. $\qquad\square$

Theorem 4.2. *Let $1 < q < 2$ and a sequence of random variables V_n with the associated densities f_{V_n} is q-convergent to a random variable V_0 with the associated density f_{V_0} and $\mathcal{F}_q[f_{V_0}](\xi)$ is continuous at $\xi = 0$. Then V_n weakly q-converges to V_0.*

Proof. Suppose that f_{V_n} converges to f_{V_0} in the sense of q-convergence. It follows from Proposition 4.2 that the corresponding sequence of induced probability measures $\mu_N = P(V_n^{-1})$ is tight. This yields relatively weak compactness of the sequence μ_n, $n = 1, 2, \dots$. Theorem 4.1 implies that each weakly convergent subsequence $\{\mu_{n_j}\}$ of the sequence μ_n converges to $\mu_0 = P(V_0^{-1})$. Hence, $\mu_n \Rightarrow \mu_0$, or equivalently, $V_n \Rightarrow V_0$. Now applying Proposition 4.1 we complete the proof. $\qquad\square$

4.3. *q*-Central Limit Theorem for $q \geq 1$

CLTs are based on the type of dependency of random variables. Therefore, in fact, there exist a series of q-versions of the classical central limit theorem. All these theorems recover the classical CLT if $q = q' = q'' = 1$.

In this section, we will consider three types of dependencies. An important distinction from the classic CLT is the fact that a complete formulation of q-CLT is not possible within only one given q. The parameter q is connected with two other numbers, $q_* = u^{-1}(q)$ and $q^* = u(q)$, where $u(q) = (1 + q)/(3 - q)$. We will see that q_* identifies an attractor, while q^* determines the scaling rate in the q-CLT. In a more general setting, the q-generalization of the CLT formulated in this section, is connected with a suitable triplet (q_{k-1}, q_k, q_{k+1}) determined by a given $q \in [1, 2)$. The sequence q_k is defined in (3.38) and (3.39). For q_k-independent identically distributed random variables the index q_{k-1} determines the attracting q-Gaussian, while the index q_{k+1} determines the scaling rate. Note that, if $q = 1$, then the entire family of q_k-central limit theorems collupses to one element, thus recovering the classic CLT.

For identically distributed random variables $X_1, X_2, \ldots, X_n, \ldots$, we will study limits of sums

$$Z_n = \frac{1}{\kappa_n(q)} (X_1 + \cdots + X_n - n\mu_q), \quad n = 1, 2, \ldots \quad (4.11)$$

as $n \to \infty$, in the sense of weak convergence. Here

$$\kappa_n(q) = (\sqrt{n\nu_{2q-1}}\sigma_{2q-1})^{\frac{1}{2-q}}, \quad n = 1, 2, \ldots, \quad (4.12)$$

is the scaling coefficient.

Two important questions we are seeking in this section are

- *Is there a random variable distributed as a q-Gaussian distribution that serves as the limit of the sequence Z_n in the sense of weak convergence?*
- *Is the limiting random variable uniquely defined?*

For $q = 1$ the answers for both questions are well known to be affirmative and constitute the content of the classical CLT.

The three types of q-independence which will be used in theorems of this section are defined as follows.

Definition 4.3. The sequence of identically distributed random variables $\{X_n\}_{n=1}^{\infty}$ are called q-independent of Type I, if

$$F_q[X_1 + \cdots + X_n](\xi) = F_q[X_1](\xi) \otimes_q \cdots \otimes_q F_q[X_n](\xi), \qquad (4.13)$$

of Type II, if

$$F_q[X_1 + \cdots + X_n](\xi) = F_q[X_1](\xi) \otimes_{q_1} \cdots \otimes_{q_1} F_q[X_n](\xi), \qquad (4.14)$$

of Type III, if

$$F_q[X_1 + \cdots + X_n](\xi) = F_{q_1}[X_1](\xi) \otimes_{q_1} \cdots \otimes_{q_1} F_{q_1}[X_n](\xi), \qquad (4.15)$$

hold for all $n \geq 2$ and $\xi \in (-\infty, \infty)$. Here, $q_1 = u(q) = \frac{1+q}{3-q}$, and F_q is the q-Fourier transform.

The following lemma will be used in the proof of the q-CLT.

Lemma 4.3. *Let* $a \in \mathbb{R}$. *The following formulas hold:*

(1) $\mu_q(aX) = a\mu_q(X)$;
(2) $\mu_q(X - \mu_q(X)) = 0$;
(3) $\sigma_q^2(aX) = a^2 \sigma_q^2(X)$.

Proof. Using the fact that the density function $f_{aX}(x)$ of the random variable aX is related to the density $f_X(x)$ of the random variable X via

$$f_{aX}(x) = \frac{1}{a} f_X\left(\frac{x}{a}\right),$$

we have

$$\mu_q(aX) = \frac{\int\limits_{\text{supp } [f_{aX}]} x \, [f_{aX}(x)]^q \, dx}{\int\limits_{\text{supp } [f_{aX}]} [f_{aX}(x)]^q \, dx}$$

$$= \frac{a \int\limits_{\text{supp } [f_X]} x \, [f_X(x)]^q \, d}{\int\limits_{\text{supp } [f_X]} [f_X(x)]^q \, dx} = a\mu_q(X).$$

Similarly one can show the equality $\sigma_q^2(aX) = a^2\sigma_q^2(X)$. To prove Part (2), we use the fact $f_{X-b}(x) = f_X(x+b)$, valid for any $b \in \mathbb{R}$. We have

$$\mu_q(X - \mu_q) = \frac{\int\limits_{\text{supp } [f_{X-\mu_q}]} x \, [f_X(x+\mu_q)]^q \, dx}{\int\limits_{\text{supp } [f_{X-\mu_q}]} [f_X(x+\mu_q)]^q \, dx}$$

$$= \frac{\int\limits_{\text{supp } [f_X]} (x-\mu_q) \, [f_X(x)]^q \, dx}{\int\limits_{\text{supp } [f_X]} [f_X(x)]^q \, dx}$$

$$= \frac{\int\limits_{\text{supp } [f_X]} x \, [f_X(x)]^q \, dx}{\int\limits_{\text{supp } [f_X]} [f_X(x)]^q \, dx} - \mu_q = 0.$$

$\square$

As we noted above, the formulation of the q-CLT depends on the type of q-independence. The theorem below is the q-generalization of the CLT under the condition of Type I q-independence.

Theorem 4.4. *Let the sequence $q_k, k \in \mathcal{Z}$, be defined in (3.40) with $q_k \in [1,2)$. Let X_j, $j = 1, 2, \ldots$, be a sequence of q_k-independent of type I and identically distributed random variables with a finite q_k-mean μ_{q_k} and a finite second $(2q_k - 1)$-moment $\sigma_{2q_k-1}^2$.*

Then the sequence

$$Z_n = \frac{X_1 + \cdots + X_n - n\mu_{q_k}}{\kappa_n(q_k)}, \quad n = 1, 2, \ldots,$$

is q_k-convergent as $n \to \infty$ to the q_{k-1}-Gaussian distribution $G_{q_{k-1}}(\beta_k; x)$ with

$$\beta_k = \left(\frac{3 - q_{k-1}}{4q_k C_{q_{k-1}}^{2q_{k-1}-2}}\right)^{\frac{1}{2-q_{k-1}}}. \tag{4.16}$$

The proof of this theorem follows from Theorem 4.5 proved below and Lemma 3.10. Theorem 4.5 represents one element ($k = 0$) in the series of assertions contained in Theorem 4.4.

Theorem 4.5. *Assume that $1 \leq q < 2$. Let X_j, $j = 1, 2, \ldots$, be a sequence of q-independent of type I and identically distributed random variables with a finite q-mean μ_q and a finite second $(2q-1)$-moment σ_{2q-1}^2.*
Then the sequence Z_n defined in (4.11) is q-convergent as $n \to \infty$ to a q_{-1}-Gaussian distribution $G_{q_{-1}}(\beta; x)$, with

$$\beta = \left(\frac{3 - q_{-1}}{4q C_{q_{-1}}^{2q_{-1}-2}}\right)^{\frac{1}{2-q_{-1}}}.$$

Proof. Due to Lemma (4.3), and changing X_1 to $X_1 - \mu_q$ if necessary, we can assume that $\mu_q = 0$. Let $f(x)$ be the density function of the random variable X_1. First we evaluate $F_q[f](\xi)$, which is the same as the q-characteristic function of the random variable X_1 (see Definition 3.3), that is $F_q[f](\xi) = \varphi_{q,X_1}(\xi)$. Due to Proposition 3.12,

$$F_q[f](\xi) = 1 - \frac{q}{2}\xi^2 \sigma_{2q-1}^2 \nu_{2q-1} + o(\xi^2), \quad \xi \to 0. \tag{4.17}$$

Denote

$$Y_j = \frac{X_j - \mu_q}{\kappa_n(q)}, \quad j = 1, 2, \ldots,$$

where $\kappa_n(q)$ is defined in (4.12). Then $Z_N = Y_1 + \cdots + Y_N$. Further, it is readily seen that, for a given random variable X and real $a > 0$, the q-characteristic function of aX satisfies the equality $\varphi_{q,aX}(\xi) = \varphi_{q,X}(a^{2-q}\xi)$. Hence, for the density function $f_{aX_1}(x)$ of aX_1 we have

$$F_q[f_{aX_1}](\xi) = F_q[f](a^{2-q}\xi). \tag{4.18}$$

It follows from this relation that

$$F_q[f_{Y_1}](\xi) = F_q[f]\left(\frac{\xi}{\sqrt{n\nu_{2q-1}\sigma_{2q-1}}}\right). \tag{4.19}$$

Moreover, it follows from the q-independence of $X_1, X_2, \ldots$ and the associativity of the q-product that

$$F_q[f_{Z_n}](\xi) = \prod_{j=1}^{n}{}_q F_q[f]\left(\frac{\xi}{\sqrt{n\nu_{2q-1}\sigma_{2q-1}}}\right) \tag{4.20}$$

Hence, making use of properties of the q-logarithm, from (4.20) we obtain

$$\ln_q F_q[Z_n](\xi) = n\ln_q F_q[f]\left(\frac{\xi}{\sqrt{n\nu_{2q-1}\sigma_{2q-1}}}\right)$$

$$= n\ln_q\left(1 - \frac{q}{2}\frac{\xi^2}{n} + o\left(\frac{\xi^2}{n}\right)\right)$$

$$= -\frac{q}{2}\xi^2 + o(1), \quad n \to \infty, \tag{4.21}$$

locally uniformly with respect to ξ. Consequently,

$$\lim_{n\to\infty} F_q(Z_n) = e_q^{-\frac{q}{2}\xi^2}. \tag{4.22}$$

Thus, Z_n is q-convergent as $n \to \infty$ to the random variable Z whose q-Fourier transform is $e_q^{-\frac{q}{2}\xi^2} \in \mathcal{G}_q$.

In accordance with Theorem 3.4 for q and some β there exists a q-Gaussian density $G_{q-1}(\beta; x)$ with

$$q_{-1} = v(q) = \frac{3q-1}{1+q},$$

such that

$$F_{q-1}\left(G_{q-1}(\beta; x)\right) = e_q^{-(q/2)\xi^2}.$$

Let us now find β. It follows from Corollary 3.1 (see (3.22)) that $\beta_*(q_{-1}) = q/2$. Solving this equation with respect to β we obtain

$$\beta = \left(\frac{3 - q_{-1}}{4qC_{q_{-1}}^{2(q_{-1}-1)}}\right)^{\frac{1}{2-q_{-1}}}, \qquad (4.23)$$

where $C_{q_{-1}}$ is the normalizing constant corresponding to the density $G_{q_{-1}}(\beta;x)$. The explicit form of the corresponding q_{-1}-Gaussian reads as

$$G_{q_{-1}}(\beta;x) = C_s^{-1}\left(\frac{\sqrt{3-s}}{2C_s^{s-1}\sqrt{z(s)}}\right)^{\frac{1}{2-s}} e_s^{-\left(\frac{3-s}{4z(s)C_s^{2(s-1)}}\right)^{\frac{1}{2-s}}x^2}, \qquad (4.24)$$

where $s = q_{-1}$. $\qquad\qquad\qquad\square$

Analogously, the q-CLT can be proved for q_k-i.i.d. of the second and third types. The formulations of the corresponding theorems are given below. The reader can verify their validity through comparison with the proof of Theorem 1.

Theorem 4.6. *Assume a sequence $q_k, k \in \mathcal{Z}$, is given as (3.40) with $q_k \in [1,2)$. Let X_j, $j = 1,2,\ldots$, be a sequence of q_k-independent of second type and identically distributed random variables with a finite q_{k-1}-mean $\mu_{q_{k-1}}$ and a finite second $(2q_{k-1} - 1)$-moment $\sigma_{2q_{k-1}-1}^2$.*
Then the sequence

$$Z_n = \frac{X_1 + \cdots + X_n - n\mu_{q_{k-1}}}{\kappa_n(q_{k-1})}, \quad n = 1,2,\ldots,$$

is q_{k-1}-convergent to a q_{k-1}-normal distribution as $n \to \infty$. The parameter β_k of the corresponding attractor $G_{q_{k-1}}(\beta_k;x)$ is

$$\beta_k = \left(\frac{3 - q_{k-1}}{4q_{k-1}C_{q_{k-1}}^{2q_{k-1}-2}}\right)^{\frac{1}{2-q_{k-1}}}. \qquad (4.25)$$

Theorem 4.7. *Assume a sequence $q_k, k \in \mathcal{Z}$, is given as (3.40) with $q_k \in [1,2)$. Let X_j, $j = 1,2,\ldots$, be a sequence of q_k-independent of third type and identically distributed random variables with a finite q_k-mean μ_{q_k} and a finite second $(2q_k - 1)$-moment $\sigma_{2q_k-1}^2$.*

Then the sequence

$$Z_n = \frac{X_1 + \cdots + X_n - n\mu_{q_k}}{\kappa_n(q_k)}, \quad n = 1, 2, \ldots,$$

is q_{k-1}-convergent to a q_{k-1}-normal distribution as $n \to \infty$. Moreover, the corresponding attractor $G_{q_{k-1}}(\beta_k; x)$ in this case is the same as in Theorem 4.4 with β_k given in (4.16).

Obviously, $\frac{q+1}{3-q} = 1$ if and only if $q = 1$. This fact yields the following corollary.

Corollary 4.8. *Let X_j, $j = 1, 2, \ldots$, be a given sequence of q_k-independent (of any type) random variables satisfying the corresponding conditions of Theorems 4.4–4.7. Then the sequence Z_n, $n = 1, 2, \ldots$, converges weakly to a normal distribution if and only if $q_k = 1$, that is, in the classic case.*

Concluding, we note that the formulation of a q-generalization of the CLT depends on the type of q-independence. Table 4.1 summarizes the q-CLT for three types of the q-independence.

We notice that in all three cases the limiting random variable is distributed according to a q_{k-1}-Gaussian distribution, where

$$q_{k-1} = v(q_k) = \frac{3q_k - 1}{1 + q_k}.$$

Obviously $q_k \neq q_{k-1}$ if $q_k \neq 1$. In the classic case both $q_k = 1$ and $v(q_k) = v(1) = 1$, so that the corresponding Gaussians do not

Table 4.1: Interrelation between the type of q-independence, conditions for q-mean and q-variance, type of convergence, and parameter β_k of the attractor.

q-Independence	Conditions	Convergence	q-Gaussian parameter
1st type	$\mu_{q_k} < \infty$, $\sigma^2_{2q_k-1} < \infty$	q_k-convergence	$\beta_k = \left(\dfrac{3-q_{k-1}}{4q_k C_{q_{k-1}}^{2q_k-1-2}}\right)^{\frac{1}{2-q_{k-1}}}$
2nd type	$\mu_{q_{k-1}} < \infty$, $\sigma^2_{2q_{k-1}-1} < \infty$	q_{k-1}-convergence	$\beta_k = \left(\dfrac{3-q_{k-1}}{4q_{k-1} C_{q_{k-1}}^{2q_k-1}}\right)^{\frac{1}{2-q_{k-1}}}$
3rd type	$\mu_{q_k} < \infty$, $\sigma^2_{2q_k-1} < \infty$	q_{k-1}-convergence	$\beta_k = \left(\dfrac{3-q_{k-1}}{4q_k C_{q_{k-1}}^{2q_k-1-2}}\right)^{\frac{1}{2-q_{k-1}}}$

differ. So, Corollary 4.8 emphasizes that such a duality is a specific feature of nonextensive statistical mechanics and is an implication of q-independence of random variables X_j.

4.4. Uniqueness of the Limiting Distribution

In this section, we prove the uniqueness of the limiting distribution in the q-CLTs formulated in the previous section. The uniqueness question does not appear in the case of the classic CLT, since the Fourier transform is a one-to-one transform of $L_1(\mathbb{R})$ to $C_0(\mathbb{R})$, or unitary operator in the space $L_2(\mathbb{R})$. However, as discussed in Section 3.10, the q-Fourier transform does not possess the one-to-one property. To prove the uniqueness we need some auxiliary facts, which are provided in subsections below.

4.4.1. *On the support of the limit distribution*

The limiting distributions obtained in Theorems 4.4–4.6, i.e., q-Gaussians, are symmetric. Any of their deformations coming from the invariance principle are also symmetric. Therefore, we can restrict our discussion of the uniqueness problem to symmetric distributions. For the sake of simplicity, we also consider continuous density functions. Other cases can be considered in a similar manner with appropriate care.

Let f be a continuous and symmetric about the origin density function. Denote $\lambda(x) = x[f(x)]^{q-1}$, where $1 \leq q < 2$. Since f is symmetric, it suffices to consider $\lambda(x)$ only for positive x. Suppose the maximum value of λ is m and $x_m > 0$ is the rightmost point where λ attains its maximum, i.e., $m = \max_{0 < x \leq a}\{x[f(x)]^{q-1}\} = x_m[f(x_m)]^{q-1}$. Let

$$\tau_* = \begin{cases} \dfrac{1}{m(q-1)}, & \text{if } 1 < q < 2, \\ \infty, & \text{if } q = 1. \end{cases}$$

Theorem 4.9. *Let f be a continuous symmetric density function with supp $f \subseteq [-a, a]$. Then the q-Fourier transform of f satisfies* *the following estimate*

$$\left|F_q[f](\eta - i\tau)\right| \le \exp_q(x_m M_q \tau), \tag{4.26}$$

where $\eta \in (-\infty, \infty)$, $\tau < \tau_*$, $M_q = \max_{[0,a]}\{[f(x)]^{q-1}\}$, *and* x_m *is the rightmost point at which the function* $x f^{q-1}(x)$ *attains its maximum* m.

Proof. For $q = 1$ this statement is the Paley–Wiener theorem. Assume $1 < q < 2$. For the density function f with the support $supp\, f \subseteq [-a, a]$, the q-Fourier transform takes the form

$$F_q[f](\xi) = \int_{-a}^{a} \frac{f(x)dx}{[1 + i(1-q)x\xi f^{q-1}(x)]^{\frac{1}{q-1}}}. \tag{4.27}$$

Let $\xi = \eta + i\tau$ where $\eta = \Re(\xi)$ is the real part of ξ and $\tau = \Im(\xi)$ its imaginary part. We assume that $\eta \in (-\infty, \infty)$ and $|\tau| < \frac{1}{m(q-1)}$. Then for the denominator of the integrand in (4.27) one has

$$\left[1 + i(1-q)x(\eta - i\tau)f^{q-1}(x)\right]^{\frac{1}{q-1}}$$

$$= \left[1 + i(1-q)\eta f^{q-1}(x) + (1-q)\tau x f^{q-1}(x)\right]^{\frac{1}{q-1}}$$

$$= \left[1 + (1-q)\tau x f^{q-1}(x)\right]^{\frac{1}{q-1}} \left[1 + i\frac{(1-q)\eta f^{q-1}(x)}{1 - (1-q)\tau x f^{q-1}(x)}\right]^{\frac{1}{q-1}}$$

$$= \left(\exp_q(\tau x f^{q-1}(x))\right)^{-1} \left(\exp_q\left(i\frac{(1-q)\eta f^{q-1}(x)}{1 - (1-q)\tau x f^{q-1}(x)}\right)\right)^{-1}. \tag{4.28}$$

Using the inequality $|\exp(iy)| \le 1$ valid for all $y \in (-\infty, \infty)$ if $q > 1$, it follows from (4.28) that

$$\left|1 + i(1-q)x(\eta + i\tau)f^{q-1}(x)\right|^{\frac{1}{q-1}} \ge \left(\exp_q(\tau x f^{q-1}(x))\right)^{-1},$$

or

$$\left|\frac{1}{[1 + i(1-q)x(\eta + i\tau)f^{q-1}(x)]^{\frac{1}{q-1}}}\right| \le \left(\exp_q(\tau x f^{q-1}(x))\right). \tag{4.29}$$

Now, (4.27) together with (4.29) and $f(x)$ being a density function, yield (4.26).   □

Remark 4.2. Theorem 4.9 can be viewed as a generalization of the well known Paley–Wiener theorem. Indeed, if $q = 1$ then (4.26) takes the form

$$|\tilde{f}(\eta - i\tau)| \leq \exp(a\tau), \quad \eta + i\tau \in \mathcal{C}, \tag{4.30}$$

which represents the Paley–Wiener theorem for continuous density functions f with the support $supp\,[f] \in [-a, a]$.

Inequality (4.30) can be used for estimation of the size of the support of f. Consider an example with $f(x) = (2a)^{-1}\mathcal{I}_{[-a,a]}(x)$, where $\mathcal{I}_{[-a,a]}(x)$ is the indicator function of the interval $[-a, a]$. The Fourier transform of this function is $\tilde{f}(\xi) = (a\xi)^{-1}\sin(a\xi)$, $M_q = M_1 = 1$, and $x_m = a$. Therefore, we have

$$\left|\tilde{f}(-i\tau)\right| \leq \frac{\sinh a\tau}{a\tau} \leq e^{a\tau}, \quad \tau > 0.$$

The latter yields

$$2a \geq 2\sup_{\tau>0} \frac{\ln|\tilde{f}(-i\tau)|}{\tau},$$

which gives an estimate from below for the size $d(f) = 2a$ of the support of f. This idea can be used to estimate the size of the support of f using the q-Fourier transform and Theorem 4.9.

Namely, inequality (4.26) implies the following estimate for the size of support of any density.

Lemma 4.1. *Let f be a density function satisfying conditions of Theorem 4.9. Then for the size $d(f)$ of its support the following estimate holds:*

$$d(f) \geq \frac{2}{M_q} \sup_{0<\tau<\tau_*} \frac{\ln_q\left|F_q[f](-i\tau)\right|}{\tau}. \tag{4.31}$$

Proof. Using inequality (4.26) with $\eta = 0$, we have

$$d(f) = 2a \geq 2x_m \geq \frac{2}{M_q} \sup_{0 < \tau < \tau_*} \frac{\ln_q \left| F_q[f](-i\tau) \right|}{\tau}. \tag{4.32}$$

□

Let $f_n(x) = f_{S_n}(x)$ be the density function of $S_n = X_1 + \cdots + X_n$, where $X_1, \ldots, X_n$ are q-independent random variables with the same density function $f = f_{X_1}$ whose support is $[-a, a]$. We show that the q-independence condition can not reduce the support of f_n to an interval independent of n. More precisely, $d(f_n)$ increases at least proportionally to n when $n \to \infty$.

Theorem 4.10. *Let X_j, $j = 1, 2, \ldots$, be q-independent of any Types I–III random variables, all having the same density function f with supp $f \subseteq [-a, a]$. Then, for the size $d(f_n)$ of the support of the density f_n of S_n, there exists a constant $C = C(q, f) > 0$, such that the estimate*

$$d(f_n) \geq Cn \tag{4.33}$$

holds.

Proof. Using formula (4.32) one has

$$d(f_n) \geq \frac{2}{M_{q,n}} \sup_{0 < \tau < \tau_*} \frac{\ln_q \left| F_q[f_n](-i\tau) \right|}{\tau}, \tag{4.34}$$

where

$$M_{q,n} = \max_{x \in [-na, na]} f_n^{q-1}(x).$$

It is clear from probabilistic arguments that $M_{q,n} \leq M_q$ for all $n \geq 2$. Therefore, it follows from (4.34) that

$$d(f_n) \geq \frac{2}{M_q} \sup_{0 < \tau < \tau_*} \frac{\ln_q \left| F_q[f_n](-i\tau) \right|}{\tau}, \tag{4.35}$$

Let X_n be a sequence of random variables q-independent of Type I (see (4.13)). Making use of the inequality $|z - r| \geq |z| - r$, which holds true for any complex z and positive integer number r, one has

$$\left|\widetilde{(f_n)}_q(-i\tau)\right| = \left|\tilde{f}_q(-i\tau) \otimes_q \cdots \otimes_q \tilde{f}_q(-i\tau)\right|$$

$$= \left|\left[n\big(\tilde{f}_q(-i\tau)\big)^{1-q} - (n-1)\right]^{\frac{1}{1-q}}\right|$$

$$\geq \left[n\left|\tilde{f}_q(-i\tau)\right|^{1-q} - (n-1)\right]^{\frac{1}{1-q}}$$

$$= \left|\tilde{f}_q(-i\tau)\right| \otimes_q \cdots \otimes_q \left|\tilde{f}_q(-i\tau)\right|.$$

Taking q-logarithm of both sides in this inequality and using the property $\ln_q(g \otimes_q h) = \ln_q g + \ln_q h$, one obtains

$$\ln_q\left|F_q[f_n](-i\tau)\right| \geq n \ln_q\left|F_q[f](-i\tau)\right|, \quad n = 1, 2, \ldots. \qquad (4.36)$$

Now using estimates (4.35) and (4.36) we obtain (4.33) with the constant

$$C = \frac{2}{M_q} \sup_{0 < \tau < \tau_*} \frac{\ln_q\left|F_q[f](-i\tau)\right|}{\tau}.$$

For random variables independent of Type II, equation (4.36) takes the form

$$\ln_{q_1}\left|F_q[f_n](-i\tau)\right| \geq n \ln_{q_1}\left|F_q[f](-i\tau)\right|. \qquad (4.37)$$

For $1 < q < 2$ one can easily verify that

$$q_1 = \frac{1+q}{3-q} = q + \frac{(q-1)^2}{3-q}.$$

Hence, $q_1 > q$. Therefore, due to Proposition 2.18, Part 1, we have

$$\ln_q\left|F_q[f_n](-i\tau)\right| \geq \ln_{q_1}\left|F_q[f_n](-i\tau)\right|,$$

consequently

$$\ln_q\left|F_q[f_n](-i\tau)\right| \geq n \ln_{q_1}\left|F_q[f](-i\tau)\right|. \qquad (4.38)$$

Using estimates (4.35) and (4.38) we obtain (4.33) with the constant

$$C = \frac{2}{M_q} \sup_{0 < \tau < \tau_*} \frac{\ln_{q_1}\left|F_q[f](-i\tau)\right|}{\tau}.$$

Similarly, for random variables independent of Type III, we have

$$\ln_q \left| F_q[f_n](-i\tau) \right| \geq n \ln_{q_1} \left| F_{qq}[f](-i\tau) \right|. \tag{4.39}$$

Again estimates (4.35) and (4.39), similarly to previous cases, imply estimate (4.33) with the constant

$$C = \frac{2}{M_q} \sup_{0 < \tau < \tau_*} \frac{\ln_{q_1} \left| F_{q_1}[f](-i\tau) \right|}{\tau}. \qquad \square$$

Corollary 4.11. *Let the sequence X_j, $j = 1, 2, \ldots$, be q-independent of any Types I–III random variables all having the same density function f with supp$f \subseteq [-a, a]$. If the sequence Z_N (in Theorems 4.4–4.7) has a limit random variable in some sense, then this random variable can not have a density function with a compact support. Moreover, due to the scaling present in Z_n, the support of the limit variable is the entire set of real numbers.*

The proof obviously follows immediately from (4.33) upon letting $n \to \infty$.

4.4.2. *On the variance and quasi-variance of a limit distribution*

Let $1 \leq q < 2$ and a random variable X with a density function $f(x)$ having zero q-mean, i.e., $\mu_q(X) = 0$.

Definition 4.4. The quasi-variance (or quasi $(2q - 1)$-variance) of order $2q - 1$ of the random variable X is defined by

$$QV(X) = QV_{2q-1}(X) = \nu_{2q-1}(X)\sigma^2_{2q-1}(X)$$

$$= \int_{\mathbb{R}} (x - \mu_q)^2 [f(x)]^{2q-1} dx. \tag{4.40}$$

Remark 4.3. The quasi-variance of X, of course, depends on q. We will omit the index q in the notation of the quasi-variance in cases when there is no confusion.

Recall that the escort normalization constant $\nu_q(X)$, the q-mean μ_q, and $(2q-1)$-variance $\sigma_{2q-1}(X)$ are defined respectively as

$$\nu_q = \nu_q(X) = \int_{\mathbb{R}} [f(x)]^q dx, \tag{4.41}$$

$$\mu_q = \mu_q(X) = \int_{\mathbb{R}} x \frac{[f(x)]^q}{\nu_q} dx, \tag{4.42}$$

and

$$\sigma_{2q-1}^2 = \sigma_{2q-1}^2(X) = \int_{\mathbb{R}} (x - \mu_q)^2 \frac{[f(x)]^{2q-1}}{\nu_{2q-1}} dx. \tag{4.43}$$

Note that if $q = 1$ then $\nu_q = 1$, $\mu_q = \mathbb{E}[X]$, and $\sigma_{2q-1}^2 = \text{Var}(X)$, the variance of X, implying the equality $QV(X) = \text{Var}(X)$. As was noted above (see Lemma 4.3), the q-mean of $X - \mu_q$ is zero. Therefore, without loss of generality, we can assume that $\mu_q = 0$. It follows from Proposition 3.12 that for a random variable X with zero q-mean the equality

$$\varphi_{q,X}(\xi) = 1 - \frac{q}{2} QV[X] \, \xi^2 + o(|\xi|^2), \quad |\xi| \to 0,$$

holds. This, in turn, in accordance with Corollary 2.7, implies

$$\ln_q \left(F_q[f_X](\xi) \right) = -\frac{q}{2} QV[X] \, \xi^2 + o(|\xi|^2), \quad |\xi| \to 0, \tag{4.44}$$

The proposition below generalizes the property $\text{Var}(X + Y) = \text{Var}(X) + \text{Var}(Y)$ of variances of independent random variables X and Y to the case of quasi-variances of q-independent random variables.

Proposition 4.4. *If X and Y are q-independent (of any Types I–III) random variables with $\mu_q(X) = 0$ and $\mu_q(Y) = 0$, then for their quasi-variances (of order $2q-1$) the relation*

$$QV(X + Y) = QV(X) + QV(Y) \tag{4.45}$$

holds.

Proof. To prove the validity of this fact we use Proposition 3.11 with $n = 1$ and $n = 2$. Then, we have

$$\frac{dF_q[f_X](0)}{d\xi} = i\mu_q(X), \quad \frac{dF_q[f_Y](0)}{d\xi} = i\mu_q(Y), \qquad (4.46)$$

and taking into account (4.40),

$$\frac{d^2 F_q[f_X](0)}{d\xi^2} = -qQV(X), \quad \frac{d^2 F_q[f_Y](0)}{d\xi^2} = -qQV(Y). \qquad (4.47)$$

Further, making use of the definition of q-independence of Type I, i.e., (4.13), we obtain

$$QV(X + Y) = -\frac{d^2 F_q[f_{X+Y}](0)}{q \, d\xi^2}$$

$$= -\frac{1}{q} \frac{d^2}{d\xi^2} \left(F_q[f_X](\xi) \otimes_q F_q[f_Y](\xi) \right)\big|_{\xi=0}.$$

For a random variable Z introduce the following notations:

$$A_q(Z; \xi) = \frac{\frac{dF_q[f_Z](\xi)}{d\xi}}{[F_q[f_Z](\xi)]^q},$$

and

$$B_q(Z; \xi) = \frac{F_q[f_Z](\xi)\frac{d^2 F_q[f_Z](\xi)}{d\xi^2} - q[\frac{dF_q[f_Z](\xi)}{d\xi}]^2}{[F_q[f_Z](\xi)]^{q+1}}.$$

One can verify by direct calculation that

$$\frac{d^2}{d\xi^2} \left(F_q[f_X](\xi) \otimes_q F_q[f_Y](\xi) \right)$$

$$= q \left[F_q[f_{X+Y}](\xi) \right]^{2q-1} \left(A_q(X; \xi) + A_q(Y; \xi) \right)^2$$

$$+ \left[F_q[f_{X+Y}](\xi) \right]^q \left(B_q(X; \xi) + B_q(Y; \xi) \right).$$

Now taking into account

$$F_q[f_X](0) = F_q[f_Y](0) = F_q[f_{X+Y}](0) = 1,$$

$$\frac{dF_q[f_X](0)}{d\xi} = \frac{dF_q[f_Y](0)}{d\xi} = 0,$$

and equalities (4.47), we arrive at (4.45).

Similarly, one can show (4.45) for random variables X and Y, which are q-independent of Type II or Type III.   □

Consider a sequence of identically distributed q-independent (of any Types I–III) random variables X_n, $n = 1, 2, \ldots$, with zero q-mean. Then it follows from Proposition 4.4 that

$$QV(S_n) = nQV(X_1),$$

where $S_n = X_1 + \cdots + X_n$. The latter implies

$$QV(S_n) = O(n), \quad n \to \infty.$$

Moreover, it follows from the definition of the quasi-variance that

$$\sigma^2_{2q-1}(S_n) = \frac{n}{\nu_{2q-1}(S_n)} QV(X_1),$$

or

$$\nu_{2q-1}(S_n) = \frac{n}{\sigma^2_{2q-1}(S_n)} QV(X_1).$$

We note that for the case $q = 1$, i.e., for independent random variables X_n the expression $\nu_{2q-1}(S_n) = 1$ does not depend on n. In this case we have

$$\sigma_1^2(S_n) = \mathrm{Var}(S_n) = n\mathrm{Var}(X_1) = n\sigma_1^2(X_1). \tag{4.48}$$

The crucial difference of the case $q > 1$ is that $\nu_{2q-1}(S_n)$ does depend on n. We assume that this dependence of $\nu_{2q-1}(S_n)$ on n can be expressed by

$$\nu_{2q-1}(S_n) = O(n^\gamma), \quad n \to \infty, \tag{4.49}$$

where $\gamma = \gamma(q)$ is a real number dependent on q.

Definition 4.5. A sequence of identically distributed random variables X_n is said to belong to the class $\mathcal{A}(\gamma)$ if $S_n = X_1 + \cdots + X_n$ satisfies the condition (4.49).

Proposition 4.5. *Let for a sequence of identically distributed random variables X_n the relation*

$$\sigma^2_{2q-1}(S_n) = O(n^\lambda), \quad n \to \infty,$$

where $\lambda > 0$, holds. Then the sequence X_n belongs to $\mathcal{A}(\gamma)$ with $\gamma = 1 - \lambda$.

Proof. Making use of relationship (4.45), we have

$$\begin{aligned}
\sigma^2_{2q-1}(S_n) &= \frac{QV(X_1 + \cdots + X_n)}{\nu_{2q-1}(S_n)} \\
&= \frac{QV(X_1) + \cdots + QV(X_n)}{\nu_{2q-1}(S_n)} \\
&= \frac{nQV(X_1)}{\nu_{2q-1}(S_n)} \\
&= n\frac{\nu_{2q-1}(X_1)\sigma^2_{2q-1}(X_1)}{\nu_{2q-1}(S_n)} = O(n^{1-\gamma}), \quad n \to \infty.
\end{aligned}$$
$\qquad\qquad\square$

Consider some examples of random variables $X_n \in \mathcal{A}(\gamma)$.

Example 4.2.

(1) If $q = 1$, then any independent identically distributed random variables obviously belong to $\mathcal{A}(0)$. That is in this case $\gamma = 0$.
(2) Let X_1 be a q-Gaussian random variable with the density function

$$G_q(\beta, x) = \frac{\sqrt{\beta}}{C_q}e_q^{-\beta x^2}, \quad \beta > 0,$$

introduced in Chapter 2. We have seen (Corollary 2.9 in Chapter 2) that $(2q-1)$-variance of X_1 is

$$\sigma^2_{2q-1}(X_1) = \frac{1}{\beta(q+1)}. \tag{4.50}$$

and its q-Fourier transform is

$$F_q[f_{X_1}](\xi) = e_{q_1}^{-\beta_* \xi^2},$$

where

$$q_1 = u(q) = \frac{1+q}{3-q} \quad \text{and} \quad \beta_* = \frac{3-q}{8\beta^{2-q}C_q^{2(q-1)}}.$$

Suppose that the sequence $X_1, \dots, X_N$ is q-independent of Type II and identically distributed. We show that the density of S_n in this case is the q-Gaussian with the density function $G_q(n^{-\frac{1}{2-q}}\beta; x)$. Making use of the definition of q-independence of Type II, we have

$$F_q[f_{S_n}](\xi) = \prod_{k=1}^{n} {}_q F_q[f_{X_k}](\xi) = e_{q_1}^{-n\beta_*\xi^2}. \tag{4.51}$$

On the other hand, the q-Fourier transform of $G_q(n^{-\frac{1}{2-q}}\beta; x)$ is

$$F[G_q(n^{-\frac{1}{2-q}}\beta; \cdot)](\xi) = e_{q_1}^{-\bar{\beta}_*\xi^2},$$

where

$$\bar{\beta}_* = \frac{3-q}{8(n^{-\frac{1}{2-q}}\beta)^{2-q}C_q^{2(q-1)}} = n\beta_*.$$

Therefore, the density function in the family of q-Gaussians with the q-Fourier transform on the right-hand side of (4.51) is

$$f_{S_n}(x) = G_q(n^{-\frac{1}{2-q}}\beta; x). \tag{4.52}$$

Using this fact one can calculate $\nu_{2q-1}(S_n)$. Indeed, due to (4.50)

$$\sigma_{2q-1}^2(S_n) = \frac{n^{\frac{1}{2-q}}}{\beta(1+q)} = n^{\frac{1}{2-q}}\sigma_{2q-1}^2(X_1). \tag{4.53}$$

Therefore it follows from (4.48) that

$$\nu_{2q-1}(S_n) = \frac{nQV(X_1)}{\sigma_{2q-1}^2(S_n)}$$

$$= \frac{n\nu_{2q-1}(X_1)\sigma_{2q-1}^2(X_1)}{n^{\frac{1}{2-q}}\sigma_{2q-1}^2(X_1)}$$

$$= n^{\frac{1-q}{2-q}}\nu_{2q-1}(X_1).$$

Thus, $X_n \in \mathcal{A}(\gamma)$ with $\gamma = \frac{1-q}{2-q}$.

(3) The sequence of q-Gaussian random variables X_n identically distributed and q-independent of Type I also belongs to the class $\mathcal{A}(\gamma)$ with $\gamma = \frac{1-q}{2-q}$. To see this fact, first we notice that for any positive real number r, its n times q-product can be expressed as

$$r \otimes_q \cdots \otimes_q r = e_q^{\ln_q r} \otimes_q \cdots \otimes_q e_q^{\ln_q r} = e_q^{n \ln_q r}.$$

Therefore, due to q-independence of Type I of the sequence X_n, we have

$$F_q[f_{S_n}](\xi) = F_q[G_q(\beta; \cdot)](\xi) \otimes_q \cdots \otimes_q F_q[G_q(\beta; \cdot)](\xi)$$
$$= e_{q_1}^{-\beta_* \xi^2} \otimes_q \cdots \otimes_q e_{q_1}^{-\beta_* \xi^2}$$
$$= e_q^{n \ln_q(e_{q_1}^{-\beta_* \xi^2})}.$$

Now using asymptotic relations $e_q^x = 1 + x + O(x^2)$, $x \to 0$, (Corollary 2.4), which we write as $e_q^x \sim 1 + x$, $x \to 0$, and $\ln_q(1+x) = x + O(x^2)$, $x \to 0$, (Corollary 2.7), which we write as $\ln_q(1+x) \sim x$, $x \to 0$, for large n (if necessary taking ξ small enough), we have

$$F_q[f_{S_n}](\xi) = e_q^{n \ln_q(e_{q_1}^{-\beta_* \xi^2})} \sim 1 + n \ln_q(e_{q_1}^{-\beta_* \xi^2})$$
$$\sim 1 + n(e_{q_1}^{-\beta_* \xi^2} - 1)$$
$$\sim 1 + n((1 - \beta_* \xi^2) - 1)$$
$$\sim 1 - n \beta_* \xi^2$$
$$\sim e_{q_1}^{-n \beta_* \xi^2}, \quad n \to \infty.$$

The latter implies similarly to the previous case $\nu_{2q-1}(S_n) \sim n^{\frac{1-q}{2-q}}$.

(4) Similarly to the case (3), the sequence of identically distributed and q-independent of Type III q-Gaussian random variables belongs to $\mathcal{A}(\gamma)$ as well with the same $\gamma = \frac{1-q}{2-q}$.

(5) Random variables studied in Vignat and Plastino (2007b) also belongs to $\mathcal{A}(\gamma)$ with $\gamma = \frac{1-q}{2-q}$, since they are asymptotically equivalent to q-indpendent random variables (see Vignat and Plastino, 2007b). As shown in Jiang *et al.* (2010) random

variables considered in Vignat and Plastino (2007b) are variance mixtures of normal densities.

Remark 4.4. Example 4.2 (5) gives a strong evidence of the fact that the subclass of variance mixtures of normal densities leading to q-Gaussians will also belong to $\mathcal{A}(\gamma)$ with $\gamma = \frac{1-q}{2-q}$. In our further considerations we assume this condition for q-i.i.d. sequences of random variables.

4.4.3. *Again on Hilhorst's invariance principle*

In Section 3.10, we discussed Hilhorst's invariance principle which describes classes of density functions with the same q-Fourier transform. As we have seen, the class of density functions with the same q-Fourier transform is equivalent to the class of continuous functions $H(\lambda)$, such that functions $X_\pm(\lambda)$ defined by

$$X_+(\lambda) = \left[x_+^{\frac{q-2}{q-1}}(\lambda) + H(\lambda)\right]^{\frac{q-1}{q-2}} = \frac{1}{\left[\frac{1}{x_+^{\frac{2-q}{q-1}}(\lambda)} + H(\lambda)\right]^{\frac{q-1}{2-q}}}, \quad (4.54)$$

$$X_-(\lambda) = \left[x_-^{\frac{q-2}{q-1}}(\lambda) + H(\lambda)\right]^{\frac{q-1}{q-2}} = \frac{1}{\left[\frac{1}{x_-^{\frac{2-q}{q-1}}(\lambda)} + H(\lambda)\right]^{\frac{q-1}{2-q}}}, \quad (4.55)$$

are invertible (see for these and other notations in Section 3.10). Like in Section 3.10, we assume that $\lambda(x)$ has only one maximum on the interval $(0,\infty)$. The proofs provided below can easily be modified to the general case. On the other hand, we saw in Section 4.4.1, that the density of the limiting distribution in the q-CLT can not have a compact support. From this point of view it is important to determine for which functions $H(\lambda)$ the support of the deformed density function $G_{q,H}(x)$ is the whole set of real numbers $\mathbb{R}$.

Proposition 4.6. *Let $H(0) > 0$. Then the support of $G_{q,H}(x)$ is compact, and*

$$supp\, G_{q,H} = \left[-(H(0))^{\frac{q-1}{q-2}}, (H(0))^{\frac{q-1}{q-2}}\right].$$

Proof. Since the function $x_-(\lambda)$ satisfies (see Fig. 3.1)

$$\lim_{\lambda \to 0} x_-(\lambda) = +\infty,$$

the largest value of $X_-(\lambda)$ is equal to $\lim_{\lambda \to 0} X_-(\lambda) = [H(0)]^{\frac{q-1}{q-2}}$. Therefore, the inverse of X_- is defined on the interval $\left[x_0, [H(0)]^{\frac{q-1}{q-2}}\right]$, where $x_0 > 0$ is some number obtained by shifting of x_m depending on $H(m)$. On the other hand the smallest value of x_+ is zero, taken at $\lambda = 0$. Therefore, the inverse of X_+ is defined on the interval $[0, x_0]$. Hence, by symmetry, $G_{q,H}$ has the support $\left[-[H(0)]^{\frac{q-1}{q-2}}, [H(0)]^{\frac{q-1}{q-2}}\right]$.

$\square$

Remark 4.5.

(1) Note that $H(0)$ can not be negative. In fact, if $H(0) < 0$, then either $X_{\pm}$ is not invertible or, if it is invertible, its inverse does not define a density function.
(2) Proposition 4.6 implies that if $H(0) > 0$ then, due to Corollary 4.11, $G_{q,H}(x)$ can not be the density function of the limit distribution in the q-CLT. Thus none of the densities in Hilhorst's counterexamples,[1] except the q-Gaussian, can serve as an attractor in the q-CLT.

Now let us discuss the case when for the deforming function $H(\lambda)$ the condition $H(0) = 0$ holds. The next proposition establishes that, in this case, any deformation $G_{q,H}(x)$ of the q-Gaussian density is asymptotically equivalent to the q-Gaussian density $G_q(x) \equiv G_{q,0}(x)$.

Proposition 4.7. *Let $H(0) = 0$. Then*

$$\lim_{|x| \to \infty} \frac{G_{q,H}(x)}{G_q(x)} = 1.$$

---

[1] See Examples 2 and 3 in Hilhorst (2010). Example 4 is not relevant to the q-CLT, since in this case, $(2q - 1)$-variance of the 2-Gaussian does not exist, and consequently the q-CLT is not applicable.

Proof. Let $H(0) = 0$. Then we have

$$\lim_{\lambda \to 0} \frac{X_-(\lambda)}{x_-(\lambda)} = \frac{\left[x_-^{\frac{q-2}{q-1}}(\lambda) + H(\lambda) \right]^{\frac{q-1}{q-2}}}{\left[x_-^{\frac{q-2}{q-1}}(\lambda) \right]^{\frac{q-1}{q-2}}}$$

$$= \lim_{\lambda \to 0} \left[1 + \frac{H(\lambda)}{x_-(\lambda)} \right]^{\frac{q-1}{q-2}} = 1.$$

Therefore, for inverses we obtain

$$\lim_{x \to +\infty} \frac{X_-^{-1}(x)}{x_-^{-1}(x)} = 1.$$

This implies

$$\lim_{x \to +\infty} \frac{G_{q,H}(x)}{G_q(x)} = \lim_{x \to +\infty} \left[\frac{\frac{X_-^{-1}(x)}{x}}{\frac{x_-^{-1}(x)}{x}} \right]^{\frac{1}{q-1}} = 1,$$

which proves the statement. $\qquad\square$

Remark 4.6.

(1) Propositions 4.6 and 4.7 establish that $G_{q,H}$ can identify a limiting distribution in the q-CLT only if $H(0) = 0$. However, in this case, independently from other values of $H(\xi)$, the density $G_{q,H}(x)$ is asymptotically equivalent to the q-Gaussian, i.e., $G_{q,H}(x) \sim G_q(x)$ as $|x| \to \infty$.

(2) In the next section, we prove that if the sequence X_n belongs to the class $\mathcal{A}(\gamma)$ with certain γ, then the limiting distribution in the q-CLT is unique.

The statement of the following proposition can be proved similarly to Proposition 4.7, replacing $X_-(\lambda)$, $x_-(\lambda)$ by functions $X_+(\lambda)$, $x_+(\lambda)$, respectively.

Proposition 4.8. *Let* $H(0) = 0$. *Then*

$$\lim_{x \to 0} \frac{G_{q,H}(x)}{G_q(x)} = 1.$$

Finally, let us prove the following lemma, which will be used in the next section.

Lemma 4.12. *Let $\omega(x)$ be a continuous function defined on $[0, \infty)$ such that*

(a) $\omega(1) = 0$,
(b) $\omega(x) > 0$ *on* $(0, 1)$, *and* $\omega(x) < 0$ *on* $(1, \infty)$,
(c) $\int\limits_0^1 \omega(x)dx = \int\limits_1^\infty |\omega(x)|dx.$

Then

$$\int\limits_0^1 x^2\omega(x)dx < \int\limits_1^\infty x^2|\omega(x)|dx. \tag{4.56}$$

Proof. Since $x^2 < 1$ for $x \in (0, 1)$, one has

$$\int\limits_0^1 x^2\omega(x)dx < \int\limits_0^1 \omega(x)dx. \tag{4.57}$$

Similarly, for $x > 1$,

$$\int\limits_1^\infty |\omega(x)|dx < \int\limits_1^\infty x^2|\omega(x)|dx. \tag{4.58}$$

Now, condition (c) and estimates (4.57) and (4.58) imply (4.56). $\quad\square$

4.4.4. *Uniqueness of the limit distribution in the q-CLT*

Let X be a random variable with a symmetric density function G and let G_H be the density function obtained from G by H-deformation, where $H(\xi)$ is a continuous function such that $H(0) = 0$ and does not change its sign on the interval $(0, x_m)$. Denote by X_H the random variable corresponding to the density function G_H.

Lemma 4.13. *Let X and X_H be random variables with the respective densities G and G_H, and let $QV(X) = QV(X_H)$. Then $\sigma^2_{2q-1}(X) = \sigma^2_{2q-1}(X_H)$ if and only if $H(\xi)$ is identically zero.*

Proof. *Sufficiency.* Let $\sigma^2_{2q-1}(X) = \sigma^2_{2q-1}(X_H)$ and assume that $H(\xi)$ is not identically zero. This equality together with $QV(X) = QV(X_H)$ implies that $\nu_{2q-1}(X) = \nu_{2q-1}(X_H)$. Due to conditions on $H(\xi)$ both densities, G and G_H, are symmetric, decreasing on the positive semiaxis. Propositions 4.6 and 4.8 imply that $G(0) = G_H(0)$ since $H(0) = 0$. Moreover, since both G and G_H are densities there is a point $a > 0$ such that $G(a) = G_H(a)$. Depending on the sign of $H(\xi)$ the following two cases may occur:

Case (A): $G(x) > G_H(x)$ *on the interval* $(0, a)$ *and* $G(x) < G_H(x)$ *on the interval* (a, ∞),

or

Case (B): $G(x) < G_H(x)$ *on the interval* $(0, a)$ *and* $G(x) > G_H(x)$ *on the interval* (a, ∞).

If necessary, switching the order of G and G_H we can always assume that the case (A) holds. Notice, that the case $G(x) \equiv G_H(x)$ is obviously excluded, since $H(\xi)$ is not identically zero. Further, due to Proposition 4.7, G and G_H share the same asymptotic behavior at infinity: $G(x) \sim G_q(x)$, $x \to \infty$. Since G and G_H are symmetric about the origin, it suffices to consider these functions only for $x \geq 0$. Furthermore, it follows from the case (A) that $G^{2q-1}(x) > G_H^{2q-1}(x)$ on the interval $[0, a)$, and $G^{2q-1}(x) < G_H^{2q-1}(x)$ on the interval (a, ∞).

Consider the function $\omega(x) = a \left[G^{2q-1}(ax) - G_H^{2q-1}(ax) \right]$. This function ω is continuous by construction. Moreover, $\omega(1) = 0$, $\omega(x) > 0$ if $x \in (0, 1)$, and $\omega(x) < 0$ if $x > 1$. The existence of finite $(2q - 1)$-variances of X and X_H implies that $\int_1^\infty x^2 |\omega(x)| dx < \infty$. The calculations below, where the symmetry of densities is taken into account, show that $\omega(x)$ satisfies condition (c) of Lemma 4.12 as well:

$$2 \int_0^1 \omega(x) dx = a \int_{-1}^1 \left(G^{2q-1}(ax) - G_H^{2q-1}(ax) \right) dx$$

$$= \int_{-a}^a \left(G^{2q-1}(x) - G_H^{2q-1}(x) \right) dx$$

$$= \int_{-a}^a G^{2q-1}(x) dx - \int_{-a}^a G_H^{2q-1}(x) dx. \tag{4.59}$$

Due to the condition $\nu_{2q-1}(X) = \nu_{2q-1}(X_H)$ of the lemma, the equality (4.59) implies that

$$2 \int_0^1 \omega(x) dx$$

$$= \nu_{2q-1}(X) - \int_{|x| \geq a} G^{2q-1}(x) dx - \left[\nu_{2q-1}(X_H) - \int_{|x| \geq a} G_H^{2q-1}(x) dx \right]$$

$$= \int_{|x| \geq a} \left| G^{2q-1}(x) - G_H^{2q-1}(x) \right| dx = 2 \int_1^\infty |\omega(x)| dx.$$

Now, it follows from Lemma 4.12 that

$$\int_0^1 x^2 \omega(x) dx - \int_1^\infty x^2 |\omega(x)| dx < 0, \tag{4.60}$$

which is equivalent to

$$\int_0^a x^2 \left(G^{2q-1}(x) - G_H^{2q-1}(x) \right) dx$$

$$- \int_a^\infty x^2 \left(G_H^{2q-1}(x) - G^{2q-1}(x) \right) dx < 0. \tag{4.61}$$

Inequality (4.61) is the same as $QV(X) < QV(X_H)$. Switching the order of G and G_H in the above analysis one can see that the case (B) implies $QV(X) > QV(X_H)$. Both obtained relations contradict to equality $QV(X) = QV(X_H)$. Hence, our assumption on $H(\xi)$ is wrong. Thus, we conclude that $H(\xi) \equiv 0$.

Necessity. The necessity is obvious, since $H(\xi) \equiv 0$ immediately implies $G_H = G$, consequently yielding $\sigma^2_{2q-1}(X) = \sigma^2_{2q-1}(X_H)$. $\quad\square$

Theorem 4.14. *Let X_n be q-independent (of any Types I–III) and identically distributed random variables with zero q-mean and finite quasi-variance. Moreover, let $X_n \in \mathcal{A}(\gamma)$ with $\gamma = \frac{1-q}{2-q}$. Then the sequence Z_n defined in (4.11) (with corresponding $\kappa_n(\cdot)$) has a unique limit distribution.*

Proof. The existence of a limit distribution depending on the type of q-independence was proved in Theorems 4.4–4.7. Suppose that there are two limit distributions Z_∞ and Z_H of the sequence Z_n with respective distinct densities $G(x)$ and $G_H(x)$. Obviously, both distributions have the same quasi-variance (compare (4.44) and (4.21))

$$QV(Z_\infty) = QV(Z_H) = 1. \qquad (4.62)$$

Moreover, due to the condition $\nu_{2q-1}(S_n) \sim O(n^{\frac{1-q}{2-q}})$,

$$\sigma^2_{2q-1}(Z_N) = \sigma^2_{2q-1}\left(\frac{X_1 + \cdots + X_N}{N^{\frac{1}{2(2-q)}}}\right)$$

$$= \frac{1}{N^{\frac{1}{2-q}}}\sigma^2_{2q-1}(X_1 + \cdots + X_N)$$

$$= \frac{1}{N^{\frac{1}{2-q}}}\frac{NQV(X_1)}{\nu_{2q-1}(X_1 + \cdots + X_N)}$$

$$= \frac{1}{N^{\frac{q-1}{2-q}}}\frac{QV(X_1)}{\nu_{2q-1}(X_1 + \cdots + X_N)} \to CQV(X_1),$$

as $N \to \infty$, where C is a positive constant. This yields that

$$\sigma^2_{2q-1}(Z_\infty) = \sigma^2_{2q-1}(Z_H) = CQV(X_1).$$

Hence, all the conditions of Lemma 4.13 are satisfied. Thus, $H(\xi) \equiv 0$, which implies $Z_\infty = Z_H$, that is the uniqueness of the limit distribution. $\qquad\square$

Remark 4.7. Theorem 4.14 immediately implies the uniqueness of limiting distribution in Theorem 4.5. Setting $q = q_k$, one obtains the uniqueness of limiting distributions in Theorems 4.4, 4.6, and 4.7, as well.

4.5. Multivariate *q*-Central Limit Theorems

In this section, we prove a multivariate version of the *q*-CLT. The proof of existence of the uniqueness of the limiting distribution is similar to the univariate case. Therefore, we will be brief in the discussion of the uniqueness. However, we provide enough details in the proof of the *q*-CLT itself, due to the circumstance that the related notions in multivariate case require explanation.

We start with the fact that, like for the univariate case, random vectors can be *q*-centered by a translation. This means that if a *d*-dimensional random vector $X = (X^{(1)}, \ldots, X^{(d)})$ has the *q*-mean $\mu_q = \mu_q(X) = (\mu_q(X^{(1)}), \ldots, \mu_q(X^{(d)}))$, then the translated random vector $X^c = X - \mu_q$ has zero *q*-mean: $\mu_q(X^c) = (0, \ldots, 0)$. This fact can be shown similarly to the one-dimensional case.

Definition 4.6. Let $d \geq 1$ be an integer number and q a number such that $1 \leq q < 1 + 2/d$. A sequence of *d*-dimensional random vectors

$$X_N = (X_N^{(1)}, \ldots, X_N^{(d)}), \quad N = 1, 2, \ldots,$$

is said to be (q^d_{-1}, q, q)-*independent* if, for every $N = 2, 3, \ldots$,

$$\mathcal{F}_{q^d_{-1}}[f_N^c](\xi) = \mathcal{F}_q[f_{X_1^c}](\xi) \otimes_q \cdots \otimes_q \mathcal{F}_q[f_{X_N^c}](\xi), \qquad (4.63)$$

where (see (3.157))

$$q^d_{-1} = 1 + \frac{2(q-1)}{2 + dn(q-1)},$$

$f_{X_j^c}$, $j = 1, \ldots, N$, are respective density functions of centered random vectors $X_j^c = X_j - \mu_q(X_j)$, $j = 1, \ldots, N$, and f_N^c is the density function of $X_1^c + \cdots + X_N^c$.

For $q = 1$ the condition (4.63) turns into the usual independence of random vectors $X_1, \ldots, X_N$. If $q \neq 1$, then (q_{-1}^d, q, q)-independence describes a special type of strong dependence.

Remark 4.8. In Definition 4.6 of (q_{-1}^d, q, q)-independence the first number q_{-1}^d indicates the parameter of the q-Fourier transform on the left-hand side of (4.63), the second number q indicates the parameter of the q-Fourier transform on the right, and the third number q indicates the parameter of the q-product. Similarly, one can introduce (q_{-1}^d, q_{-1}^d, q) and (q, q, q)-independences, as well.

Theorem 4.15. *Assume $1 < q \leq 1 + 2/d$. Let $X_1, \ldots, X_N, \ldots$ be a sequence of (q_{-1}^d, q, q)-i.i.d. d-dimensional random vectors with a finite q-mean*

$$\mu_q = \left(\mu_q(X_1^{(1)}), \ldots, \mu_q(X_1^{(d)})\right),$$

and a finite $(2q - 1)$-covariance matrix Σ_{2q-1} with entries

$$\sigma_{q,i,j} = \mu_{2q-1}\left[\left(X_1^{(i)} - \mu_q(X_1^{(i)})\right)\left(X_1^{(j)} - \mu_q(X_1^{(j)})\right)\right], \quad i,j = 1, \ldots, d.$$

Then the sequence of random vectors

$$Z_N = \frac{X_1 + \cdots + X_N - N\mu_q}{(\nu_{2q-1}N)^{\frac{1}{2(1+d-dq)}}}, \quad N = 1, 2, \ldots, \tag{4.64}$$

is q_{-1}^d-convergent as $N \to \infty$ to a d-dimensional q_{-1}^d-Gaussian distribution with the covariance matrix $\beta\Sigma_{2q-1}^{-1}$, where

$$\beta = \left(\frac{[2 - d(q_{-1}^d - 1)]C_{q_{-1}^d}^{2(q_{-1}^d - 1)}}{4q}\right)^{\frac{1}{1+d-dq_{-1}^d}}. \tag{4.65}$$

Proof. First we assume that X_1 is centered, that is $\mu_q(X_1) = (\mu_{q,1}, \ldots, \mu_{q,d}) = (0, \ldots, 0)$, where $\mu_{q,j}$ is the q-mean of the jth component of the random vector X_1. Let f_{X_1} be the density function of the random vector X_1. Using the asymptotic expansion of q-exponential, namely (see Corollary 2.4)

$$\exp_q(z) = 1 + z + \frac{q}{2}z^2 + o(z^2), \quad z \to 0,$$

and finiteness of the q-mean and $(2q-1)$-covariance matrix of X_1, we obtain

$$
\begin{aligned}
\mathcal{F}_q[f_{X_1}](\xi) &= \int_{\mathbb{R}^d} f(x)\exp_q\left(i(x,\xi)[f(x)]^{q-1}\right)dx \\
&= \int_{\mathbb{R}^d} f_{X_1}(x)\big(1 + i(x,\xi)[f_{X_1}(x)]^{q-1} \\
&\quad - \frac{q}{2}(x,\xi)^2[f_{X_1}(x)]^{2(q-1)} + o((x,\xi)^2[f_{X_1}(x)]^{2(q-1)})\big)dx \\
&= 1 + i\nu_q(X_1)(\xi,\mu_q) - (q/2)\nu_{2q-1}(X_1)(\Sigma_{2q-1}\xi,\xi) \\
&\quad + o(|\xi|^2), \quad |\xi| \to 0,
\end{aligned}
\tag{4.66}
$$

where $(x,\xi) = x_1\xi_1 + \cdots + x_d\xi_d$ and

$$(\xi,\mu_q) = \sum_{j=1}^{d} \xi_j\mu_{q,j} = 0 \quad \text{and} \quad (\Sigma_{2q-1}\xi,\xi) = \sum_{j,k=1}^{d} \sigma^d_{2q-1,j,k}\xi_j\xi_k.$$

Hence, for $\mathcal{F}_q[f_{X_1}](\xi)$, we have

$$\mathcal{F}_q[f_{X_1}](\xi) = 1 - (q/2)\nu_{2q-1}(X_1)(\Sigma_{2q-1}\xi,\xi) + o(|\xi|^2), \quad |\xi| \to 0.$$

Further, one can show that, for a given random variable X and a real $\rho > 0$, the equality

$$\mathcal{F}_q[f_{\rho X}](\xi) = \mathcal{F}_q[f_X](\rho^{1+d-dq}\xi), \quad \xi \in \mathbb{R}^d, \tag{4.67}$$

holds. Indeed, the density function $f_{\rho X}(x)$ of ρX is given by

$$f_{\rho X}(x) = \rho^{-d}f_X(\rho^{-1}x).$$

Due to Proposition 3.21,

$$\mathcal{F}_q[f_\rho x](\xi) = \mathcal{F}_q[\rho^{-d} f_X(\rho^{-1}x)](\xi)$$
$$= \rho^{-d} \mathcal{F}_q[f_X(\rho^{-1}x)](\rho^{-d(q-1)}\xi)$$
$$= \mathcal{F}_q[f_X](\rho^{1+d-dq}\xi),$$

thus proving relation (4.67). This fact implies that the q-Fourier transform of the density function of

$$Y_1 = \frac{X_1}{\left[\nu_{2q-1}(X_1)N\right]^{-\frac{1}{2(1+d-dq)}}}$$

takes the form

$$\mathcal{F}_q[f_{Y_1}](\xi) = \mathcal{F}_q[f_{X_1}]\left(\frac{\xi}{\sqrt{N\nu_{2q-1}(X_1)}}\right).$$

Moreover, it follows from the (q_{-1}, q, q)-independence of $X_1, X_2, \ldots$ and the associativity of the q-product that

$$\mathcal{F}_{q-1}[f_{Y_N}](\xi)$$
$$= \mathcal{F}_q[f_{X_1}]\left(\frac{\xi}{\sqrt{N\nu_{2q-1}(X_1)}}\right) \otimes_q \cdots \otimes_q \mathcal{F}_q[f_{X_1}]\left(\frac{\xi}{\sqrt{N\nu_{2q-1}(X_1)}}\right)$$

$$\tag{4.68}$$

with N factors, where

$$Y_N = \frac{X_1 + \cdots + X_N}{\left[\nu_{2q-1}(X_1)N\right]^{\frac{1}{2(1+d-dq)}}}$$

Hence, making use of properties of the q-logarithm, from (4.68) we obtain

$$\ln_q \mathcal{F}_{q_{-1}^d}[f_{Y_N}](\xi) = N \ln_q \mathcal{F}_q[f_{X_1}]\left(\frac{\xi}{\sqrt{N\nu_{2q-1}(X_1)}}\right)$$

$$= N \ln_q \left(1 - \frac{q}{2}\frac{(\Sigma_{2q-1}\xi, \xi)}{N} + o\left(\frac{|\xi|^2}{N}\right)\right)$$

$$= -\frac{q}{2}(\Sigma_{2q-1}\xi, \xi) + o(1), \quad N \to \infty, \tag{4.69}$$

locally uniformly by ξ. It follows that

$$\lim_{N \to \infty} \mathcal{F}_{q^d_{-1}}[f_{Y_N}](\xi) = e_q^{-(q/2)(\Sigma_{2q-1}\xi,\xi)} \in \mathcal{G}_q. \tag{4.70}$$

Thus, Y_N is q-convergent to the random vector Z, whose q^d_{-1}-Fourier transform is

$$G_q \left(\frac{q}{2}\Sigma_{2q-1}; \xi \right) = e_q^{-(q/2)(\Sigma_{2q-1}\xi,\xi)}.$$

In accordance with Theorem 3.21, there exists a q^d_{-1}-Gaussian, whose q^d_{-1}-Fourier transform is $G_q(\frac{q}{2}\Sigma_{2q-1}; \xi)$. Hence, its density $f_Z(x)$ is a q^d_{-1}-Gaussian $G_{q^d_{-1}}(A; x)$ with a covariance matrix A, i.e.,

$$\mathcal{F}_{q^d_{-1}}[f_Z](\xi) = \mathcal{F}_{q^d_{-1}}[G_{q^d_{-1}}(A; x)] = e_q^{-(\frac{q}{2}\Sigma_{2q-1}\xi,\xi)}.$$

Let us now find the matrix A. Again, it follows from Theorem 3.21 that A has the form $\beta\Sigma_{2q-1}^{-1}$, where β satisfies the equation (see (3.146))

$$\frac{q}{2} = \frac{[2 - d(q^d_{-1} - 1)] \left(C_{q^d_{-1}}(\beta) \right)^{2(q^d_{-1}-1)}}{8\beta}.$$

Solving this equation with respect to β we obtain (4.65).

Finally, if $\mu_q(X_1) \neq 0$, then as it was noted above the random vector $X_k^c = X_k - \mu_q(X_k)$ has zero q-mean. Therefore, the sequence of random vectors defined in (4.64) is q-convergent to the random vector Z, thus proving the theorem. $\square$

Theorem 4.16. *Assume a sequence $q_k^d, k = 0, \pm 1, \ldots$ is given as (3.157). Let X_N, $N = 1, 2, \ldots$, be a sequence of $(q_{k-1}^d, q_k^d, q_k^d)$-independent for some $k \in \mathcal{Z}$ and identically distributed d-dimensional random vectors with a finite q_k^d-mean and a finite second $(2q_k^d - 1)$-covariance matrix $\Sigma_{2q_k^d-1}$.*

Then the sequence of random vectors

$$Z_N = D_N(q_k^d)\left(X_1 + \cdots + X_N - N\mu_{q_k^d}\right), \quad N = 1, 2, \ldots,$$

with

$$D_N(q_k^d) = \left[N\nu_{2q_k^d-1}(X_1) \right]^{-\frac{1}{2(1+d-dq_k^d)}}$$

is q_{k-1}^d-convergent to the multivariate q_{k-1}^d-Gaussian distribution with the covariance matrix $\beta_k \Sigma_{2q_k^d-1}^{-1}$ where

$$\beta_k = \left(\frac{[2 - d(q_{k-1}^d - 1)] C_{q_{k-1}^d}^{2(q_{k-1}^d-1)}}{4q_k^d} \right)^{\frac{1}{1+d-dq_{k-1}^d}}. \tag{4.71}$$

Remark 4.9. The formulation of Theorem 4.16 changes if we change the definition of q-independence. Similarly to the univariate case, in the multivariate case one can prove the q-CLTs for the other two versions of q-independence. Below we exhibit the corresponding results in Table 4.2. The current formulation, where the definition of $(q_{k-1}^d, q_k^d, q_k^d)$-independence is given by (4.63), corresponds to the first raw of Table 4.2. The second and third rows reflect the cases when q-independence is replaced by (q_k^d, q_k^d, q_k^d)- and $(q_{k-1}^d, q_{k-1}^d, q_k^d)$-independence, respectively. Consequently, the outcome, that is, the type of convergence and the parameter β_k of the attractor are also changed. In all cases the attractor is q_{k-1}^d-Gaussian with different values of parameter $\beta = \beta_k$ (see Table 4.2). Note that all three cases generalize the classic multivariate CLT, recovering it if $q_k = 1$.

Table 4.2:   Interrelation between the type of q-independence, convergence and the parameter of the attractor in the multivariate q-CLT.

	q-Independence	Convergence	Parameter β_k
1	$(q_{k-1}^d, q_k^d, q_k^d)$	q_{k-1}^d-convergence	$\left(\dfrac{[2-d(q_{k-1}^d-1)]C_{q_{k-1}^d}^{2(q_{k-1}^d-1)}}{4q_k^d} \right)^{\frac{1}{1+d-dq_{k-1}^d}}$
2	(q_k^d, q_k^d, q_k^d)	q_k^d-convergence	$\left(\dfrac{[2-d(q_{k-1}^d-1)]C_{q_{k-1}^d}^{2(q_{k-1}^d-1)}}{4q_k^d} \right)^{\frac{1}{1+d-dq_{k-1}^d}}$
3	$(q_{k-1}^d, q_{k-1}^d, q_k^d)$	q_{k-1}^d-convergence	$\left(\dfrac{[2-d(q_{k-1}^d-1)]C_{q_{k-1}^d}^{2(q_{k-1}^d-1)}}{4q_k^d} \right)^{\frac{1}{1+d-dq_{k-1}^d}}$

4.5.1. *Sequential multivariate q-CLT*

In this section, we consider random vectors X_N with symmetric densities. By definition a sequence of random vectors X_N, $N = 1, 2, \ldots$, is said to be (Q_k)-independent if for all $N = 2, 3, \ldots$ the relations

$$\mathcal{F}_{Q_k}[f_N](\xi) = \mathcal{F}_{Q_k}[f_{X_1}](\xi) \otimes_{q_{k+d}} \cdots \otimes_{q_{k+d}} \mathcal{F}_{Q_k}[f_{X_N}](\xi), \qquad (4.72)$$

hold for all $\xi \in \mathbb{R}^d$, where f_N is the joint density function of the sum $X_1 + \cdots + X_N$, f_{X_j} is the density function of X_j, and $\mathcal{F}_{Q_k}$ is d-dimensional Q_k-sequential Fourier transform with $Q_k = \{q_k, q_{k+1}, \ldots, q_{k+d-1}\}$.

For further analysis we need the following notations. Let $f(x)$ be a density function and $x_{(j)} = (x_1, \ldots, x_j)$, $\xi_{(d-j)} = (\xi_{j+1}, \ldots, \xi_d)$. Identify consequently the functions

$$g_0(x_1, \ldots, x_d), \quad \ldots, \quad g_j(x_{(d-j)}, \xi_{(j)}), \quad \ldots, \quad g_d(\xi_1, \ldots, \xi_d)$$

by

$$g_0(x) = f(x),$$

$$g_1(x_{(d-1)}, \xi_d) = \int_{\mathbb{R}^1} f(x)\, e_{q_{k+d-1}}^{i x_d \xi_d [g_0]^{q_{k+d-1}-1}}\, dx_d,$$

$$g_j(x_{(d-j)}, \xi_{(j)}) = \int_{\mathbb{R}^j} f(x) e_{q_{k+d-1}}^{i x_d \xi_d [g_0]^{q_{k+d-1}-1}} \cdots e_{q_{k+d-j}}^{i x_{d-j+1} \xi_{d-j+1} [g_{j-1}]^{q_{k+d-j}-1}}$$

$$\times\, dx_{d-j+1} \ldots dx_d, \qquad (4.73)$$

for $j = 2, \ldots, d$.

It is not hard to see that the last of these coincides with the Q_k-sequential Fourier transform, namely

$$g_d(\xi_1, \ldots, \xi_d) = \mathcal{F}_{Q_k}[f](\xi)$$

$$= \int_{\mathbb{R}^d} f(x) e_{q_{k+d-1}}^{i x_d \xi_d [g_0]^{q_{k+d-1}-1}} \cdots e_{q_k}^{i x_1 \xi_1 [g_{d-1}]^{q_k-1}}\, dx_1 \ldots dx_d.$$

$$(4.74)$$

The functions $g_j(x_{(d-j)}, \xi_{(j)})$, $j = 1, \ldots, d$, in (4.73) are also connected with the sequential Fourier transforms. To see that let us introduce the following notations:

Denote by $f_j(x_{(j)})$ the marginal density of $f(x)$ defined by

$$f_j(x_{(j)}) = \int_{\mathbb{R}^{d-j}} f(x_{(j)}, x_{j+1}, \ldots, x_d) dx_{j+1} \ldots dx_d.$$

It follows from (4.73) that

$$g_j(x_1, \ldots, x_{d-j}, \xi_{d-j+1}, \ldots, \xi_d)$$
$$= f_{d-j}(x_1, \ldots, x_{d-j}) + o(|\xi_{d-j+1}| + \cdots + |\xi_d|), \qquad (4.75)$$
$$|\xi_{d-j+1}| + \cdots + |\xi_d| \to 0.$$

Further, we introduce the notation

$$\sigma^2_{j,q_{k+j-1}-1} = \int_{\mathbb{R}^d} x^2_{d-j} f(x) \prod_{l=1}^{j} [f_{d-l}(x_1, \ldots, x_{d-l})]^{q_{k+d-j}-1} dx, \quad (4.76)$$

where $j = 1, \ldots, d$. Lastly, for a d-dimensional random vector $V = (V_1, \ldots, V_d)$ and d-dimensional (non-random) vector $\rho = (\rho_1, \ldots, \rho)$ we use the notation $[\rho, V]$ for the random vector $(\rho_1 V_1, \ldots, \rho_d V_d)$.

Theorem 4.17. *Assume* $1 < q \leq 1 + 2/d$. *Let* $X_1, \ldots, X_N, \ldots$ *be a sequence of* Q_k-*independent and identically distributed d-dimensional random vectors with a symmetric density* $f(x)$ *and a finite* $\sigma^2_{j,2q_{k+j-1}-1}$, $j = 1, \ldots, d$.

Then the sequence of random vectors $Z_N = [a_N, Y_N]$, *where*

$$Y_N = X_1 + \cdots + X_N, \quad N = 1, 2, \ldots,$$

and

$$a_N = (a_1, \ldots, a_d), \quad a_j = (\sqrt{q_{k+j-1}N} \sigma_{j,q_{k+j-1}-1})^{-\frac{1}{2-q_{k+j-1}}},$$
$$j = 1, \ldots, d, \quad N = 1, 2, \ldots, \qquad (4.77)$$

is q_k-*convergent as* $N \to \infty$ *to a d-dimensional* q_{k-1}-*Gaussian distribution with a diagonal covariance matrix.*

Proof. It follows from (4.74) and (4.75) that

$$\mathcal{F}_{Q_k}[f](\xi) = 1 - \frac{1}{2}\sum_{j=1}^{d} q_{k+j-1}\sigma^2_{j,q_{k+j-1}-1}\xi_j^2 + o(|\xi|^2). \qquad (4.78)$$

Further, it is readily seen that, for a given random variable V with the density function f_V and a real d-dimensional vector ρ with positive components ρ_j, $j = 1, \ldots, d$, the Q_k-Fourier transform of the density function f_W of the random vector $W = [\rho, V]$ is

$$\mathcal{F}_{Q_k}[f_W](\xi_1, \ldots, \xi_d) = \mathcal{F}_{Q_k}[f_V](\rho_1^{2-q_{k+d-1}}\xi_1, \ldots, \rho_d^{2-q_k}\xi_d).$$

It follows from this relation and (4.77) that

$$\mathcal{F}_{Q_k}\left[f_{[a_N, X_1]}\right] = \mathcal{F}_{Q_k}[f_{X_1}]\left(\frac{\xi_1}{\sqrt{q_k N}\sigma_{1,q_k-1}}, \ldots, \frac{\xi_d}{\sqrt{q_{k+d-1}N}\sigma_{d,q_{k+d-1}-1}}\right).$$

Moreover, it follows from the Q_k-independence of $X_1, X_2, \ldots$ and the associativity of the q-product that

$$\mathcal{F}_{Q_k}[f_{Z_N}](\xi) = \mathcal{F}_{Q_k}[f_{X_1}]\left(\frac{\xi}{\sqrt{N}}\right) \otimes_{q_{k+d}} \cdots \otimes_{q_{k+d}} \mathcal{F}_{Q_k}[f_{X_1}]\left(\frac{\xi}{\sqrt{N}}\right), \qquad (4.79)$$

with N factors. Hence, making use of properties of the q-logarithm, from (4.79) we obtain

$$\ln_{q_{k+d}}\mathcal{F}_{Q_k}[f_{Z_N}](\xi)$$

$$= N \ln_{q_{k+d}}\mathcal{F}_{Q_k}[f_{X_1}]\left(\frac{\xi}{\sqrt{N}}\right)$$

$$= N \ln_{q_{k+d}}\left(1 - \sum_{j=1}^{d} \frac{q_{k+j-1}\sigma^2_{j,q_{k+j-1}-1}\xi_j^2}{2N} + o\left(\frac{|\xi|^2}{N}\right)\right)$$

$$= -\sum_{j=1}^{d} \frac{q_{k+j-1}\sigma^2_{j,q_{k+j-1}-1}}{2}\xi_j^2 + o(1), \quad N \to \infty, \qquad (4.80)$$

locally uniformly by ξ.

It follows from (4.80) that locally uniformly by ξ,

$$\lim_{N \to \infty} \mathcal{F}_{Q_k}(f_{Z_N}) = e_{q_k+d}^{-\sum_{j=1}^{d} \frac{q_{k+j-1}\sigma_{j,q_{k+j-1}-1}^2}{2}\xi_j^2}. \tag{4.81}$$

Thus, Z_N is Q_k-convergent to the random vector Z, whose Q_k-Fourier transform is $e_{q_k+d}^{-\sum_{j=1}^{d} \frac{q_{k+j-1}\sigma_{j,q_{k+j-1}-1}^2}{2}\xi_j^2}$.

Now we show that for the density function of Z there exists a d-dimensional q_k-Gaussian with a diagonal covariance matrix Σ, such that

$$\mathcal{F}_{Q_k}[f_Z](\xi) = \mathcal{F}_{Q_k}[G_{q_k}(\Sigma; x)] = e_{q_k+d}^{-\sum_{j=1}^{d} \frac{q_{k+j-1}\sigma_{j,q_{k+j-1}-1}^2}{2}\xi_j^2}.$$

In accordance with Theorem 3.23 the covariance matrix Σ is diagonal with the diagonal entries β_j, $j = 1, \ldots, d$, uniquely determined from the system of equations (see (3.156))

$$\frac{\left[_{d-j+1}C_{q_{k+j-1}} \right]^{2(q_{k+j-1}-1)}}{4\beta_j} \frac{\prod_{n=k+j-1}^{d+k-1} \frac{3-q_n}{2}}{\prod_{n=k}^{k+j-2} \frac{3-q_n}{2}} = \frac{q_{k+j-1}\sigma_{j,q_{k+j-1}-1}^2}{2},$$

$$j = 1, \ldots, d.$$

where $_{d-m}C_{q_{k+m}}$, $m = 0, \ldots, d-1$, are defined in (3.148). The proof of the theorem is complete. $\qquad\square$

Remark 4.10.

(1) The sequential version of the q-CLT formulated above can be extended for the general case of non-diagonal covariance matrices, as well as for other types of q-i.i.d. random vectors X_j, $j = 1, 2, \ldots$.

(2) Theorem 4.17 reduces to Theorem 4.4 if $d = 1$. Indeed, in this case $\beta_1 = \beta_k$ and it is easy to verify that $\sigma_{1,q_k}^2 = \sigma_{2q_k-1}^2$,

$_1C_{q_k} = \sqrt{\beta}C_{q_{k-1}}$, and

$$\beta_k = \left(\frac{3 - q_{k-1}}{4q_k C_{q_{k-1}}^{2q_{k-1}-2}} \right)^{\frac{1}{2-q_{k-1}}}.$$

4.5.2. On uniqueness in multivariate q-central limit theorems

In Section 4.4, we discussed the uniqueness of limiting distribution in detail in the case of univariate central limit theorems. Similar arguments work in the multivariate case as well. Therefore, in this section we briefly discuss the uniqueness question, without going into details. However, we emphasize important statements, which do not follow immediately from the univariate case, and which are leading to the uniqueness.

We start with the multi-dimensional version of Theorem 4.9, which represents a q-generalization of the celebrated Paley–Wiener theorem on Fourier transform of functions with compact support. Let f be a continuous and symmetric density function. The exact meaning of symmetry will be clarified below. Denote $\lambda_j(x) = x_j[f(x)]^{q-1}$, $j = 1, \ldots, d$, where $1 \leq q < 2$. These functions form a continuous mapping $\Lambda : \mathbb{R}^d \to \mathbb{R}^d$. Since f is symmetric, it suffices to consider $\Lambda(x)$ only for points x with positive coordinates $x_j \geq 0, j = 1, \ldots, d$. Suppose the maximum value of $\lambda_j(x)$ in the variable x_j, subject to all the other variables are fixed, is M_j and $x_{*,j} > 0$ is the rightmost point where the maximum is attained. We denote $x_* = (x_{*_1}, \ldots, x_{*_d})$. Further, let

$$\tau_* = \begin{cases} \dfrac{1}{m(q-1)}, & \text{if } 1 < q < 2, \\ \infty, & \text{if } q = 1. \end{cases}$$

and $\Pi_a = [-a, a] \times \cdots \times [-a, a]$ be a d-dimensional hypercube. We also use the notations $\mathcal{T} = \{\tau \in \mathbb{R}^d : \tau_j < \tau_*, \ j = 1, \ldots, d\}$, and $\mathcal{T}_+ = \{\tau \in \mathbb{R}^d : 0 < \tau_j < \tau_*, \ j = 1, \ldots, d\}$.

Theorem 4.18. *Let f be a continuous symmetric about coordinate hyperplanes density function with $\operatorname{supp} f \subseteq \Pi_a$. Then the q-Fourier* *transform of f satisfies the following estimate*

$$\left| F_q[f](\eta - i\tau) \right| \le e_q^{M_q(x_*, \tau)}, \qquad (4.82)$$

where $\eta \in \mathbb{R}^d$, $\tau \in \mathcal{T}$, $M_q = \max\limits_{x \in \Pi_a}\{[f(x)]^{q-1}\}$, *and* $x_* = (x_{*1},$ $\ldots, x_{*d}) \in \Pi_a$ *is the point with positive coordinates* x_{*j} *at which the function* $x_j f^{q-1}(x)$ *attains its maximum as a function of* x_j.

The case $d = 1$ was proved in Theorem 4.9. The proof in the multidimensional case is similar to the one-dimensional case. Inequality (4.82) implies an estimate from below for the volume of the support of any density function with compact support.

Lemma 4.2. *Let* f *be a density function satisfying conditions of Theorem 4.18. Then, for the volume* $V(\mathrm{supp}[f])$ *of its support the following estimate holds*

$$V(\mathrm{supp}[f]) \ge (\sqrt{2}x_*)^d \ge \frac{2^{d/2}}{M_q^d}\left[\sup_{\tau \in \mathcal{T}_+} \frac{\ln_q \left| F_q[f](-i\tau)\right|}{|\tau|}\right]^d. \qquad (4.83)$$

Proof. Using inequality (4.82) with $\eta = 0$, we have

$$|x_*||\tau| \ge |(x_*, \tau)| \ge \frac{1}{M_q}\ln_q \left| F_q[f](-i\tau)\right|, \quad \tau \in \mathcal{T}.$$

The latter implies the following estimate for the volume of the support:

$$V(\mathrm{supp}[f]) = (2a)^d \ge \left(\frac{2}{\sqrt{d}}|x_*|\right)^d$$

$$\ge \frac{2^d}{(\sqrt{d}M_q)^d}\left[\sup_{\tau \in \mathcal{T}_+}\frac{\ln_q \left| F_q[f](-i\tau)\right|}{|\tau|}\right]^d, \qquad (4.84)$$

proving the statement. □

Lemma 4.2 helps to estimate the support of the density function $f_N(x) = f_{S_N}(x)$, $x \in \mathbb{R}^d$, of the sum $S_N = X_1 + \cdots + X_N$ of identically distributed q-independent random vectors $X_1, \ldots, X_n$ with the (same) density function $f = f_{X_1}$ whose support is a subset of Π_a, $a > 0$. Namely, the following statement holds.

Theorem 4.19. *Let $X_1, \ldots, X_n$ be q-independent of any type random vectors all having the same density function f with $\operatorname{supp} f \subseteq \Pi_a$. Then, for the volume $V(\operatorname{supp}[f_N])$ of the support of the density function f_N of S_N, there exists a constant $C = C(q, f, d) > 0$, such that the estimate*

$$V(\operatorname{supp}[f_N]) \geq CN^d \tag{4.85}$$

holds.

The proof repeats the proof of Theorem 4.10. Therefore, we omit it. Theorem 4.19 implies the following important statement.

Theorem 4.20. *Let $X_1, \ldots, X_N$ be q-independent of any type random vectors all having the same density function f with $\operatorname{supp} f \subseteq \Pi_a$. If the sequence Z_N has a distributional limit random variable in some sense, then this random variable can not have a density with compact support. Moreover, due to the scaling present in Z_N, the support of the limit variable is the entire space $\mathbb{R}^d$.*

The proof obviously follows immediately from (4.85) upon letting $N \to \infty$.

Similarly to the one-dimensional case, due to estimate (4.85), no distribution with compact support can be a limiting distribution in multivariate q-CLTs. In the one-dimensional case we used some asymptotic relationships between limiting distributions and their quasi-variances. Similar arguments hold in the multivariate case as well, leading to uniqueness of the limiting distribution.

4.6. Other *q*-Central Limit Theorems

4.6.1. *Vignat–Plastino q-central limit theorem*

Vignat and Plastino (2007b) proved a multivariate version of the q-central limit theorem without using the q-Fourier transform. The dependence assumed is not exactly the same as in Theorem 4.16, however they are asymptotically equivalent. The obtained q-CLT can be interpreted as a version of a limiting process for a sequence of time-changed random vectors, in which the random time is distributed according to the positive square root of the χ^2-distribution with m degrees of freedom, also called a χ-*distribution*. Let a denote the χ-distribution. Its density function is [Wiki-Chi]

$$f_a(t) = \frac{t^{m-1}e^{-\frac{t^2}{2}}}{\Gamma(m/2)2^{m/2-1}}, \quad t > 0.$$

However, we need the scaled version of the χ-distribution. Namely, the distribution $\chi = \sigma a$, where σ is a positive real number (a scale parameter), which is referred to as a scaled χ-distribution. The density function of the scaled χ-distribution χ with m degrees of freedom and the scale parameter σ is given by

$$f_\chi(t) = \frac{t^{m-1}e^{-\frac{t^2}{2\sigma^2}}}{\Gamma(m/2)2^{m/2-1}\sigma^m}, \quad t > 0.$$

Vignat and Plastino proved their theorem for both cases: $q > 1$ and $q < 1$.

Theorem 4.21 (Vignat and Plastino, 2007b). *Let $q > 1$ and $X_1, \ldots, X_n$ be a sequence of d-dimensional independent and identically distributed random vectors with zero mean and covariance matrix Σ, and assume that the scaled χ-distribution χ is independent of X_1, has m degrees of freedom, and the scale parameter $\sigma = 1/\sqrt{m-2}$. Then the sequence*

$$Z_n = \frac{1}{\chi\sqrt{n}} \sum_{j=1}^{n} X_j \tag{4.86}$$

weakly converges to a multivariate q-Gaussian random vector Z with covariance matrix Σ, and

$$q = \frac{m+d+2}{m+d}. \tag{4.87}$$

Theorem 4.22 (Vignat and Plastino, 2007b). *Let $q < 1$ and $X_1, \ldots, X_n$ be a sequence of d-dimensional i.i.d. random vectors with zero mean and covariance matrix Σ, and assume that the scaled*

χ-*distribution* χ *is independent of* X_1, *has* m *degrees of freedom, and the scale parameter* $\sigma = \sqrt{m-2}$. *Then the sequence*

$$Z_n = \frac{\frac{1}{\sqrt{n}} \sum_{j=1}^{n} X_j}{\sqrt{\chi^2 + \left(\frac{1}{\sqrt{n}} \sum_{j=1}^{n} X_j\right)^T \Sigma^{-1} \left(\frac{1}{\sqrt{n}} \sum_{j=1}^{n} X_j\right)}} \tag{4.88}$$

weakly converges to a multivariate q-*Gaussian random vector* Z *with covariance matrix* Σ, *and*

$$q = \frac{m-4}{m-2}. \tag{4.89}$$

In these two theorems, though X_j, $j = 1, \ldots, n$, are independent, nevertheless random variables $Y_j = X_j/\chi$, $j = 1, \ldots, n$, are dependent through the scaled χ-distribution. Can this dependence be described analytically? Perhaps not. However, as it is proved in Vignat and Plastino (2007b), the dependence indirectly assumed in these two theorems, is asymptotically equivalent to the q-independence assumed in Theorems 4.15 and 4.16.

Further, the indices q shown in (4.87) and (4.89) of the limiting q-Gaussian distributions of Theorems 4.21 and 4.22 depend on the degrees of freedom m of the scaled χ-distribution. Relation between q and m as indicated in paper Vignat and Plastino (2007b) is (see also Vignat and Plastino, 2005)

$$m = \frac{2}{q-1} - d, \tag{4.90}$$

if $q > 1$. At the same time the index of limiting distributions in Theorem 4.16 is q_{k-1}^d defined in (3.157), that is

$$q_{k-1}^d = \frac{2q - d(k-1)(q-1)}{2 - d(k-1)(q-1)}.$$

Since Theorem 4.16 is valid only for $q_k > 1$, we can compare the latter with q in (4.87). We have

$$\frac{m+d+2}{m+d} = \frac{2q - d(k-1)(q-1)}{2 - d(k-1)(q-1)}.$$

Solving this equation for m we have

$$m = \frac{2}{q-1} - (k-1)d,$$

which coincides with (4.90) if $k = 2$. Thus, Theorem 4.21 of Vignat and Plastino is consistent with the $k = 2$ particular case of Theorem 4.16.

Finally, we note that Theorem 4.21 of Vignat and Plastino is also a special case of Theorem 4.25, which represents the central limit theorem for the sequence of exchangeable random variables.

4.6.2. *q-Central limit theorem for exchangeable random variables*

The results provided in this section were published in Jiang *et al.* (2010). We introduce some necessary notions. For details of these notions and their relevant properties we refer the reader to the following sources (Chow and Teicher, 1978; Keilson and Steutel, 1974; Gneiting, 1997; Jiang *et al.*, 2010).

The first notion we are going to introduce is the exchangeability of random variables.

Definition 4.7. A sequence of random variables $\{X_j, \ j = 1, 2, \ldots\}$ defined on some probability space is called *exchangeable* if for each n,

$$P(X_1 \leq x_1, \ldots, X_n \leq x_n) = P(X_{\pi(1)} \leq x_1, \ldots, X_{\pi(n)} \leq x_n)$$

for any permutation π of $\{1, 2, \ldots, n\}$ and any $x_j \in \mathbb{R}, \ j = 1, \ldots, n$.

Bruno de Finetti proved a theorem (see, e.g., Chow and Teicher, 1978), according to which an infinite sequence of exchangeable random variables is conditionally i.i.d., given a σ-field $\mathcal{G}$ of permutable events. Coordinate random variables $\{\xi_n \equiv \xi_n^\omega = X_n(\omega) | \mathcal{G}, n \geq 1\}$, given $\mathcal{G}$, are called *mixands*, and due to de Finetti's theorem are i.i.d. for each $\omega \in \Omega$. Moreover, there exists an empirical conditional distribution P^ω for X_j such that for each natural number n, any

Borel function $f : \mathbf{R}^n \to \mathbf{R}$, and any Borel subset B of $\mathbf{R}$,

$$P(f(X_1, \ldots, X_n) \in B) = \int_\Omega P(f(X_1, \ldots, X_n) \in B | \mathcal{G}) dP$$

$$= \int_\Omega P^\omega(f(\xi_1, \ldots, \xi_n) \in B) dP.$$

An important fact is that for exchangeable sequences of random variables, the dependence never dies in contrast to weakly dependent sequences. Hence if the covariance exists, it does not change along the sequence. Moreover, for an infinite exchangeable sequence the covariance is always non-negative.

Another notion we need in this section is a variance mixture of normal random variables.

Definition 4.8. A *variance (or scale) mixture of normal distributions (VMON for short)* by definition is a random variable whose characteristic function has the form

$$\phi(t) = \int_0^\infty \exp(-t^2 u/2) dH(u),$$

with H a distribution function on $[0, \infty)$, called the *mixing distribution*. The corresponding density is

$$f(x) = \int_0^\infty (2\pi u)^{-1/2} \exp(-x^2/(2u)) dH(u).$$

Each VMON has a representation in the form of product $V \cdot N$ of random variables, where N is the standard normal random variable, and $V > 0$ and is independent of N. Examples of VMONs include many commonly used distributions such as the symmetric stable distributions, the Cauchy, Laplace, double exponential, logistic, hyperbolic, and Student distributions and their mixtures, and many others. See Keilson and Steutel (1974) and Gneiting (1997), for details.

The connection between the q-Gaussians and the VMON can be established using completely monotone functions. By definition, a function $h(x)$, $x > 0$, is called *completely monotone* on $(0, \infty)$ if

$$(-1)^n h^{(n)}(x) \geq 0, \quad x > 0, \tag{4.91}$$

for all $n = 0, 1, 2, \ldots$. One can show that for the q-Gaussian density function $G_q(x)$ the function $G_q(\sqrt{x})$ is completely monotone for $q > 1$ and not completely monotone for $q < 1$. Indeed, for $q > 1$

$$(-1)^n G_q^{(n)}(\sqrt{x}) = \frac{A_n}{C_q[1 + (q-1)x]^{\frac{nq-n+1}{q-1}}} > 0, \quad x > 0,$$

for all $n = 0, 1, \ldots$, where A_n is defined in (2.38), which implies validity of (4.91). For $q < 1$, one can easily verify that (4.91) is not verified for all $n \geq 0$.

It is known (see Andrew and Mallows, 1974), that a symmetric density function $f(x)$ is a variance mixture of normals if and only if $f(\sqrt{x})$ is completely monotone. This fact leads to the following statement.

Theorem 4.23. *q-Gaussians are variance mixtures of normals when $1 \leq q < 3$, and not variance mixtures of normals when $q < 1$.*

We note that not every VMON is a q-Gaussian. Therefore, it is important to identify the mixing distributions that yield the q-Gaussians for $1 < q < 3$. Namely, the following statement holds. Due to its importance in terms of representation of the q-Gaussian density, and also for completeness, we provide the proof, as well.

Theorem 4.24 (Jiang *et al.*, 2010). *Let V_q be a positive random variable with the corresponding density function*

$$f_{V_q}(v) = C_{V_q} v^{-\frac{2}{q-1}} \exp\left(-\frac{1}{2(q-1)v^2}\right), \quad v > 0, \tag{4.92}$$

where the normalizing constant is

$$C_{V_q} = \frac{2}{[2(q-1)]^{\frac{3-q}{2(q-1)}} \Gamma\left(\frac{3-q}{2(q-1)}\right)}.$$

Then

(1) *the q-Gaussian random variable N_q can be represented as the variance mixture of normals $N_q = V_q \cdot N$, where N is normal; and*

(2) *for the corresponding q-Gaussian density function $G_q(x)$, $1 < q < 3$, the following representation holds:*

$$G_q(x) = \int_0^\infty \frac{1}{\sqrt{2\pi}v} \exp\left(\frac{-x^2}{2v^2}\right) f_V(v)\, dv, \quad x \in \mathbb{R}. \qquad (4.93)$$

Proof. It suffices to show only (4.93), since Part (1) is a simple implication of Part(2). Let L denote the Laplace transform. Using the formula

$$L_{t \to s}\left[\exp(-bt) \cdot t^\alpha\right] = \frac{\Gamma(\alpha+1)}{(s+b)^{\alpha+1}}$$

valid for $\alpha > -1$, and taking

$$\alpha = \frac{1}{q-1} - 1 = \frac{2-q}{q-1}, \quad b = \frac{1}{q-1}$$

one has

$$L_{t \to s}\left[\exp\left(-\frac{t}{q-1}\right) \cdot t^{\frac{2-q}{q-1}}\right] = \frac{\Gamma\left(\frac{1}{q-1}\right)}{\left(s + \frac{1}{q-1}\right)^{\frac{1}{q-1}}}$$

$$= \frac{\Gamma\left(\frac{1}{q-1}\right)(q-1)^{\frac{1}{q-1}}}{[1 + (q-1)s]^{\frac{1}{q-1}}}.$$

Changing s to x^2, the latter can be written as

$$\frac{a_q}{[1 - (1-q)x^2]^{1/(q-1)}} = \int_0^\infty \exp(-x^2 t)\, dH(t),$$

where

$$dH(t) = \exp\left(-\frac{t}{q-1}\right) \cdot t^{\frac{2-q}{q-1}}\, dt$$

and

$$a_q = \Gamma\left(\frac{1}{q-1}\right)(q-1)^{\frac{1}{1-q}}.$$

Routine calculations verify the claim where the following substitutions are used here below: $v^2 \to v$ in the second equality, $1/v \to u$ in the third equality, and $\frac{1}{2}\left(x^2 + \frac{1}{q-1}\right)u \to v$ in the fourth equality.

$$\int_0^\infty \frac{1}{\sqrt{2\pi}v}\exp\left(\frac{-x^2}{2v^2}\right)f_V(v)\,dv$$

$$= C_{V_q}\int_0^\infty \frac{1}{\sqrt{2\pi}v}\exp\left(\frac{-x^2}{2v^2}\right)\exp\left(-\frac{1}{2(q-1)v^2}\right)v^{-\frac{2}{q-1}}\,dv$$

$$= \frac{1}{2\sqrt{2\pi}}C_{V_q}\int_0^\infty \exp\left(-\frac{1}{2}\left(x^2+\frac{1}{q-1}\right)\cdot\frac{1}{v}\right)\left(\frac{1}{v}\right)^{\frac{q}{q-1}}\,dv$$

$$= \frac{1}{2\sqrt{2\pi}}C_{V_q}\int_0^\infty \exp\left(-\frac{1}{2}\left(x^2+\frac{1}{q-1}\right)u\right)u^{\frac{1}{q-1}-1}\,du$$

$$= \frac{1}{2\sqrt{2\pi}}C_{V_q}\int_0^\infty \exp(-v)\cdot v^{\frac{1}{q-1}-1}\cdot\left(\frac{1}{2}\left(x^2+\frac{1}{q-1}\right)\right)^{\frac{-1}{q-1}}\,dv$$

$$= \frac{1}{2\sqrt{2\pi}}C_{V_q}\Gamma\left(\frac{1}{q-1}\right)\left(\frac{1}{2}\left(x^2+\frac{1}{q-1}\right)\right)^{\frac{-1}{q-1}}$$

$$= \frac{1}{\sqrt{2\pi}}C_{V_q}2^{\frac{2-q}{q-1}}\Gamma\left(\frac{1}{q-1}\right)(q-1)^{\frac{1}{q-1}}\frac{1}{[1+(q-1)x^2]^{\frac{1}{q-1}}}.$$

It is not hard to verify that

$$\frac{1}{\sqrt{2\pi}}C_{V_q}2^{\frac{2-q}{q-1}}\Gamma\left(\frac{1}{q-1}\right)(q-1)^{\frac{1}{q-1}} = C_q^{-1},$$

where C_q is the normalizing constant of the q-Gaussian density $G_q(x)$ defined in (2.68), thus completing the proof. $\qquad\square$

Remark 4.11. In fact, in Theorem 4.24, the random variable V_q is 1 over χ-distribution with the number of degrees of freedom being $\frac{2}{q-1} - 1$. In other words, Theorem 4.24 is a 1-D case of Vignat–Plastino's Theorem 4.21.

Now we turn our attention to the central limit theorem for exchangeable random variables (see also Jiang an Hahn, 2003).

Theorem 4.25 (Jiang *et al.*, 2010). *Let* $\{X_j, \ j = 1, 2, \dots \}$ *be an infinite sequence of exchangeable random variables where* X_1 *has mean 0 if the mean exists or is symmetric otherwise. Assume that coordinate mixands satisfy* $0 < E[\xi_1^2] < \infty$ *almost surely (a.s.) Then either*

$$\frac{1}{\sqrt{n}}\sum_{j=1}^{n} X_j \to V_* \cdot N, \quad \text{in distribution,} \tag{4.94}$$

or

$$\frac{1}{n}\sum_{j=1}^{n} X_j \to V_{**}, \quad \text{a.s.} \tag{4.95}$$

where $V_* = \sqrt{\mathrm{Var}(\xi_1)} = \sqrt{\mathrm{Var}(X_1 | \mathcal{G})}$ *and* $V_{**} = E[\xi_1] = E[X_1 | \mathcal{G}]$.

Remark 4.12. Case (4.94) of Theorem 4.25 shows that q-Gaussians with $1 < q < 3$ are among the possible limits in the central limit theorem for exchangeable sequences of random variables. However, they are not the only ones. It seems more natural to consider q-Gaussians as limit distributions in case (4.94) since theoretically the range of distributions in case (4.95) is vast. For example, let ϵ_i's be i.i.d. with mean 0 and standard deviation 1, and Y be any random variable that is independent of all ϵ_i's. Then $\{Y + \epsilon_i, i \geq 1\}$ is a sequence of exchangeable random variables and

$$\sum_{i=1}^{n}(Y + \epsilon_i)/n \to Y, \quad \text{a.s.}$$

In case (4.94), $E[X_1 X_2] = 0$, if it exists, which means that q-Gaussians can be attractors of uncorrelated but not independent exchangeable sequences of random variables. In case (4.95),

$E[X_1 X_2] \neq 0$, if it exists, and in this case the central limit theorem is really the strong law of large numbers for exchangeable random variables. This case explains why the examples in the papers Rodriguez *et al.* (2008) and Hanel *et al.* (2009) achieve q-Gaussian limits with normalizers at the rate of n (see Section 4.11).

Remark 4.13. Returning to the Vignat–Plastino CLT (Theorem 4.21), we note that for one-dimensional case it is a special instance of Theorem 4.25 with $X_j = V_q \cdot Y_j$, where Y_j are i.i.d. with finite variance, and $V_q = \frac{1}{\chi}$, where the χ-distribution has $2/(q-1) - 1$ degrees of freedom. Note that, in general, a sequence of exchangeable random variables does not necessarily have the form $V_* \cdot Y_j$.

Lastly, we formulate a triangular array version of Theorem 4.25.

Theorem 4.26 (Jiang *et al.*, 2010). *Let* $\{X_{n,j}, j \leq n \leq n, n = 1, 2, \ldots\}$ *be a triangular array of rowwise exchangeable random variables that can be embedded into infinite exchangeable sequences. Assume* $X_{n,i}$*'s are centered or symmetric with* $0 < E[\xi_{n,1}^2] < \infty$ *in probability (i.e., with probability one) for each* n*, and* $\mathrm{Var}(\xi_{n,1}) \to V^{*2}$ *in probability when* $n \to \infty$*, with* $V^* > 0$ *a.s. Then*

(1) *the convergence*

$$\frac{1}{\sqrt{n}} \sum_{j=1}^{n} X_{n,j} \to V^* \cdot N + V^{**}, \quad \text{in distribution}$$

holds when $\sqrt{n} E[\xi_{n,1}] \to V^{**}$ *as* $n \to \infty$*; and*
(2) *the convergence*

$$\frac{1}{n^\alpha} \sum_{j=1}^{n} X_{n,j} \to V^{***}, \quad \text{in probability}$$

with $\alpha > 1/2$ *holds when* $n^{1-\alpha} E[\xi_{n,1}] \to V^{***}$ *in probability as* $n \to \infty$.

Remark 4.14. Note that when $X_{n,j}$'s in Theorem 4.26 are identically distributed, then $V^{**} = 0$ and $\alpha = 1$, which is consistent with Theorem 4.25.

4.7. *q*-Central Limit Theorem and *q*-Triplets

We recall that the density of the classical Gaussian distribution N_t, describing the normal diffusion process with the diffusion coefficient κ and evolving in time, can be calculated via the inverse Fourier transform of $e^{-\kappa t|\xi|^2}$, namely,

$$G(t, x) = \frac{1}{2\pi} \int_{\mathbb{R}} e^{-\kappa t|\xi|^2 - ix\xi} d\xi$$

$$= \frac{1}{2\sqrt{\pi\kappa t}} e^{-\frac{x^2}{4\kappa t}} \quad t > 0, \ x \in \mathbb{R}.$$

It follows from this equation that the mean square displacement (or, the variance) of N_t is related to time as

$$\langle N_t^2 \rangle = O(t), \quad t \to \infty,$$

provided the variance $\langle N_t^2 \rangle$ exists. It is known (for details, see, e.g., Metzler and Klafter, 2000; Zaslavsky, 2002) that, for a possibly anomalous diffusion process X_t, the mean square displacement behaves as

$$\langle X_t^2 \rangle = O(t^\delta), \quad t \to \infty,$$

with $\delta < 1$ for subdiffusive processes, $\delta > 1$ for superdiffusive processes, and $\delta = 1$ for normal diffusion.

If the variance of X_t does not exist, then we must focus on how the square displacement asymptotically grows in time. In the case of evolving in time limiting q-Gaussians Z_t, obtained in the q-CLTs (in all three types of q_k-independence), one can show that the scaling rate is

$$\delta = \frac{1}{2 - q_{k-1}} = q_{k+1}.$$

Indeed, it follows from relationship (4.71) and Theorem 3.21 that the q_{k-1}-Fourier preimage of $e_{q_k^d}^{-\kappa_q t|\xi|^2}$ can be represented in the form

$$At^{-\frac{\gamma}{2}} e_{q_{k-1}}^{-B\frac{x^2}{t^\gamma}},$$

where A and B are some constants, and

$$\gamma = \frac{1}{2 - q_{k-1}}.$$

Assume $\langle Z_1^2 \rangle$ is finite. Then we have

$$\langle Z_t^2 \rangle = At^{-\gamma/2} \int_{\mathbb{R}} x^2 \exp_{q_{k-1}} \left(-B \frac{x^2}{t^\gamma} \right) dx$$

$$= At^\gamma \int_{\mathbb{R}} x^2 \exp_{q_{k-1}} \left(-Bx^2 \right) dx$$

$$= At^\gamma \langle X_1^2 \rangle = O(t^\gamma), \quad t \to \infty,$$

which implies

$$\delta = \gamma = 1/(2 - q_{k-1}) = q_{k+1}.$$

On the other hand, for all three types of q-independence the limiting random variable in the q-CLT is distributed as q-Gaussian with $q = q_{k-1}$ (see Theorems 4.5–4.7). Hence, three consecutive members of

$$q_k = \frac{2q + k(1 - q)}{2 + k(1 - q)}, \quad q \in [1, 2), \quad k = 0, \pm 1, \ldots,$$

namely, the triplet (q_{k-1}, q_k, q_{k+1}), play an important role in the description of non-extensive phenomena. Namely, if a strong dependence (q-independence) is given by q_k, then the corresponding attractor is a q_{k-1}-Gaussian and, in turn, the scaling rate is q_{k+1}. Thus, processes corresponding to the q-CLT, have a triple property (*attractor, strong dependence, scaling rate*), which can be described by the triplet

$$(q_{k-1}, \ q_k, \ q_{k+1}).$$

The same pattern can be obtained for multivariate processes as well. The direct version of the multivariate q-CLT (Theorem 4.16) provides information about the attractor, the type of strong dependence (or, global correlation in physics literature) described by q-independence, and the scaling rate. More precisely, if the strong dependence is identified by the index q_k^d, then the attractor is a q_{k-1}^d-Gaussian, while the scaling rate is

$$\delta = \frac{1}{1 + d - dq_{k-1}^d}.$$

The latter follows from Theorem 4.16 similarly to the univariate case; compare exponents in β_k (last columns) in Tables 4.1 and 4.2. It follows from Lemma 3.25 that

$$\frac{1}{1 + d - dq_{k-1}^d} = q_{k+1}^d.$$

Thus, the triplet

$$(q_{k-1}^d, q_k^d, q_{k+1}^d)$$

describes a triple property (*attractor, strong dependence, scaling rate*) in the d-dimensional case.

In non-extensive statistical mechanics, another q-triplet (q_{sen}, q_{rel}, q_{stat}) has been introduced in Tsallis (2004a), where *sen, rel* and *stat* stand for *sensitivity to the initial conditions, relaxation,* and *stationary state*, respectively. This triplet describes important features of some complex systems, such as the fluctuating magnetic field of the plasma within the solar wind, as observed in the data sent by Voyager 1 (Burlaga and Vinas, 2005). It appears in fact that there is not just one or other triplet which is relevant, but the entire infinite countable family $\{q_k\}$, each member corresponding, in principle, to some physical quantity. The one-to-one association of each member of the family to a concrete physical quantity is still an open question; some knowledge is however available. Namely, as is noted in (Tsallis et al., 2005a):

> "...an interesting mathematical structure which might well be at the basis of the q-triplet conjectured in [Tsallis (2004a)] and recently confirmed [Burlaga and Vinas (2005)] with data received from the spacecraft Voyager 1 in the distant heliosphere. The q-triplet observed in the solar wind is given by $q_{sen} \approx -0.6 \pm 0.2$, $q_{rel} \approx 3.8 \pm 0.3$, and $q_{stat} \approx 1.75 \pm 0.06$ [Burlaga and Vinas (2005)]. These values are consistent with $q_{rel} + (1/q_{sen}) = 2$ and $q_{stat} + (1/q_{rel}) = 2$, hence $1 - q_{sen} = [1 - q_{stat}]/[3 - 2q_{stat}]$. Therefore,

we expect only one q of the triplet to be independent. The most precisely determined value in ref. [Burlaga and Vinas (2005)] is $q_{stat} = 1.75 = 7/4$. It immediately follows that $q_{sen} = -1/2$ (neatly consistent with -0.6 ± 0.2) and $q_{rel} = 4$ (again neatly consistent with 3.8 ± 0.3)."

4.8. q-Central Limit Theorem and the Scale Invariance

An important concept in statistical mechanics is the concept of exact or asymptotic *scale-invariance* (or *scale-freedom*). We define the notion of scale invariance as the case where the joint distribution (depending on Nd real variables) of a system made of N particles has (either exactly, or asymptotically for large N), as marginal distribution, the joint distribution (depending on $(N-1)d$ real variables) of a system made of $(N-1)$ particles (see details in Tsallis *et al.*, 2005a; Mendes and Tsallis, 2001). We address here the study of a similar problem, but related to the dimension d. More precisely, we focus on the scale invariance problem for a rescaled process of $X_1 + \cdots + X_N$ where $X_1, \ldots, X_N, \ldots$, is a sequence of identically distributed Q_k-independent (or (q_{k-1}, q_k)-independent) d-dimensional random vectors, which satisfy the conditions of Theorem 4.17 (Theorem 4.16, respectively). Assume $q_k \geq 1$ and $H_N(x_1, x_2, \ldots, x_d)$ is the density of

$$Z_N = [a_N, Y_N],$$

where

$$a_N = (a_1, \ldots, a_d), \; a_j = \left(\sqrt{q_{k+j-1} N} \sigma_j\right)^{-\frac{1}{2-q_{k+j-1}}},$$

(or

$$Z_N = D_N(q_k^d)(X_1 + \cdots + X_N - N\mu_{q_k^d})$$

with

$$D_N(q_k^d) = (N \nu_{2q_k^d - 1})^{-\frac{1}{2(2-q_k^d)}}).$$

Then, in accordance with Theorem 4.17 (Theorem 4.16),

$$H_N(x_1, \ldots, x_d) \to G_{d, q_{k-1}}(x_1, \ldots, x_d)$$

in the sense of weak convergence. It follows from Lemma 3.22 and Corollary 3.23 that

$$\lim_{N \to \infty} \int_{-\infty}^{\infty} H_N(x_1, x_2, \ldots, x_d) dx_d$$

$$= \int_{-\infty}^{\infty} G_{d, q_{k-1}}(x_1, x_2, \ldots, x_d) dx_d$$

$$= G_{d-1, q_k}(x_1, \ldots, x_{d-1}),$$

where $G_{d-1, q_k}(x_1, \ldots, x_{d-1})$ is a $(d-1)$-dimensional q_k-Gaussian. Hence, for large N, the probability density function of the system Z_N and its $(d-1)$-dimensional marginal density have the same form of q-generalized Gaussians with the indices q_{k-1} and q_k respectively. In other words $H_N(x_1, x_2, \ldots, x_d)$ is, in this specific sense, asymptotically scale invariant. This property enlightens the applicability of the present concepts and theorems to many natural and artificial systems, where q-Gaussians are observed.

4.9. *q*-Central Limit Theorem and Nonlinear Fokker–Planck Equation

The following one-dimensional nonlinear Fokker–Planck equation was studied in Plastino and Plastino (1995), Tsallis and Bukman (1996) in connection with anomalous diffusion due to correlation,

$$\frac{\partial p(t, x)}{\partial t} = -\frac{\partial}{\partial x}[F(x)p(t, x)] + D\frac{\partial^2 [p(t, x)]^\nu}{\partial x^2}, \quad t > 0, \ x \in \mathbb{R},$$

where $p = p(t, x)$ is the probability of the state of a diffusion process described by a position x at time t, the function $F(x)$ corresponds to a drift, D is a diffusion coefficient, and $\nu > -1$ is a real number describing correlations between the states of the underlying diffusion process. The case without drift, that is $F(x) = 0$, was considered in Section 6.2.2. We note that the correlation defined in Tsallis and Bukman (1996) by the parameter ν is different from q-independence of Types I–III we used in q-CLTs. The parameter ν is connected with the parameter Q (this Q and our q in the q-CLT are not the same. Therefore, to avoid a possible confusion, we use capital Q) through the equation $\nu = 2 - Q \neq 1$. The relevant stable solutions are, for simple forms of drift, Q-Gaussians with $Q \in (-\infty, 3)$, and $\delta = 2/(3 - Q) \in [0, \infty]$, hence both superdiffusion and subdiffusion can exist in addition to normal diffusion. In the particular case of Theorem 4.4, $k = 1$, we have

$$\delta = q_2 = 1/(2 - q). \tag{4.96}$$

This coincides with the nonlinear Fokker–Planck equation mentioned above. Indeed, in the theorem $(k = 1)$ we require the finiteness of $(2q - 1)$-variance. Denoting $2q - 1 = Q$, we get $\delta = 1/(2 - q) = 2/(3 - Q)$. It should be pointed out that the attractors in these two cases are different. Indeed, as it follows from Theorem 1 $(d = 1)$, the attractor is not a Q-Gaussian, but rather a q-Gaussian, with $q = (Q + 1)/2$. This discrepancy possibly is a result of somewhat different assumptions regarding the correlations. The multivariate versions of the q-CLT and relationship (4.96) allow us to conjecture that, for the *nonlinear Fokker–Planck equation with d-dimensional spatial variable* $x \in \mathbb{R}^d$, *corresponding to anomalous d-dimensional correlated diffusion processes, the scaling rate is*

$$\delta = q_2^d = \frac{2}{2 + d - dQ}.$$

We will discuss the nonlinear Fokker–Planck equation in greater detail in Chapter 6.

4.10. Stochastic Processes Based on q-Gaussians

In this section, we discuss q-generalizations of Brownian motion based on the q-Gaussian distribution. It should be noted that there are several different versions of q-Brownian motion in the modern mathematical and physics literature; see more in Section 4.11. Our interest is strictly restricted to the version based on the q-Gaussian random variable (vector) N_q, introduced and studied in Chapter 2. Attempts to construct such q-Brownian motions were done in papers Jiang *et al.* (2010) and Borland (1998) using different approaches, thus obtaining two different versions. Both versions can be used in modeling of random processes depending on the presence of various types of correlations.

The approach used in Jiang *et al.* (2010) is based on *q*-Gaussians for $q > 1$ in terms of variance mixtures of normals and leads to the definition of a process that might naturally be called *q*-Brownian motion. However, that name is already in use in other areas (see Section 4.11). Thus, the process to be constructed will be called a *q*-VM Brownian motion, where *VM* reflects the fact that it is a variance mixture of Brownian motions.

Definition 4.9. A stochastic process is called *exchangeable* if it is continuous in probability with $X_0 = 0$ and such that the increments over disjoint intervals of equal length form an exchangeable sequence.

Definition 4.10. A stochastic process X is called *conditionally independent and with stationary increments, given some σ-field*, if *both* properties of the increments are conditionally valid for any finite collection of disjoint intervals of the same length.

These two definitions are connected by the following theorem of Bühlmann which characterizes exchangeable processes on $\mathbb{R}_+$; see, e.g., Kallenberg (2002, Theorem 9.21).

Theorem 4.27. *Let the process $(X_t)_{t \geq 0}$ be $\mathbb{R}^d$-valued and continuous in probability with $X_0 = 0$. Then X is exchangeable if and only if it has conditionally independent and stationary increments given some σ-field.*

We can now define our *q*-VM Brownian motion.

Definition 4.11. By definition, a *q-VM Brownian motion*, B_t^q, with $1 \leq q < 3$, is a stochastic process having the following properties:

(1) conditionally independent and with stationary increments;
(2) all increments are *q*-Gaussian;
(3) continuous sample paths.

Theorem 4.28 (Jiang *et al.*, 2010). (Existence) *For each $1 \leq q < 3$, a q-VM Brownian motion exists.*

Proof. Let B_t be a standard Brownian motion. When $q = 1$, $B_t^1 = B_t$ is a process satisfying all three conditions with the conditional σ-field being the trivial σ-field. For $1 < q < 3$, let V_q be a random variable with the density f_q given in Theorem 4.24. Then $B_t^q = V_q \cdot B_t$ has conditionally stationary and independent increments given the value of V_q. All increments are q-Gaussian. Furthermore, a mixture of processes with continuous sample paths, has continuous sample paths.   □

The q-VM Brownian motion, by definition, has conditionally stationary and independent increments, each of which is q-Gaussian. This is expressed in the fact that its transition density is expressed in the form

$$P(t, x \mid y) = \int_0^\infty p(t, x \mid y, v) f_{V_q}(v) dv,$$

where $p(t, x \mid y, v)$ for each fixed $v > 0$ satisfies the linear Fokker–Planck equation

$$\frac{d}{dt} p(t, x \mid y, v) = \frac{1}{2} v^2 \frac{\partial^2 p(t, x \mid y, v)}{\partial x^2}, \quad t > 0, \ x, y \in \mathbb{R},$$

with the initial condition $p(0, x \mid y, v) = \delta_y(x)$, the Dirac delta function concentrated at y, and the occurrence of the transition probability for a particular v is weighted according to the density function $f_{V_q}(v)$, $v > 0$, defined in (4.92).

4.11. Additional Notes

(1) On q-Central Limit Theorems

Non-extensive statistical mechanics deals with strongly dependent random variables. That is the strength of dependencies do not rapidly decrease with increasing "distance" (in both space and time) between random variables localized or moving in some geometrical lattice (or continuous space) on which a "distance" can be defined. We used the term *q-independence* for such strong dependencies.

This type of dependencies in physics literature is referred to as *global correlations* (see Tsallis *et al.*, 2005a for more details). In the case of strong dependencies, imposed by the q-independence, the limiting distribution is no longer a Gaussian distribution. Instead, the limiting distribution in this case is a q-Gaussian.

The q-central limit theorems established in Sections 4.3 and 4.4 of this chapter were first published in Umarov and Tsallis (2007) and Umarov *et al.* (2008), and the uniqueness of the limiting distribution in Umarov and Tsallis (2016). These theorems are consistent with non-extensive statistical mechanics. Strong dependencies are represented through the three forms of the q-independence. Obviously, q-CLTs remains valid if q-independence conditions hold asymptotically for large n. The q-independence conditions, represented analytically with the help of q-algebra, may not be simply expressed analytically if we rely on classic algebra. However, Vignat–Plastino's q-CLT and the CLT for exchangeable random variables, discussed in Sections 4.6.1 and 4.6.2, shed light on this issue. Namely, strong dependencies, imposed through variance mixtures of random variable (vectors) and leading to q-Gaussian attractors, are asymptotically equivalent to q-independence.

(2) Connection of q-CLT with non-extensive statistical mechanics and non-additive entropic forms

If we optimize the non-additive entropy S_q by appropriately constraining the mean value of the random variable, we obtain q-exponential distributions. If we similarly constrain the variance, we obtain q-Gaussians instead. These q-Gaussian forms are plethorically found in natural, artificial and social complex systems, which strongly suggests a mathematical grounding in terms of a CLT, as it happens with the Boltzmann–Gibbs entropy for usual systems. In the realm of non-extensive statistical mechanics, this grounding relies on the q-CLT. This theorem, and similar ones, constitutes therefore a pillar of the present generalization of standard statistical mechanics.

(3) Examples of models with exchangeable random variables

As we have seen in Section 4.6, exchangeable random variables can play an important role in describing q-Gaussian limits. Here are some examples of models which use exchangeable random variables. The model proposed in Marsh *et al.* (2006), considers N identical and distinguishable, but not necessarily independent binary subsystems. Let $r_{N,n}$ be the probability that there are n subsystems in state 1, which is given by the Leibnitz rule:

$$r_{N,n} + r_{N,n+1} = r_{N-1,n}.$$

Since construction of the model only considers the number of subsystems in state 1, the order of 1's and 0's does not matter, a typical property of exchangeable sequences. It is shown in Hilhorst and Schehr (2007) that the weak limit of this sequence *is not a q-Gaussian*. However, using the same Leibnitz rule, Rodriguez *et al.* (2008) and Hanel *et al.* (2009) constructed exchangeable models that *do have a q-Gaussian limit*.

Another example is Beck–Cohen's superstatistics (Beck and Cohen, 2003), which can be interpreted as a variance mixture of normals. As a test particle moves from cell to cell, its velocity $v(t)$ (changing in time t) is usually modeled through the Langevin stochastic differential equation

$$dv = -\gamma v dt + \sigma dB_t, \quad t > 0,$$

where B_t is a standard Brownian motion. Instead of γ and σ being deterministic, as in classical statistical physics, Beck and Cohen let $\beta = \gamma/\sigma^2$ be a random variable with density $f(\beta)$. In this setting, the widely used classical Boltzman factor $e^{-\beta E}$ takes a generalized form called superstatistics:

$$B(E) = \int_0^\infty k(\beta)e^{-\beta E}d\beta, \tag{4.97}$$

where E represents the energy of a microstate associated with each cell. Notice that equation (4.97) is exactly a variance mixture of normals after standardization. The special case where $k(\beta)$ (the density of $1/V_q^2$ in Theorem 4.24) is the density of a χ^2-distribution yields q-statistics (see Wilk and Wlodarczyk, 2000; Beck, 2001).

(4) On q-Brownian motion constructions

In section 4.10, we constructed the q-Gaussian stochastic process called q-VM Brownian motion. This approach was discussed in Jiang *et al.* (2010). Another approach, used in Borland (1998), for construction of q-Gaussian stochastic processes is based on a stochastic differential equation with a special correlated driving process. Namely, Borland defines a q-Gaussian process Y_t as the log returns of stock prices that follow the stochastic differential equation

$$dY_t = \mu dt + \sigma d\Omega_t, \quad t > 0,$$

where Ω_t evolves according to

$$d\Omega_t = P(\Omega_t)^{(1-q)/2} dB_t.$$

The evolution of the probability distribution P of Y_t is given by the nonlinear Fokker–Planck equation ($\mu = 0, \sigma = 1$)

$$\frac{\partial}{\partial t} P(x,t|y) = \frac{1}{2} \frac{\partial^2}{\partial x^2} \left[P^{2-q}(x,t|y) \right], \quad P(x,0|y) = \delta(x-y),$$

where $\delta(\cdot)$ is the Dirac distribution. Solutions for $P(x,t|y)$ are given by q-Gaussians for each fixed t, as established in Borland (1998) or using the q-Fourier transform as in Plastino and Plastino (1995), Tsallis and Bukman (1996) and Borland (1998), or using the q-Fourier transform in Umarov and Queiros (2010).

The Borland process clearly differs from the q-VM Brownian motion, even if both stochastic processes are non-Markovian, have continuous paths and q-Gaussian marginals. As noted above, the q-VM Brownian motion has conditionally stationary and independent increments. However, even though the Borland process also has stationary increments (see Chapter 6), they are not conditionally independent. We will discuss the Borland process in more detail in Chapter 6.

Another concept in the realm of q-Gaussian processes is based on randomized non-commutative C^*-spaces of operators. We do not go into details of this approach, since it requires a heavy background from the modern non-commutative operator algebra. The interested reader can examine it in Bozejko and Speicher (1991, 1992), Bozejko *et al.* (1997), Van Leeuwen and Maassen (1998) and Blitvić (2012). We only note that all the above mentioned concepts of q-Gaussian processes have wide applications in modern sciences and engineering.

Chapter 5

(q, α)-Stable Distributions

5.1. Introduction

The q-central limit theorem, proved in the previous chapter, grounds the ubiquity of q-Gaussian distributions. Chapter 6 discusses emergence of q-Gaussians in various processes as an attractor. The q-central limit theorem holds for q-independent (or "globally correlated" in physics terminology) random variables with finite $(2q-1)$-variance. What is the attracting variable if the latter condition, that is finiteness of the $(2q-1)$-variance fails for q-independent identically distributed random variables? The current chapter discusses this question for q-generalized symmetric α-stable distributions, hereafter referred to as (q, α)-stable distributions.

In Section 5.2, we briefly provide basic facts related to α-stable distributions. If $0 < \alpha < 2$, then α-stable distributions possess an asymptotic power-law decay at infinity. As we have seen, q-Gaussians also possess the power-law decay property. However, unlike the q-Gaussians, α-stable distributions, in general, do not have closed explicit density representations.

Distributions with asymptotic power-law decay found a huge number of applications in various practical studies (see, e.g., Montroll and Shlesinger, 1982; Mittnik and Rachev, 1993; Beck and Schloegel, 1993; McCulloch, 1996; Mandelbrot, 1997; Metzler and Klafter, 2000; Caruso *et al.*, 2007; Tsallis, 2009a; Umarov *et al.*, 2018, to mention but a few), confirming the frequent nature of these distributions. As it will become clear later on, (q, α)-stable distributions generalize both α-stables and q-Gaussians. Indeed, $(1, \alpha)$-stable distributions correspond to the α-stables, and the $(q, 2)$-stable distributions correspond to the q-Gaussians. All (q, α)-stable distributions, except Gaussians $((1, 2)$-distributions), exhibit asymptotic power-laws. In practice, the researcher is often interested in the identification of a correct attractor of correlated states, which plays a major role in the adequate modeling of the physical phenomenon itself. This motivates the study of sequences of (q, α)-stable distributions and their attractors, as the current chapter focuses on.

In Section 5.3, we introduce (q, α)-stable random variables. It should be noted that there is an essential difference between (q, α)-stability of random variables and the classic stability of α-stables. Namely, (q, α)-stability holds for q-independent random variables, while the standard stability holds for independent random variables. As noted, the q-independence exhibits a special long-range correlation between random variables. In practice this notion reflects physical states (arising, e.g., in nonextensive statistical mechanics), which are strongly correlated. Examples of such systems include earthquakes (Caruso *et al.*, 2007), cold atoms in optical dissipative lattices (Duglas *et al.*, 2006), and dusty plasma (Liu and Goree, 2008), to mention but a few. A modeling of nonextensive processes with strong correlation into independent states can not adequately reflect their evolution. Likewise, (q, α)-stable distributions can not be captured by the existing theory of α-stable distributions, which is heavily based on the concept of independence.

In Section 5.4, we prove the central limit theorem for symmetric (q, α)-stable distributions, generalizing existing central limit theorem for stable distributions. For simplicity we will consider only symmetric univariate (q, α)-stable distributions (see Additional Notes for multivariate distributions). Section 5.5 studies q-weak limits of scaled sums of q-independent identically distributed (q, α)-stable random variables. The central limit theorem for (q, α)-stable random variables and the limit theorem for their scaled sums provide two different, but complimentary classifications of (q, α)-stable random variables. These classifications are discussed in Section 5.8. Finally, Section 5.9 contains some historical notes, open problems, and other facts related to stable distributions, and in particular, (q, α)-stable distributions.

5.2. Lévy's α-Stable Distributions

X and Y are said to be random variables of the same type if $Y = cX + d$ for some real c and d. In words this means that the same type of random variables can be obtained one from the other by scaling and shifting.

Definition 5.1. A random variable X is called *stable* if for independent copies X_1, X_2 of X the random variables $a_1 X_1 + a_2 X_2$ and X are of the same type for any $a_1, a_2 \in \mathbb{R}$, i.e.,

$$a_1 X_1 + a_2 X_2 = cX + d. \tag{5.1}$$

A random variable X is called *strictly stable* if $d = 0$ in equation (5.1), i.e., $a_1 X_1 + a_2 X_2 = cX$.

Stable random variables (or distributions) can be conveniently described with the help of their characteristic functions. For the complete description of stable distributions four parameters are used:

- a stability index $\alpha \in (0, 2]$,
- a skewness parameter $\beta \in [-1, 1]$,
- a scale parameter $\sigma > 0$, and
- a location parameter $\mu \in \mathbb{R}$.

Among these parameters, α is the most important one, determining a peculiar class of stability. To emphasize this, the term α-stable distribution is used for distributions with the stability index α. We denote the class of α-stable distributions by $S_\alpha(\beta, \sigma, \mu)$. In literature there are different parameterizations used for the description of stable distributions; see, e.g., Samorodnitsky and Taqqu (1994), Meerschaert and Scheffler (2001) and Nolan (2002). For us it is convenient to adopt the parameterization used in Samorodnitsky and Taqqu (1994). Namely, the characteristic function $\varphi_X(\xi) = \mathbb{E}(e^{i\xi X})$

of α-stable distribution X has the form

$$\varphi_X(\xi) = \begin{cases} \exp\left(i\xi\mu - \sigma^\alpha|\xi|^\alpha\left(1 - i\beta\mathrm{sign}(\xi)\tan\frac{\alpha\pi}{2}\right)\right), & \text{if } \alpha \neq 1, \\ \exp\left(i\xi\mu - \sigma|\xi|\left(1 + i\beta\mathrm{sign}(\xi)\frac{2}{\pi}\ln|\xi|\right)\right), & \text{if } \alpha = 1, \end{cases}$$
(5.2)

where

$$\mathrm{sign}(\xi) = \begin{cases} 1, & \text{if } \xi \geq 0, \\ -1, & \text{if } \xi < 0. \end{cases}$$

If $\alpha = 2$, then $\varphi_X(\xi) = \exp(i\xi\mu - \sigma^2\xi^2)$, which is the characteristic function of the normal distribution with mean μ and variance $2\sigma^2$. The density function of the latter is

$$f(x) = \frac{1}{2\sigma\sqrt{\pi}}e^{-\frac{(x-\mu)^2}{4\sigma^2}}, \quad x \in \mathbb{R}.$$
(5.3)

If $\alpha < 2$, then the density $f_X(x)$ of α-stable distribution X has power law decay at infinity, and (Samorodnitsky and Taqqu, 1994)

$$f_X(x) = O(|x|^{-1-\alpha}), \quad |x| \to \infty.$$
(5.4)

In the case $1 < \alpha < 2$ the variance does not exist, but the mean exists and equals μ. The support of X in the symmetric case ($\beta = 0$) is the whole real axis $\mathbb{R}$. In the case $0 < \alpha \leq 1$ both the variance and mean do not exist. In this case if $\beta = 1$, then the support is the interval $[\mu, \infty)$; if $\beta = -1$, then the support is $(-\infty, -\mu]$. The case $0 < \alpha < 1$, $\beta = 1, \mu = 0$, provides important stable subordinators widely used in the study of anomalous diffusion processes.

As seen from (5.2), if $\beta = 0$, then skewness disappears making the α-stable distribution symmetric about μ. Note that the probability density function of α-stable distribution has a closed form only in three particular cases:

- $\alpha = 2$ (normal distribution), given by (5.3);
- $\alpha = 1$ (Cauchy distribution), given by

$$f(x) = \frac{1}{\pi}\frac{\sigma}{(x-\mu)^2 + \sigma^2};$$

and

- $\alpha = 1/2, \beta = 1$ (Lévy distribution), given by

$$f(x) = \sqrt{\frac{\sigma}{2\pi}} \frac{e^{-\frac{\sigma}{2(x-\mu)}}}{(x-\mu)^{3/2}}, \quad x > \mu.$$

It follows from (5.4) and Proposition 2.28 that when $\alpha < 2$ the α-stable distribution is asymptotically equivalent to the q-Gaussian with

$$q = \frac{\alpha + 3}{\alpha + 1}, \quad 0 < \alpha < 2.$$

Notice that when α changes in the interval $0 < \alpha < 2$, the parameter q changes in the interval $5/3 < q < 3$. The significant difference between q-Gaussians and α-stables is that q-Gaussian densities have closed forms for all q, while α-stable densities have closed form only in the three particular cases mentioned above.

The central limit theorem for α-stable random variables is formulated as follows; see Nolan (2002).

Theorem 5.1. *A non-degenerate random variable Z is α-stable for some $0 < \alpha \leq 2$ if and only if there is an independent, identically distributed sequence of random variables $X_1, X_2, \ldots$, and constants $a_n > 0$, $b_n \in \mathbb{R}$, such that*

$$a_n(X_1 + \cdots + X_n) - b_n \to Z, \quad \text{in distribution.} \tag{5.5}$$

Remark 5.1. Theorem 5.1 is about description of α-stable distributions. Namely, if Z is an α-stable distribution, then one can construct a sequence of independent random variables X_n, and sequences $a_n > 0, b_n$, such that the limit (5.5) holds. On the other hand if there is a sequence of random variables X_n and numbers $a_n > 0, b_n$, such that the limit (5.5) holds, then Z is an α-stable random variable. Similar description is valid in the case of q-generalized α-stable distributions; see Theorem 5.4.

5.3. Generalized (q, α)-Stable Random Variables for $q > 1$

Replacing the exponential function in (5.2) by the q-exponential we can introduce a q-version of α-stable distributions.

Definition 5.2. Let $1 \leq q < 2$ and $0 < \alpha \leq 2$. A random variable X is called (q, α)-stable, if its q-characteristic function $\varphi_{q,X}(\xi) = F_q[f_X](\xi)$ has the representation

$$F_q[f_X](\xi) = \begin{cases} \exp_q\left(i\xi\mu - \sigma^\alpha|\xi|^\alpha\left(1 - i\beta\text{sign}(\xi)\tan\frac{\alpha\pi}{2}\right)\right), & \text{if } \alpha \neq 1, \\ \exp_q\left(i\xi\mu - \sigma|\xi|\left(1 + i\beta\text{sign}(\xi)\frac{2}{\pi}\ln|\xi|\right)\right), & \text{if } \alpha = 1. \end{cases}$$
$$(5.6)$$

The meaning of (q, α)-stability differs from the α-stability introduced in Definition 5.1. Namely, let X_1 and X_2 be two q-independent (of Type I) copies of an (q, α)-stable random variable X. That is X_1 and X_2 satisfy the condition

$$F_q[f_{X_1+X_2}](\xi) = F_q[f_X](\xi) \otimes_q F_q[f_X](\xi), \quad \xi \in \mathbb{R}.$$

Then, the characteristic function of $a_1 X_1 + a_2 X_2$ for any positive real numbers a_1, a_2 is also of the form (5.6), with the same α. Indeed, applying the q-independence and the equality (see (4.18)),

$$F_q[f_{aX}](\xi) = F_q[f_X](a^{2-q}\xi),$$

we have

$$F_q[f_{a_1 X_1 + a_2 X_2}](\xi)$$

$$= F_q[f_{a_1 X_1}](\xi) \otimes_q F_q[f_{a_2 X_2}](\xi)$$

$$= F_q[f_X](a_1^{2-q}\xi) \otimes_q F_q[f_X](a_2^{2-q}\xi)$$

$$= \exp_q\left(i\xi a_1^{2-q}\mu - a_1^{(2-q)\alpha}\sigma^\alpha|\xi|^\alpha\left(1 - i\beta\text{sign}(\xi)\tan\frac{\alpha\pi}{2}\right)\right)$$

$$\otimes_q \exp_q\left(i\xi a_2^{2-q}\mu - a_2^{(2-q)\alpha}\sigma^\alpha|\xi|^\alpha\left(1 - i\beta\text{sign}(\xi)\tan\frac{\alpha\pi}{2}\right)\right)$$

$$= \exp_q\left(i\xi\bar{\mu} - \bar{\sigma}^\alpha|\xi|^\alpha\left(1 - i\beta\text{sign}(\xi)\tan\frac{\alpha\pi}{2}\right)\right),$$

where

$$\bar{\mu} = (a_1^{2-q} + a_2^{2-q})\mu, \quad \bar{\sigma}^\alpha = (a_1^{(2-q)\alpha} + a_2^{(2-q)\alpha})\sigma^\alpha.$$

Thus, the meaning of the (q, α)-stability is that the random variable $a_1 X_1 + a_2 X_2$ with two q-independent copies of X is again of the same type as X. By induction, one can extend this to any finite number of q-independent copies of (q, α)-stable random variables.

Remark 5.2. Similarly, one can define (q, α)-stable distributions adapted to q-independence of Types II and III. In the next section we will establish a q-generalization of the central limit theorem for α-stables (Theorem 5.1). For simplicity we will prove the theorem only for symmetric (q, α)-stables.

Definition 5.3. Let $1 \leq q < 2$ and $0 < \alpha \leq 2$. A random variable X is said to have a symmetric (q, α)-stable distribution if its q-Fourier transform is represented in the form

$$F_q[f_X](\xi) = e_q^{-\beta|\xi|^\alpha}, \quad \xi \in \mathbb{R}, \tag{5.7}$$

with $\beta > 0$. We denote the set of random variables with (q, α)-stable distributions by $\mathcal{L}_q[\alpha]$.

Introduce the set of functions

$$\mathcal{G}_q[\alpha] = \{b\, e_q^{-\beta|\xi|^\alpha}, \ b > 0, \ \beta > 0\}. \tag{5.8}$$

Due to Definition 5.3, a random variable $X \in \mathcal{L}_q[\alpha]$ if its q-Fourier transform $F_q[f_X](\xi) \in \mathcal{G}_q[\alpha]$ with $b = 1$. Note that if $\alpha = 2$, then $\mathcal{G}_q[2]$ represents the set of q-Gaussians and $\mathcal{L}_q[2]$ is the set of random variables whose densities are q_*-Gaussians, where $q_* = (3q - 1)/(1 + q)$.

In Chapter 3, we introduced the set of functions $H_{q,\alpha}(\mathbb{R})$; see (3.61). Denote by $SH_{q,\alpha}$ the set of symmetric density functions in $H_{q,\alpha}(\mathbb{R})$, that is

$$SH_{q,\alpha} = \{f(x) \in H_{q,\alpha}(\mathbb{R}) : f(x) \geq 0, \ f(-x) = f(x),$$

$$\text{and} \int_{\mathbb{R}} f(x)dx = 1\}. \tag{5.9}$$

Further, it follows from the asymptotic relation (2.39) that

$$e_q^{-\beta|\xi|^\alpha} = 1 - \beta|\xi|^\alpha + o(|\xi|^\alpha). \tag{5.10}$$

Proposition 5.1. *Let X be a random variable with a density function f_X. Then*

(1) *if $f_X \in H_{q,\alpha}$, then $X \in \mathcal{L}_q[\alpha]$,*
(2) *if $X \in \mathcal{L}_q[\alpha]$ and its support is $\mathbb{R}$, then $f_X \in H_{q,\alpha}$.*

Proof. The proof of (1) immediately follows from Proposition 3.15 and (5.10). The proof of (2) is an implication of the fact that, due to Hilhorst's invariance principle, densities supported on $\mathbb{R}$ and having the same Fourier transform are asymptotically equivalent; see Proposition 4.7.   □

5.4. Central Limit Theorem for (q, α)-Stable Random Variables

In this section, we establish a q-generalization of the central limit theorem for stable random variables. Let $Q = 2q - 1$. We notice that if $1 \le q < 2$, then $1 \le Q < 3$. We need the following notations:

$$\mathcal{Q}_1 = \{(Q, \alpha) : 1 \le Q < 3, \ \alpha = 2\},$$

$$\mathcal{Q}_2 = \{(Q, \alpha) : 1 \le Q < 3, \ 0 < \alpha < 2\},$$

$$\mathcal{Q} = \mathcal{Q}_1 \cup \mathcal{Q}_2.$$

Proposition 5.2. *Let q-independent random variables $X_j \in \mathcal{L}_q[\alpha], j = 1, \ldots, m$. Then for constants $a_1, \ldots, a_m$,*

$$\sum_{j=1}^m a_j X_j \in \mathcal{L}_q[\alpha].$$

Proof. Let

$$F_q[X_j](\xi) = e_q^{-\beta_j}|\xi|^\alpha, \quad j = 1, \ldots, m.$$

Using the properties $e_q^x \otimes_q e_q^y = e_q^{x+y}$ and $F_q[aX](\xi) = F_q[X](a^{2-q}\xi)$, it follows from the definition of q-independence that

$$F_q\left[\sum_{j=1}^m a_j X_j\right] = e_q^{-\beta|\xi|^\alpha}, \quad \beta = \sum_{j=1}^m \beta_j |a|^{\alpha(2-q)} > 0.$$

□

Remark 5.3. Proposition 5.2 implies that a random variable $X \in \mathcal{L}_q[\alpha]$ and $\sum_{j=1}^m a_j X_j$ with m q-independent copies X_j, $j = 1, \dots, m$, of X are the same type, justifying the stability of distributions in $\mathcal{L}_q[\alpha]$. Recall that if $q = 1$ then q-independent random variables are independent in the usual sense. Thus, if $q = 1$, $0 < \alpha < 2$, then $\mathcal{L}_1[\alpha] \equiv \mathcal{L}_{\text{sym}}[\alpha]$, where $\mathcal{L}_{\text{sym}}[\alpha]$ is the set of symmetric α-stable Lévy distributions.

Further, the appropriately scaling limit of sequences of q-independent random variables with (q, α)-stable distributions has again a (q, α)-stable distribution. To show this consider the sum

$$Z_N = \frac{1}{s_N(q, \alpha)}(X_1 + \cdots + X_N), \quad N = 1, 2, \dots \qquad (5.11)$$

where $s_N(q, \alpha)$ is the scaling coefficient specified below. First we prove a preliminary result.

Theorem 5.2. *Assume* $(2q - 1, \alpha) \in \mathcal{Q}_2$. *Let* X_j, $j = 1, 2, \dots$, *be symmetric q-independent random variables all having the same probability density function* $f(x) \in H_{q,\alpha}$. *Then* Z_N *defined in* (5.11) *with* $s_N(q, \alpha) = N^{\frac{1}{\alpha(2-q)}}$ *is q-convergent as* $N \to \infty$ *to a (q, α)-stable distribution.*

Proof. Assume $(Q, \alpha) \in \mathcal{Q}_2$. Let f be the density associated with X_1. First, we evaluate $F_q[X_1] = F_q[f(x)]$. Using asymptotic relation (2.39) we have

$$F_q[f](\xi) = 1 - \mu_{q,\alpha}|\xi|^\alpha + o(|\xi|^\alpha), \xi \to 0. \qquad (5.12)$$

Denote $Y_j = N^{-\frac{1}{\alpha}} X_j, j = 1, 2, \dots$. Then $Z_N = Y_1 + \cdots + Y_N$. Further, it is readily seen that for a given random variable X and real $a > 0$, the equality $F_q[aX](\xi) = F_q[X](a^{2-q}\xi)$ holds. It follows from this relation that $F_q[Y_j] = F_q[f](\frac{\xi}{N^{1/\alpha}})$, $j = 1, 2, \ldots$ Moreover, it follows from the q-independence of $X_1, X_2, \ldots$, and the associativity of the q-product that

$$F_q[Z_N](\xi) = F_q[f]\Big(\frac{\xi}{N^{\frac{1}{\alpha}}}\Big) \underbrace{\otimes_q \cdots \otimes_q}_{N \text{ factors}} F_q[f]\Big(\frac{\xi}{N^{\frac{1}{\alpha}}}\Big). \qquad (5.13)$$

Further, making use of asymptotic relation for the q-logarithm obtained in Corollary 2.7, equation (5.13) implies

$$\ln_q F_q[Z_N](\xi) = N \ln_q F_q[f]\Big(\frac{\xi}{N^{-\frac{1}{\alpha}}}\Big)$$

$$= N \ln_q \Big(1 - \frac{\mu_{q,\alpha}|\xi|^\alpha}{N} + o\Big(\frac{|\xi|^\alpha}{N}\Big)\Big)$$

$$= -\mu_{q,\alpha}|\xi|^\alpha + o(1), \quad N \to \infty, \qquad (5.14)$$

locally uniformly with respect to ξ. Hence,

$$\lim_{N\to\infty} F_q[Z_N] = e_q^{-\mu_{q,\alpha}|\xi|^\alpha} \in \mathcal{G}_q[\alpha]. \qquad (5.15)$$

Thus, Z_N is q-convergent to a random variable Z with (q, α)-stable distribution, as $N \to \infty$.   □

Since the density of $X \in \mathcal{L}_q[\alpha]$ is in $H_q[\alpha]$ it follows immediately the following corollary from Theorem 5.2.

Corollary 5.3. *Assume $(2q - 1, \alpha) \in \mathcal{Q}_2$. Let X_N be a sequence of symmetric q-independent (q, α)-stable random variables. Then Z_N, with the same $s_N(q, \alpha)$ in Theorem 5.2, q-weakly converges to a (q, α)-stable distribution.*

Note that $\alpha = 2$ is not included in $\mathcal{Q}_2$ in Theorem 5.2. The case $\alpha = 2$ corresponds to Theorem 4.5. Recall that in this case $\mathcal{L}_q[2]$ consists of random variables whose densities are in $\mathcal{G}_{q^*}[2]$, where $q^* = \frac{3q-1}{q+1}$.

Theorem 5.4 (Central limit theorem for (q, α)-stables). *A nondegenerate random variable Z belongs to $\mathcal{L}_q[\alpha]$ for some $(q, \alpha) \in \mathcal{Q}$ if and only if there is a q-independent, identically distributed* *sequence of random variables* $X_1, X_2, \ldots$, *and constants* $a_n > 0$ *such that*

$$a_n(X_1 + \cdots + X_n) \to Z, \qquad (5.16)$$

in the sense of q-convergence.

Proof. Let $Z \in \mathcal{L}_q[\alpha]$ for some $(q, \alpha) \in \mathcal{Q}$. If $(q, \alpha) \in \mathcal{Q}_1$, then due to the q-CLT (Theorem 4.5), there exists a sequence of q-independent identically distributed random variables X_j, $j = 1, 2, \ldots$, with finite Q-variance and scaling $\kappa_n(q)$, such that $\kappa_n(q)(X_1 + \cdots + X_n)$ q-converges to a q-Gaussian, which belongs to $\mathcal{L}_q[2]$. If $(Q, \alpha) \in \mathcal{Q}_2$, then the result follows from Theorem (5.5), if one takes as X_j, $j = 1, 2, \ldots$, q-independent copies of Z and as a_n the numbers $n^{-1/\alpha(2-q)}$. Thus, the "if" part of the theorem is valid for all $(q, \alpha) \in \mathcal{Q}$.

Conversely, let Z be a non-degenerate random variable and there exist a q-independent identically distributed random variables X_n and a sequence a_n, such that

$$S_n = a_n(X_1 + \cdots + X_n) \to Z$$

in the sense of q-convergence, that is

$$\lim_{n \to \infty} F_q[f_{S_n}](\xi) = F_q[f_Z](\xi), \quad \forall\, \xi \in \mathbb{R}.$$

We show that Z is (q, α)-stable. Indeed, let $S_n^{(1)}, \ldots, S_n^{(m)}$ be m q-independent copies of S_n. Then

$$\lim_{n \to \infty} F_q[f_{S_n^{(1)} + \cdots + S_n^{(m)}}](\xi) = \lim_{n \to \infty} F_q[f_{S_n^{(1)}}](\xi) \underbrace{\otimes_q \cdots \otimes_q}_{m\,\text{factors}} F_q[f_{S_n^{(m)}}](\xi)$$

$$= F_q[f_{Z^{(1)}}](\xi) \underbrace{\otimes_q \cdots \otimes_q}_{m\,\text{factors}} F_q[f_{Z^{(m)}}](\xi)$$

$$= F_q[f_{Z^{(1)} + \cdots + Z^{(m)}}](\xi),$$

where $Z^{(1)}, \ldots, Z^{(m)}$ are m q-independent copies of Z. This implies

$$S_n^{(1)} + \cdots + S_n^{(m)} \to Z^{(1)} + \cdots + Z^{(m)}, \quad n \to \infty,$$

in the sense of q-convergence. Further, the sum $S_n^{(1)} + \cdots + S_n^{(m)}$ can be written as $\frac{a_n}{a_{nm}} S_{nm}$. This sequence, by assumption, is q-convergent to the random variable $c_m Z$ as $n \to \infty$, where c_m is the limit of a_n/a_{nm} as $n \to \infty$. Hence, we have

$$c_m Z = Z^{(1)} + \cdots + Z^{(m)},$$

or

$$Z = c_m^{-1}\left(Z^{(1)} + \cdots + Z^{(m)}\right).$$

The latter means that Z is (q, α)-stable.   □

Theorem 5.4 allows to establish a connection between the classic Lévy distributions and q_α^L-Gaussians (here L indicates the correspondence to Lévy stable distributions). Indeed, in accordance with Proposition 5.1, the density function of (q, α)-stable random variable X has the asymptote

$$f_X(x) \sim C|x|^{-\frac{1+\alpha}{1+\alpha(q-1)}}, \quad |x| \to \infty, \tag{5.17}$$

where C is a positive constant. Now comparing the latter with (5.4), we notice that the density of a (q, α)-stable random variable shares the same asymptotic behavior as the density of the symmetric stable random variable centered at 0 and with the stability index

$$\gamma = \gamma(q, \alpha) = \frac{\alpha(2 - q)}{1 + \alpha(q - 1)}.$$

However, as noted above, there is an essential difference between (q, α)-stability of q-independent random variables and the classic stability of α-stable distributions. In modeling, a decomposition of processes with strong correlation into independent states can not adequately reflect their evolution. Likewise, (q, α)-stable distributions can not be described by the existing theory of α-stable distributions, which is heavily based on the concept of independence, unless $q = 1$. If $q = 1$, then correlation disappears, that is q-independence becomes usual probabilistic independence, and $\gamma(1, \alpha) = \alpha$, implying $\mathcal{L}_1[\alpha] = \mathcal{L}_{\text{sym}}[\alpha]$, where $\mathcal{L}_{\text{sym}}[\alpha]$ stands for symmetric α-stable distributions (centered at 0).

It is not hard to verify that there exists a q_α^L-Gaussian, which is asymptotically equivalent to f_X (equation (5.17)). Let us now find q_α^L. Denote by η the decay rate of the q_α^L-Gaussian. Due to Proposition 2.28, any q_α^L-Gaussian behaves asymptotically like $C/|x|^{2/(q_\alpha^L-1)}$, i.e., $\eta = 2/(q_\alpha^L - 1)$. Hence, we obtain the relation

$$\frac{\alpha + 1}{1 + \alpha(q - 1)} = \frac{2}{q_\alpha^L - 1}. \tag{5.18}$$

Solving this equation with respect to q_α^L, we have

$$q_\alpha^L = \frac{3 + Q\alpha}{\alpha + 1}, \quad Q = 2q - 1, \tag{5.19}$$

linking three parameters: α, the parameter of the α-stable Lévy distributions, q, the parameter of correlation, and q_α^L, the parameter of attractors in terms of q_α^L-Gaussians. The surface

$$\{(q_\alpha^L, Q, \alpha) \in \mathbb{R}^3 : (q_\alpha^L, Q, \alpha) \text{ satisfy equation (5.19)}\}$$

is depicted in Fig. 5.1.

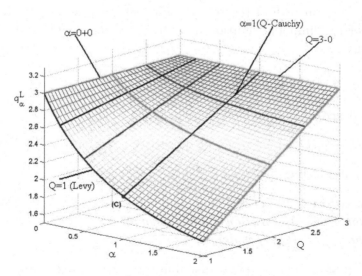

Fig. 5.1: $q_\alpha^L \equiv q^*$ as the function of (Q, α).

Equation (5.19) identifies all (Q, α)-stable distributions with the same index of attractor $G_{q_\alpha^L}$, proving the following proposition.

Proposition 5.3. *Let* $1 \leq Q < 3 \, (Q = 2q - 1)$, $0 < \alpha < 2$, *and*

$$\frac{3 + Q\alpha}{\alpha + 1} = q_\alpha^L, \tag{5.20}$$

Then the density of $X \in \mathcal{L}_q[\alpha]$ *is asymptotically equivalent to* q_α^L-*Gaussian.*

In the particular case $Q = 1$, we recover the known connection between the classical Lévy distributions $(q = Q = 1)$ and corresponding q_α^L-Gaussians. In fact, putting $Q = 1$ in equation (5.20), we obtain

$$q_\alpha^L = \frac{3 + \alpha}{1 + \alpha}, \quad 0 < \alpha < 2. \tag{5.21}$$

When α increases between 0 and 2 (i.e., $0 < \alpha < 2$), q_α^L decreases between 3 and 5/3 (i.e., $5/3 < q_\alpha^L < 3$); we remind that q-Gaussians with $q < 5/3$ have a finite variance, and therefore, the limiting distribution is a Gaussian (see Fig. 5.2).

It is useful to find the relationship between the decay rate $\eta = \frac{2}{q_\alpha^L - 1}$ of the limit distribution and (α, Q). Using formula (5.18), we obtain

$$\eta = \frac{2(\alpha + 1)}{2 + \alpha(Q - 1)}. \tag{5.22}$$

The surface

$$\{(\eta, Q, \alpha) \in \mathbb{R}^3 : (\eta, Q, \alpha) \text{ satisfy equation (5.22)}\}$$

is depicted in Fig. 5.3.

Proposition 5.4. *Let* $X \in \mathcal{L}_Q[\alpha]$, $1 \leq Q < 3$, $0 < \alpha < 2$. *Then the associated density function* f_X *has asymptotics* $f_X(x) \sim |x|^\eta$, $|x| \to \infty$, *where* $\eta = \eta(Q, \alpha)$ *is defined in* (5.22).

If $Q = 1$ (classic Lévy distributions), then (5.22) implies the well-known relation $\eta = \alpha + 1$.

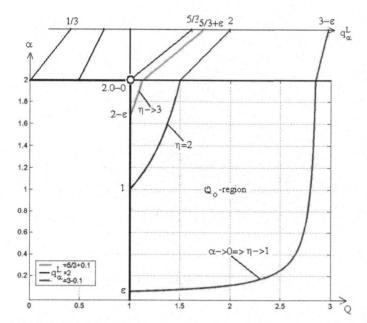

Fig. 5.2: All pairs of (Q, α) on the indicated curves are associated with the same q_α^L-Gaussian. Two curves corresponding to two different values of q_α^L do not intersect. In this sense these curves represent the constant levels of q_α^L or $\eta = 2/(q_\alpha^L - 1)$. The line $\eta = 1$ joins the points $(Q, \alpha) = (1, 0.0 - 0)$ and $(3 - 0, 2)$; the line $\eta = 2$ joins the Cauchy distribution (noted C) with itself at $(Q, \alpha) = (1, 1)$ and at $(2, 2)$; the $\eta = 3$ line joins the points $(Q, \alpha) = (1, 2.0 - 0)$ and $(5/3, 2)$ (by ϵ we simply mean to give an indication, and not that both infinitesimals coincide). The entire line at $Q = 1$ and $0 < \alpha < 2$ is mapped into the line at $\alpha = 2$ and $5/3 \leq q^* < 3$.

Analogous relationships can be obtained for other values of Q. We call, for convenience, a (Q, α)-stable distribution a Q-Cauchy distribution, if $\alpha = 1$. We obtain the classic Cauchy–Poisson–Lorentz distribution if $Q = 1$. The corresponding line can be obtained cutting the surface in Fig. 5.3 along the line $\alpha = 1$. For Q-Cauchy distributions (5.19) and (5.22) imply

$$q_1^L(Q) = \frac{3 + Q}{2} \quad \text{and} \quad \eta = \frac{4}{Q + 1}, \tag{5.23}$$

respectively (see Figs. 5.4 and 5.5).

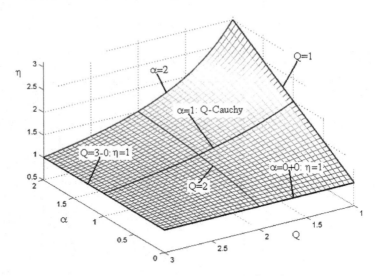

Fig. 5.3: η as the function of (Q, α).

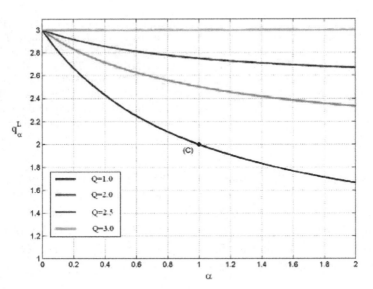

Fig. 5.4:   The relationship between α and q_α^L for typical fixed values of Q; cross-sections of the surface in Fig. 5.1.

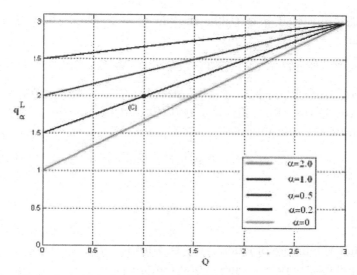

Fig. 5.5: The relationship between Q and q_α^L for typical fixed values of α; cross-sections of the surface in Fig. 5.1.

The relationship between α and q_α^L for typical fixed values of Q are given in Fig. 5.4. In this figure, we can also see that $\alpha = 1$ (Cauchy) corresponds to $q_1^L = 2$ (in the $Q = 1$ curve). In Fig. 5.5 the relationships between Q ($Q = 2q - 1$) and q_α^L are represented for typical fixed values of α.

5.5. Sums of (q, α)-Stable Random Variables and their Scaling Limits

In this section, we study sums of (q, α)-stable distributions and their scaling limits obtaing a wide generalization of the q-central limit theorem proved in Chapter 4.

Let $1 < q < 2$, $0 < \alpha \le 2$, and $f \in \mathcal{G}_q[\alpha]$, that is

$$f(x) = b e_q^{-\beta |x|^\alpha}, \quad x \in \mathbb{R},$$

with some $b > 0$, $\beta > 0$. It follows from the definition of the q-exponential function that for $f \in \mathcal{G}_q[\alpha]$ the asymptotic relation

$$f(x) \sim C_f |x|^{\frac{-\alpha}{q-1}}, \quad C_f > 0, \quad \text{as } |x| \to \infty \qquad (5.24)$$

holds. Similarly, if $g \in \mathcal{G}_q[2]$, that is $\alpha = 2$, then the asymptotic relation

$$g(x) \sim C_g \, |x|^{\frac{-2}{q-1}}, \quad C_g > 0, \text{ as } |x| \to \infty \qquad (5.25)$$

holds. Comparing orders of asymptotes in (5.24) and (5.25), we can easily verify that for a fixed $\alpha \in (0,2]$ and for any $q \in (1,2)$ there exists a one-to-one mapping

$$\mathcal{M}_{q,q^*} : \mathcal{G}_q[\alpha] \to \mathcal{G}_{q^*}[2], \qquad (5.26)$$

where

$$q^* = \frac{\alpha + 2(q-1)}{\alpha}, \qquad (5.27)$$

such that the image of a density $f \in \mathcal{G}_q[\alpha]$ is again a density in $\mathcal{G}_{q^*}[2]$. Analogously, there is a one-to-one mapping

$$\mathcal{K}_{q,q^*} : \mathcal{G}_q[\alpha] \to \mathcal{G}_{q^*}[2],$$

with the same q^*, such that it transfers a function $h(x) = e_q^{-\beta|x|^\alpha}$, an element of $\mathcal{G}_q[\alpha]$ with the coefficient $b = 1$ to the element $g(x) = e_{q^*}^{-\frac{\alpha\beta}{2}|x|^2}$ of $\mathcal{G}_{q^*}[2]$ with coefficient $b = 1$. We notice that if $\alpha = 2$, then $q^* = q$ and both operators coincide with the identity operator. As a one-to-one operator, $\mathcal{K}_{q,q^*}$ has the following inverse

$$\mathcal{K}_{q,q^*}^{-1} : e_{q^*}^{-\beta|x|^2} \to e_q^{-\frac{2\beta}{\alpha}|x|^\alpha}.$$

Definition 5.4. The operator $\mathbb{F}_q$ defined by

$$\mathbb{F}_q = \mathcal{K}_{q^*, u(q^*)}^{-1} F_{q^*} \mathcal{M}_{q,q^*}, \qquad (5.28)$$

where

$$u(q^*) = \frac{1 + q^*}{3 - q^*}, \qquad (5.29)$$

$$q_* = \frac{\alpha q + (q-1)}{\alpha + (q-1)}, \qquad (5.30)$$

is called a generalized q-Fourier transform. It is readily seen that if $\alpha = 2$, then $\mathbb{F}_q = F_q$.

Definition 5.4 schematically can be represented through the commutative diagram

$$
\begin{array}{ccc}
\mathcal{G}_q[\alpha] & \xrightarrow{\;\mathbb{F}_q\;} & \mathcal{G}_{q_*}[\alpha] \\
\mathcal{M}_{q,q^*} \downarrow & & \uparrow \mathcal{K}^{-1}_{q_*,u(q^*)} \\
\mathcal{G}_{q^*}[2] & \xrightarrow{\;\mathbb{F}_{q^*}\;} & \mathcal{G}_{u(q^*)}[2]
\end{array}
\tag{5.31}
$$

Proposition 5.5. *Assume* $1 \le q < 2$, $0 < \alpha \le 2$ *and the numbers* q^* *and* q_* *are defined as in* (5.27) *and* (5.30), *respectively. Then the mapping*

$$
\mathbb{F}_q : \mathcal{G}_q[\alpha] \to \mathcal{G}_{q_*}[\alpha] \tag{5.32}
$$

holds.

Proof. We use the commutative diagram (5.31) for the proof. Let a density $f \in \mathcal{G}_q[\alpha]$, i.e., asymptotically

$$
f(x) \sim C_f \, |x|^{-\alpha/(q-1)}, \quad x \to \infty,
$$

with some $C_f > 0$. Its image $\mathcal{M}_{q,q^*}[f](x)$, a q^*-Gaussian $G_{q^*}(\beta; x)$, in order to be asymptotically equivalent to f, necessarily

$$
G_{q^*}(\beta; x) \sim \frac{C_1}{|x|^{\frac{2}{q^*-1}}} \sim \frac{C_f}{|x|^{\frac{\alpha}{q-1}}}, \quad |x| \to \infty.
$$

Hence,

$$
q^* = \frac{\alpha + 2(q-1)}{\alpha} = 1 + \frac{2(q-1)}{\alpha}.
$$

Further, it follows from Umarov et al. (2008, Corollary 2.10), that

$$
\mathbb{F}_{q^*} : \mathcal{G}_{q^*}[2] \to \mathcal{G}_{u(q^*)}[2],
$$

where $u(q^*)$ is defined in equation (5.29). Now taking into account the asymptotic equality (the right vertical line in diagram (5.31))

$$
G_{q_1}(\beta_1; x) \sim \frac{C_2}{|x|^{\frac{2}{q_1-1}}} \sim \frac{C_3}{|x|^{\frac{\alpha}{q_*-1}}}, \quad |x| \to \infty,
$$

we obtain

$$q_* = \frac{\alpha q - (q - 1)}{\alpha - (q - 1)} = 1 + \frac{\alpha(q - 1)}{\alpha - (q - 1)}.$$

Thus, the mapping (5.32) holds with q^* and q_* defined in equations (5.27) and (5.30), respectively. □

Let us now introduce two functions that are important for our further analysis:

$$u_\alpha(s) = \frac{\alpha s - (s - 1)}{\alpha - (s - 1)} = 1 + \frac{\alpha(s - 1)}{\alpha - (s - 1)}, \tag{5.33}$$

where $0 < \alpha \le 2$, $s < \alpha + 1$, and

$$w_\alpha(s) = \frac{\alpha + 2(s - 1)}{\alpha} = 1 + \frac{2(s - 1)}{\alpha}, \quad 0 < \alpha \le 2. \tag{5.34}$$

It can be easily verified that $w_\alpha(s) = s$, if $\alpha = 2$.

The inverse, $u_\alpha^{-1}(t)$, $t \in (1 - \alpha, \infty)$, of the function in (5.33) reads

$$u_\alpha^{-1}(t) = \frac{\alpha t + (t - 1)}{\alpha + (t - 1)} = 1 + \frac{\alpha(t - 1)}{\alpha + (t - 1)}. \tag{5.35}$$

The function $u(s)$ possesses the following properties:

$$u_\alpha\left(\frac{1}{u_\alpha(s)}\right) = \frac{1}{s} \quad \text{and} \quad u_\alpha\left(\frac{1}{s}\right) = \frac{1}{u_\alpha^{-1}(s)}.$$

If we denote $q_{\alpha,1} = u_\alpha(q)$ and $q_{\alpha,-1} = u_\alpha^{-1}(q)$, then

$$u_\alpha\left(\frac{1}{q_{\alpha,1}}\right) = \frac{1}{q} \quad \text{and} \quad u_\alpha\left(\frac{1}{q}\right) = \frac{1}{q_{\alpha,-1}}. \tag{5.36}$$

Proposition 5.5 implies that for $0 < \alpha \le 2$ and $1 \le q < \min\{2, 1 + \alpha\}$ the following mappings hold:

(i) $\mathbb{F}_q : \mathcal{G}_q[\alpha] \to \mathcal{G}_{u_\alpha(q)}[\alpha]$,
(ii) $\mathbb{F}_q^{-1} : \mathcal{G}_{u_\alpha(q)}[\alpha] \to \mathcal{G}_q[\alpha]$,

where $\mathbb{F}_q^{-1}$ is the inverse to $\mathbb{F}_q$. The existence of the inverse of the q-Fourier transform F_q in the set of q-Gaussians was established in Theorem 3.5. Since mappings $\mathcal{M}_{q,q^*}$ and $\mathcal{K}_{q,q^*}$ are one-to-one, relationship (5.31) yields invertibility of $\mathbb{F}_q$ in $\mathcal{G}_{z_\alpha(q)}[\alpha]$ and validity of property (ii).

Further, we introduce the sequence

$$q_{\alpha,n} = u_{\alpha,n}(q) = u(u_{\alpha,n-1}(q)), \quad n = 1, 2, \ldots,$$

with a given $q = u_0(q)$, $q < 1 + \alpha$. We can extend the sequence $q_{\alpha,n}$ for negative integers $n = -1, -2, \ldots$ as well, setting

$$q_{\alpha,-n} = u_{\alpha,-n}(q) = u_\alpha^{-1}(u_{\alpha,1-n}(q)), \quad n = 1, 2, \ldots.$$

It is not hard to verify that

$$q_{\alpha,n} = 1 + \frac{\alpha(q-1)}{\alpha - n(q-1)} = \frac{\alpha q - n(q-1)}{\alpha - n(q-1)}, \tag{5.37}$$

for all integer n satisfying $-\infty < n \leq [\frac{\alpha}{q-1}]$. The restriction $n \leq [\alpha/(q-1)]$ implies the necessary condition $q_{\alpha,n} > 1$, since q-Fourier transform is defined for $q \geq 1$. Note that $q_{\alpha,n}$ is a function of $(q, n/\alpha)$, that $q_{\alpha,n} \equiv 1$ for all $n = 0, \pm1, \pm2, \ldots$, if $q = 1$, and that $\lim_{n \to \pm\infty} z_{\alpha,n}(q) = 1$ for all $q \neq 1$; see Fig. 5.6 for $q_{\alpha,n}$ with some typical values of α and n. Equation (5.37) can be rewritten as follows:

$$\frac{\alpha}{q_{\alpha,n} - 1} - n = \frac{\alpha}{q-1}, \quad n = 0, \pm1, \pm2, \ldots \tag{5.38}$$

We note that the latter coincides with equation (13) of Mendes and Tsallis (2001), once we identify α with the quantity z therein defined, which was obtained through a quite different approach (related to the renormalization of the index q emerging from summing a specific expression over one degree of freedom).

We also note an interesting property of $q_{\alpha,n}$. If we have a q-Gaussian in the variable $|x|^{\alpha/2}$ ($q \geq 1$), i.e., a q-exponential in the variable $|x|^\alpha$, its successive derivatives and integrations with respect to $|x|^\alpha$ precisely correspond to $q_{\alpha,n}$-exponentials in the same variable $|x|^\alpha$.

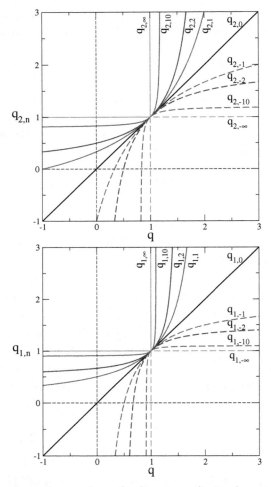

Fig. 5.6: The q-dependences of $q_{2,n}$ (*top*) and $q_{1,n}$ (*bottom*) as given by Eq. (12). We notice the tendency of all the $|n| \to \infty$ curves to collapse onto the $q_{\alpha,n} = 1$ horizontal straight line if $q \neq 1$, and onto the $q = 1$ vertical straight line if $q = 1$. This tendency is gradually intensified, $\forall n$, when α is fixed onto values decreasing from 2 towards zero.

Further, we introduce the sequence $q^*_{\alpha,n} = w(q_{\alpha,n})$, which can be written in the form

$$q^*_{\alpha,n} = 1 + \frac{2(q-1)}{\alpha - n(q-1)} = \frac{\alpha + (n-2)(1-q)}{\alpha - n(q-1)}, \qquad (5.39)$$

for $n = 0, \pm 1, \ldots$, or, equivalently,

$$\frac{2}{q^*_{\alpha,n} - 1} + n = \frac{\alpha}{q - 1}, \quad n = 0, \pm 1, \ldots. \tag{5.40}$$

It follows from Proposition 5.5 and definitions of sequences $q_{\alpha,n}$ and $q^*_{\alpha,n}$ that

$$\mathbb{F}_{q_{\alpha,n}} : \mathcal{G}_{q_{\alpha,n}}[\alpha] \to \mathcal{G}_{q_{\alpha,n+1}}[\alpha], \, -\infty < n \leq \left[\frac{\alpha}{q-1}\right]. \tag{5.41}$$

Proposition 5.6. *For all* $n = 0, \pm 1, \pm 2, \ldots$ *the following relations*

$$q^*_{\alpha,n-1} + \frac{1}{q^*_{\alpha,n+1}} = 2, \tag{5.42}$$

$$q^*_{2,n} = q_{2,n}, \tag{5.43}$$

hold.

Proof. We notice that

$$\frac{1}{q^*_{\alpha,n+1}} = 1 - \frac{2(q-1)}{\alpha - (n-1)(q-1)}.$$

On the other hand, by (5.39)

$$-\frac{2(q-1)}{\alpha - (n-1)(q-1)} = 1 - q^*_{\alpha,n-1},$$

which implies (5.42) immediately. The relation (5.43) can be checked easily. □

Remark 5.4. The property $q^*_{2,n} = q_{2,n}$ shows that the sequences (5.37) and (5.39) coincide if $\alpha = 2$. Hence, the mapping (5.41) takes the form

$$F_{q_{2,n}} : \mathcal{G}_{q_{2,n}}[2] \to \mathcal{G}_{q_{2,n+1}}[2],$$

recovering Proposition 3.8. Moreover, in this case the duality (5.42) holds for the sequence $q_{\alpha,n}$ as well. If $\alpha < 2$ then the values of $q^*_{\alpha,n}$

are distinct from the values of $q_{\alpha,n}$. The difference is given by

$$q_{\alpha,n} - q_{\alpha,n}^* = \frac{(2-\alpha)(1-q)}{\alpha + n(1-q)},$$

vanishing for $\alpha = 2, \forall q$, or for $q = 1, \forall \alpha$. In the latter case $q_{\alpha,n} = q_{\alpha,n}^* \equiv 1$.

Further, we define for $n = 0, \pm 1, \ldots,$ $k = 1, 2, \ldots,$ $n + k \leq [\frac{\alpha}{q-1}] + 1$, the operators

$$\mathbb{F}_n^k[f](\xi) = \mathbb{F}_{q_{\alpha,n+k-1}} \circ \cdots \circ \mathbb{F}_{q_{\alpha,n}}[f](\xi)$$

$$= \mathbb{F}_{q_{\alpha,n+k-1}}[\ldots \mathbb{F}_{q_{\alpha,n+1}}[\mathbb{F}_{q_{\alpha,n}}[f]] \ldots](\xi),$$

and

$$\mathbb{F}_n^{-k}[f](\xi) = \mathbb{F}_{q_{\alpha,n-k}}^{-1} \circ \cdots \circ \mathbb{F}_{q_{\alpha,n-1}}^{-1}[f](\xi)$$

$$= \mathbb{F}_{q_{\alpha,n-k}}^{-1}[\ldots \mathbb{F}_{q_{\alpha,n-2}}^{-1}[\mathbb{F}_{q_{\alpha,n-1}}^{-1}[f]] \ldots](\xi).$$

In addition, we adopt the convention $\mathbb{F}_q^k[f](\xi) = f(\xi)$ if $k = 0$ for any admissible q.

Summarizing the above mentioned relationships, we obtain the following assertions.

Proposition 5.7. *The following mappings hold:*

(1) $\mathbb{F}_{q_{\alpha,n}} : \mathcal{G}_{q_{\alpha,n}}[\alpha] \to \mathcal{G}_{q_{\alpha,n+1}}[\alpha], -\infty < n \leq [\frac{\alpha}{q-1}];$

(2) $\mathbb{F}_n^k : \mathcal{G}_{q_{\alpha,n}}[\alpha] \to \mathcal{G}_{q_{\alpha,k+n}}[\alpha],\quad k = 1, 2, \ldots,\quad n = 0, \pm 1, \ldots,$
    $-\infty < n + k \leq [\frac{\alpha}{q-1}] + 1;$

(3) $\lim_{k \to -\infty} \mathbb{F}_n^k \mathcal{G}_q[\alpha] = \mathcal{G}[\alpha],\quad n = 0, \pm 1, \ldots,$

where $\mathcal{G}[\alpha]$ is the set of densities of classic symmetric α-stable Lévy distributions.

Theorem 5.5. *Assume $(2q_{\alpha,k} - 1, \alpha) \in \mathcal{Q}$, where the sequence $q_{\alpha,k}, -\infty < k \leq [\alpha/(q-1)]$ is given as in (5.37) with $q_0 = q \in [1, \min\{2, 1+\alpha\})$. Let $X_j,\ j = 1, 2, \ldots$, be symmetric $q_{\alpha,k}$-independent (for some k) random variables all having the same probability density function $f(x) \in H_{q_{\alpha,k},\alpha}$.*

Then the sequence

$$Z_N = \frac{X_1 + \cdots + X_N}{N^{\frac{1}{\alpha(2 - q_{\alpha,k})}}}$$

is $q_{\alpha,k}$-convergent to a $(q_{\alpha,k-1}, \alpha)$-stable distribution, as $N \to \infty$.

Proof. The proof in the case $\alpha = 2$ is the same as in Theorem 4.4. If $0 < \alpha < 2$, then setting $q = q_{k,\alpha}$ in Theorem 5.5 (which corresponds to $k = 0$, that is $q_{0,\alpha} = q$) we obtain the proof in the general case. □

5.6. Scaling Rate Analysis

In Section 4.3, the formula

$$\beta_k = \left(\frac{3 - q_{k-1}}{4q_k C_{q_{k-1}}^{2q_{k-1}-2}} \right)^{\frac{1}{2 - q_{k-1}}}. \tag{5.44}$$

was obtained for the q-Gaussian parameter β of the attractor. It follows from this formula that the scaling rate in the case $\alpha = 2$ is

$$\delta = \frac{1}{2 - q_{k-1}} = q_{k+1}, \tag{5.45}$$

where q_{k-1} is the q-index of the attractor. Moreover, if we insert the "evolution parameter" t, then the translation of a q-Gaussian to a density in $\mathcal{G}_q[\alpha]$ changes t to $t^{2/\alpha}$. Hence, applying these two facts to the general case, $0 < \alpha \le 2$, and taking into account that the attractor index in our case is $q_{\alpha,k-1}^*$, we obtain the formula for the scaling rate

$$\delta = \frac{2}{\alpha(2 - q_{\alpha,k-1}^*)}. \tag{5.46}$$

In accordance with Proposition 5.6, $2 - q_{\alpha,k-1}^* = 1/q_{\alpha,k+1}^*$. Consequently,

$$\delta = \frac{2}{\alpha} q_{\alpha,k+1}^* = \frac{2}{\alpha} \frac{\alpha - (k-1)(q-1)}{\alpha - (k+1)(q-1)}. \tag{5.47}$$

Finally, in terms of $Q = 2q - 1$ the formula (5.47) takes the form

$$\delta = \frac{2}{\alpha} \frac{2\alpha - (k-1)(Q-1)}{2\alpha - (k+1)(Q-1)}. \tag{5.48}$$

In Section 4.9, it was noticed that the scaling rate in the non-linear Fokker–Planck equation can be derived from the model corresponding to the case $k = 1$. Taking this fact into account we can conjecture that the scaling rate in the *fractional generalization* of the nonlinear Fokker–Planck equation is

$$\delta = \frac{2}{\alpha + 1 - Q},$$

which can be derived from (5.48) setting $k = 1$. In the case $\alpha = 2$ we get the known result $\delta = 2/(3 - Q)$ obtained in Tsallis and Bukman (1996). The latter relation was in fact experimentally verified in Combe *et al.* (2015).

5.7. On Additive and Multiplicative Dualities

In the non-extensive statistical mechanical literature, there are two transformations that appear quite frequently in various contexts. They are sometimes referred to as *dualities*. The *multiplicative duality* is defined through

$$\mu(q) = 1/q, \tag{5.49}$$

and the *additive duality* is defined through

$$\nu(q) = 2 - q. \tag{5.50}$$

They satisfy $\mu^2 = \nu^2 = \mathbf{1}$, where $\mathbf{1}$ represents the *identity*, i.e., $\mathbf{1}(q) = q, \forall q$. We also verify that

$$(\mu\nu)^m(\nu\mu)^m = (\nu\mu)^m(\mu\nu)^m = \mathbf{1} \quad (m = 0, 1, 2, \ldots).$$

Consistently, we define $(\mu\nu)^{-m} \equiv (\nu\mu)^m$, and $(\nu\mu)^{-m} \equiv (\mu\nu)^m$.

Also, for $m = 0, \pm 1, \pm 2, \ldots$, and $\forall q$,

$$(\mu\nu)^m(q) = \frac{m - (m-1)\,q}{m + 1 - m\,q} = \frac{q + m(1-q)}{1 + m(1-q)},$$

$$\nu(\mu\nu)^m(q) = \frac{m + 2 - (m+1)\,q}{m + 1 - m\,q} = \frac{2 - q + m(1-q)}{1 + m(1-q)}, \tag{5.51}$$

and

$$(\mu\nu)^m \mu(q) = \frac{-m + 1 + m\,q}{-m + (m+1)\,q} = \frac{1 - m(1-q)}{q - m(1-q)}.$$

We can easily verify, from equations (3.40) and (5.51), that the sequences $q_{2,n}$ ($n = 0, \pm 2, \pm 4, \ldots$) and $q_{1,n}$ ($n = 0, \pm 1, \pm 2, \ldots$) coincide with the sequence $(\mu\nu)^m(q)$ ($m = 0, \pm 1, \pm, 2, \ldots$).

This structure was advanced in Tsallis *et al.* (2005a) to interpret NASA data for the solar wind. Moreover, a quite general structure was published in Gazeau *et al.* (2019), which leads to a connection between Möbius transforms and q-triplets.

5.8. (q, α)-Stables: Two Classifications and the General Framework

The central limit theorems established in Chapters 4 and 5 represent a complex interrelation of various possibilities of limit theorems of independent or strongly dependent random variables. In this section, we describe the entire picture from the angle of two classifications of (q, α)-stable distributions.

The q-central limit theorem (Theorem 4.4) states that the sequence of sums of an appropriately scaled q_k-independent and identically distributed random variables with a finite $(2q_k - 1)$-variance is q-convergent to a q_k^*-Gaussian, which is the q_k^*-Fourier preimage of a q_k-Gaussian. Here q_k and q_k^* are sequences defined as

$$q_k = \frac{2q - k(q-1)}{2 - k(q-1)}, \quad k = 0, \pm 1, \ldots,$$

and

$$q_k^* = q_{k-1}, \quad k = 0, \pm 1, \ldots.$$

Schematically this theorem can be represented as

$$\{f : \sigma_{2q_k-1}(f) < \infty\} \xrightarrow{F_{q_k}} \mathcal{G}_{q_k}[2] \xleftarrow{F_{q_k^*}} \mathcal{G}_{q_k^*}[2], \qquad (5.52)$$

where $\mathcal{G}_q[2]$ is the set of q-Gaussians. We have noted that the processes described by the q-central limit theorem can be effectively described by the q-triplet $(P_{\text{att}}, P_{\text{cor}}, P_{\text{scl}})$, where $P_{\text{att}}, P_{\text{cor}}$ and P_{scl} are parameters of the *attractor, correlation* and *scaling rate*, respectively. We found that (see details in Section 4.3)

$$(P_{\text{att}}, P_{\text{cor}}, P_{\text{scl}}) \equiv (q_{k-1}, q_k, q_{k+1}). \qquad (5.53)$$

The (q, α)-stable distributions can be classified in two different ways. Section 5.3 provides a first classification of symmetric (q, α)-stable distributions. Schematically, the corresponding theorem (Theorem 5.4) is represented as

$$\mathcal{L}(q, \alpha) \xrightarrow{F_q} \mathcal{G}_q(\alpha) \xleftarrow{F_q} \mathcal{G}_{q^L}(2), \quad 0 < \alpha < 2, \qquad (5.54)$$

where $\mathcal{L}(q, \alpha)$ is the set of (q, α)-stable distributions, $\mathcal{G}_{q^L}(2)$ is the set of q^L-Gaussians asymptotically equivalent to the densities $f \in \mathcal{L}(q, \alpha)$. The index q^L is linked with q as follows

$$q^L = q_\alpha^L(q) = \frac{3 + (2q - 1)\alpha}{1 + \alpha}.$$

Note that the case $\alpha = 2$ is peculiar in this classification and we agree to refer to the scheme (5.52) in this case.

In Section 5.5, we have studied a generalization of the q-central limit theorem to the case when the $(2q - 1)$-variance of random variables is infinite. The theorem that we have obtained generalizes the q-central limit theorem (which corresponds to $\alpha = 2$) to the full range $0 < \alpha \leq 2$ and provides a second classification of (q, α)-stable distributions. Schematically, this theorem can be represented as

$$\mathcal{L}(q_{\alpha,k}, \alpha) \xrightarrow{F_{q_{\alpha,k}}} \mathcal{G}_{q_{\alpha,k}}(\alpha) \xleftarrow{F_{q_{\alpha,k}^*}} \mathcal{G}_{q_{\alpha,k}^*}(2), \quad 0 < \alpha \leq 2, \qquad (5.55)$$

generalizing the scheme (5.52). The sequences $q_{\alpha,k}$ and $q^*_{\alpha,k}$ in this case read

$$q_{\alpha,k} = \frac{\alpha q + k(1-q)}{\alpha + k(1-q)}, \quad k = 0, \pm 1, \ldots,$$

and

$$q^*_{\alpha,k} = 1 - \frac{2(1-q)}{\alpha + k(1-q)}, \quad k = 0, \pm 1, \ldots.$$

Note that the triplet $(P_{\text{att}}, P_{\text{cor}}, P_{\text{scl}})$ mentioned above in this case takes the form

$$(P_{\text{att}}, P_{\text{cor}}, P_{\text{scl}}) \equiv (q^*_{\alpha,k-1}, \, q_{\alpha,k}, \, (2/\alpha)q^*_{\alpha,k+1}),$$

which coincides with (5.53) if $\alpha = 2$.

Finally, unifying the schemes (5.54) and (5.55) we obtain the general picture for the description of (q, α)-stable distributions:

$$\mathcal{L}[q_{\alpha,k}, \alpha] \xrightarrow{F_{q_{\alpha,k}}} \mathcal{G}_{q_{\alpha,k}}[\alpha] \xleftarrow{F_{q^*_{\alpha,k}}} \mathcal{G}_{q^*_{\alpha,k}}[2]$$
$$\updownarrow F_q \qquad\qquad (5.56)$$
$$\mathcal{G}_{q^L_{\alpha,k}}[2],$$

where

$$q^L_{\alpha,k} = q^L_\alpha(q_{\alpha,k}) = \frac{3 + (2q_{\alpha,k} - 1)\alpha}{1 + \alpha}.$$

In Fig. 5.7, connections of parameters $(Q, \alpha) \in \mathcal{Q}$ ($Q = 2q - 1$) with q^L and q^*, ($k = 0$) are represented. If $Q = 1$ and $\alpha = 2$ (the blue box in the figure), then the random variables are independent in the usual sense and have *finite* variance. The standard central limit theorem applies, and the attractors are classic Gaussians.

If Q belongs to the interval $(1, 3)$ and $\alpha = 2$ (the blue straight line on the top), the random variables are *not* independent. If the random variables have a *finite* Q-variance, then Theorem 4.4 applies, and the attractors belong to the family of q^*-Gaussians. Note that q^* runs in $[1, 5/3)$. Thus, in this case, attractors (q^*-Gaussians) have *finite* classic variance (i.e., 1-variance) in addition to *finite* q^*-variance.

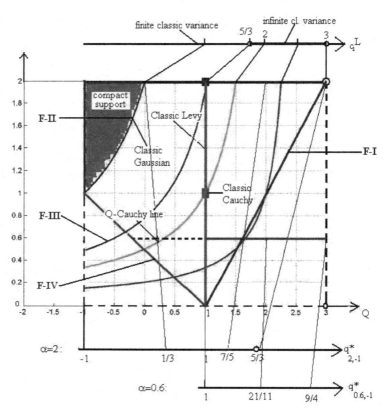

Fig. 5.7: The general picture of the q-central limit theorem for (q, α)-stable random variables. This picture represents the (Q, α)-regions, where $Q = 2q - 1$, of various appearances of the q-central limit theorem depending on values of Q and α. Four borderlines marked as F1, F2, F3, F4 separate regions of (Q, α)-distributions with significantly different properties (the details are in the text of the current section and Section 5.9).

If $Q = 1$ and $0 < \alpha < 2$ (the vertical green line in the figure), we have the classic Lévy distributions, and random variables are independent, and have *infinite* variance, i.e., their attractors belong to the family of α-stable Lévy distributions. The top q^L-line in the figure shows that, in asymptotic terms of q-Gaussians, classic symmetric α-stable distributions correspond to $\cup_{5/3 < q < 3} \mathcal{G}_q[2]$.

If $0 < \alpha < 2$ and Q belongs to the interval $(1, 3)$, we observe the rich variety of possibilities of (q, α)-stable distributions. In this case, random variables are *not* independent, have *infinite* variance and *infinite Q-variance*. The rectangle $\{1 < Q < 3;\ 0 < \alpha < 2\}$, at the right of the classic Lévy line, is covered by non-intersecting curves

$$C_{q^L} \equiv \left\{ (Q, \alpha) : \frac{3 + Q\alpha}{\alpha + 1} = q^L \right\}, \quad 5/3 < q^L < 3.$$

This family of curves describes all (Q, α)-stable distributions based on the mapping (5.54) with q-Fourier transform. The constant q^L is the index of the q^L-Gaussian attractor corresponding to the points (Q, α) on the curve C_{q^L}. For example, the green curve corresponding to $q^L = 2$ describes all Q-Cauchy distributions, recovering the classic Cauchy–Poisson–Lorentz distribution if $\alpha = 1$ (the green box in the figure). Every point (Q, α) lying on the brown curve corresponds to $q^L = 2.5$.

The second classification of (Q, α)-stable distributions based on the mapping (5.55) with q^*-Fourier transform leads to a covering of $\mathcal{Q}$ by curves distinct from C_{q^L}. Namely, in this case we have the following family of straight lines:

$$S_{q^*} \equiv \left\{ (Q, \alpha) : \frac{4\alpha}{Q + 2\alpha - 1} = 3 - q^* \right\}, \quad 1 \leq q^* < 3, \qquad (5.57)$$

which are obtained from (5.39) replacing $n = -1$ and $2q - 1 = Q$. For instance, every (Q, α) on the line F-I (the blue diagonal of the rectangle in the figure) identifies q^*-Gaussians with $q^* = 5/3$. This line is the frontier (or borderline) of points (Q, α) which separates finite and infinite classic variances. Namely, all (Q, α) above the line F-I identify distributions with *finite* variance, and points on this line and below identify distributions with *infinite* classic variance. Two bottom lines in Fig. 5.7 reflect the sets of q^* corresponding to lines $\{1 \leq Q < 3; \alpha = 2\}$ (the top boundary of the rectangle in the figure) and $\{1 \leq Q < 3; \alpha = 0.6\}$ (the brown horizontal line in the figure).

Finally, we note that Fig. 5.7 reflects the case $k = 0$ in the description (5.56). The cases $k \neq 0$ can be analyzed analogously.

5.9. Additional Notes

(1) Historical notes

The central limit theorem (CLT) and α-stable distributions have rich applications in various fields including Boltzmann–Gibbs (BG) statistical mechanics and related areas.

The classic theory of α-stable distributions was originated by Paul Lévy and further developed by Kolmogorov, Gnedenko, Feller, and many others; for details and history see, for instance, Gnedenko and Kolmogorov (1954), Feller (1966), Samorodnitsky and Taqqu (1994), Meerschaert and Scheffler (2001), Metzler and Klafter (2000), Uchaykin and Zolotarev (1999) and references therein.

Non-extensive statistical mechanics (Tsallis, 1988) (see also Prato and Tsallis, 1999; Gell-Mann and Tsallis, 2004; Boon and Tsallis, 2005; Tsallis, 2009a,b), characterized by the index q (which recovers the BG theory in the case $q = 1$) focuses on, in particular, strongly correlated random states, the mathematical models of which can be represented by specific long-range dependent random variables. The q-central limit theorem consistent with non-extensive statistical mechanics was established in Umarov *et al.* (2008). The main objective there was to study the scaling limits (attractors) of sums of q-independent random variables with a finite $(2q - 1)$-variance. The mapping

$$F_q : \mathcal{G}_q[2] \to \mathcal{G}_{z(q)}[2], \qquad (5.58)$$

where F_q is the q-Fourier transform, $z(s) = (1 + s)/(3 - s)$, and $\mathcal{G}_q[2]$ is the set of q-Gaussians up to a constant factor, was essentially used in the description of the attractors. The number 2 in the notation indicates that the mapping (5.58) holds in the class of q-Gaussians (not necessarily normalized).

(2) (q, α)-Stable distributions

The q-generalization of Lévy's α-stable distributions was studied in Umarov *et al.* (2010). The mapping (5.58) in the case of (q, α)-stables generalizes to

$$\mathcal{L}[q,\alpha] \xrightarrow{F_q} \mathcal{G}_q[\alpha] \xleftrightarrow{F_{q^*}} \mathcal{G}_{q^*}[2]$$
$$\updownarrow F_q \qquad\qquad (5.59)$$
$$\mathcal{G}_{q^L}[2],$$

opening a wide field of possible applications. For example, it would surely be interesting to study nonlinear fractional Fokker–Planck equations of the following type:

$$\frac{\partial p(x,t)}{\partial t} = D \frac{\partial^\alpha [p(x,t)]^{2-q}}{\partial |x|^\alpha}, \quad q < 3, \ 0 < \alpha \le 2. \qquad (5.60)$$

For $q = 1$, see Umarov *et al.* (2018) and for $\alpha = 2$, see Frank (2005) and references therein.

The (q, α)-stable distributions considered in the current chapter are symmetric. The theory containing the asymmetric case of (q, α)-stable distributions might be developed using the technique based on gamma distributions and suggested in Budini (2015). Another possibility is the technique used for obtaining a skewed version of the q-Gaussian, discussed in Tasaki and Koike (2017). The latter uses an abstract method for obtaining skewed distributions from a symmetric one, which was published in 1985; see Azzalini (1985, 2014).

Chapter 6

Applications and Observations of q-Gaussian Distributions

6.1. Introduction

This chapter introduces some applications of mathematical tools introduced in previous chapters. The q-Fourier transform can be used for solutions of some nonlinear partial differential equations, including the nonlinear porous medium equation. In addition to that, we briefly describe here various examples where q-Gaussians are observed in natural, artificial and social systems, consistently with the q-CLT, and related mathematical structures (such as large deviation theory). In such systems, we consistently expect the emergence of q-exponentials as well, basically when the variable takes non-negative values. Many of these results are available in the literature. For more information, a quite complete and updated bibliography is available at http://tsallis.cat.cbpf.br/biblio.htm.

6.2. Applications to Nonlinear Partial Differential Equations

6.2.1. *The Fokker–Planck and porous medium equations*

In this section, we discuss an application of the q-Fourier transform to the solution of nonlinear Fokker–Planck type (partial differential) equations. Similar to the classic Fourier transform, the q-Fourier transform also can be used to establish a relation between a nonlinear partial differential equation and a nonlinear ordinary differential equation.

The classic Fourier transform reduces the Cauchy problem for a linear partial differential equation of the form

$$u_t(t,x) = A(D_x)u(t,x), \quad t > 0, \ x \in \mathbb{R}^n,$$

$$u(0,x) = \varphi(x), \quad x \in \mathbb{R}^n,$$

where $D_x = (D_1,\ldots,D_n)$, $D_j = -i\frac{\partial}{\partial x_j}$, $j = 1,\ldots,n$, and $A(D_x)$ is an elliptic differential operator, to the Cauchy problem for associated linear ordinary differential equation

$$\hat{u}_t(t,\xi) = A(\xi)\hat{u}(t,\xi), \quad t > 0, \ \xi \in \mathbb{R}^n,$$

$$\hat{u}(0,\xi) = \hat{\varphi}(\xi), \quad \xi \in \mathbb{R}^n,$$

for every fixed $\xi \in \mathbb{R}^n$. Here the hat-notation $\hat{f}$ means the classical Fourier transform of f, that is

$$\hat{f}(\xi) = F[f](\xi) = \int_{\mathbb{R}^n} f(x)\exp(ix\xi)dx, \quad \xi \in \mathbb{R}^n.$$

In the particular case $n = 1$ and $A(D_x) = \kappa\frac{d^2}{dx^2}$, where $\kappa > 0$ is a diffusion coefficient, for the Fourier image $\hat{u}(t,\xi)$ of a solution $u(t,x)$ of the Cauchy problem

$$u_t(t,x) = \kappa\frac{\partial^2 u(t,x)}{\partial t^2}u(t,x), \quad t > 0, \ x \in \mathbb{R}^n, \tag{6.1}$$

$$u(0,x) = \varphi(x), \quad x \in \mathbb{R}^n, \tag{6.2}$$

we have the Cauchy problem for the ordinary differential equation

$$\frac{d\hat{u}(t,\xi)}{dt} = -\kappa\xi^2\hat{u}(t,\xi), \quad \xi \in \mathbb{R}, \tag{6.3}$$

$$\hat{u}(0,\xi) = \hat{\varphi}(\xi), \quad \xi \in \mathbb{R}, \tag{6.4}$$

where ξ is a parameter. The unique solution of the latter is

$$\hat{u}(t,\xi) = \hat{\varphi}(\xi)e^{-t\kappa\xi^2}, \quad t > 0, \ \xi \in \mathbb{R}.$$

Applying the inverse Fourier transform, one has a solution to Cauchy problem (6.1), (6.2) in the form

$$u(t,x) = \int_{\mathbb{R}} G(t,y)\varphi(x-y)dy, \quad t > 0, \quad x \in \mathbb{R},$$

where $G(t,y)$ is the Gaussian with mean zero and variance $2t\kappa$, namely,

$$G(t,x) = \frac{1}{2\sqrt{\pi\kappa t}}e^{-\frac{x^2}{4\kappa t}}, \quad t > 0, \quad x \in \mathbb{R}.$$

We note that if the initial condition is given by $u(0,x) = \delta(x)$, the Dirac delta function concentrated at $x = 0$ and $\kappa = 1/2$, then (6.1), (6.2) corresponds to the Fokker–Planck equation associated with Brownian motion B_t, $t \geq 0$ without drift (Risken, 1989). In this case we have

$$u(t,x) = G_1(t,x) = \frac{1}{\sqrt{2\pi t}}e^{-\frac{x^2}{2t}}, \quad t > 0, \quad x \in \mathbb{R}. \tag{6.5}$$

Now consider the following nonlinear diffusion equation with a singular diffusion coefficient,

$$\frac{\partial U}{\partial t} = (U^{1-q}U_x)_x, \quad t > 0, \quad x \in \mathbb{R},$$

where $q \geq 1$. This is the celebrated *porous medium equation* for superdiffusion regime, which ubiquitously emerged in various physical processes; see Vázquez (2007), Carillo and Toscani (2000), Otto (2001) and references therein. In the particular case $q = 1$, equation (6.6) reduces to equation (6.1) with $\kappa = 1$. Thus, the porous medium equation generalizes the diffusion equation (6.1) to a nonlinear equation describing correlated processes. The index q is responsible for correlation of states of the underlying process.

6.2.2. *Application of the q-Fourier transform to the solution of nonlinear PDEs*

In this section, we demonstrate that the role of the q-Fourier transform F_q in solving the porous medium equation is similar to the role of the classic Fourier transform $F = F_1$ in solving Cauchy problem (6.1), (6.2). Note that the monograph (Vázquez, 2007) uses a different approach for the solution of the porous medium equation.

Consider the following Cauchy problem for nonlinear diffusion equation,

$$\frac{\partial U(t,x)}{\partial t} = \frac{\partial}{\partial x}\left([U(t,x)]^{1-q}\frac{\partial U(t,x)}{\partial x}\right), \quad t > 0, \ x \in \mathbb{R}, \qquad (6.6)$$

$$U(0,x) = \varphi(x), \quad x \in \mathbb{R}. \qquad (6.7)$$

We look for a solution of this Cauchy problem in the similarity set

$$G_q^* = \{U(t,x) : U(t,x) = t^a G_q(\beta; t^b x), \ a = a(q) \in \mathbb{R},$$
$$b = b(q) \in \mathbb{R}, \ \beta = \beta(q) > 0\},$$

where a, b, and β do not depend on t and x.

Proposition 6.1. *Let* $1 \le q < 2$ *and suppose* $U(t,x) \in G_q^*$ *is a solution of Cauchy problem* (6.6), (6.7) *with* $\varphi(x) = \delta(x)$, *the Dirac delta function. Then its* q-*Fourier transform* $\hat{U}_q(t,\xi) = F_q[U(t,x)](\xi)$ *satisfies the following nonlinear ordinary differential equation with parameter* ξ

$$(\hat{U}_q)_t' = -\frac{B(\beta,q)\xi^2}{t^{\frac{q-1}{3-q}}}(\hat{U}_q)^{q_1}, \quad t > 0, \ \xi \in \mathbb{R}, \qquad (6.8)$$

where $B(\beta,q) = \frac{2-q}{4\beta^{2-q}C_q^{q-1}}$ *and* $q_1 = \frac{1+q}{3-q}$, *and the initial condition*

$$\hat{U}(0,\xi) = 1, \quad \xi \in \mathbb{R}. \qquad (6.9)$$

Proof. Let $U \in G_q^*$ be a solution of (6.6), i.e., for some $a = a(q)$ and $\beta = \beta(q)$ it has the representation

$$U(t,x) = t^a G_q(\beta; \ t^a x).$$

Then, it follows from Proposition 3.2 and equation (3.76) that,

$$\hat{U}_q(t, \xi) = F_q[U(t, x)](\xi)$$

$$= F_q[G_q(\beta; x)] \left(\frac{\xi}{t^{a(2-q)}} \right)$$

$$= \frac{1}{C_q} Y_q \left(\left(\frac{\sqrt{\beta}}{C_q} \right)^{q-1} \frac{\xi}{\sqrt{\beta} t^{a(2-q)}} \right), \quad (6.10)$$

where $Y_q(\xi)$ is a solution of equation (3.92). Computing the derivative of $\hat{U}_q(t, x)$ in variable t, taking into account that (see, e.g., Vázquez, 2007),

$$a = -\frac{1}{3-q}, \quad (6.11)$$

and using equation (3.92), we obtain

$$(\hat{U}_q)_t = -\frac{2-q}{4\beta^{2-q} C_q^{2(q-1)}} \xi^2 (\hat{U}_q)^{q_1},$$

where $q_1 = (1+q)/(3-q)$. Moreover, it follows from equality (6.10) and (6.11) that

$$\lim_{t \to 0+} \hat{U}(t, \xi) = \frac{1}{C_q} \lim_{t \to 0+} Y_q \left(\left(\frac{\sqrt{\beta}}{C_q} \right)^{q-1} \frac{\xi t^{\frac{2-q}{3-q}}}{\sqrt{\beta}} \right)$$

$$= \frac{1}{C_q} Y_q(0) = 1,$$

implying (6.9), since $Y_q(0) = C_q$ (see (3.77)).   □

The inverse statement, given in the following formulation, is also true.

Proposition 6.2. *Suppose* $V(t, \xi)$, $V(0, \xi) = 1$, *is a solution to ODE with parameter* ξ

$$V' = -\frac{B(\beta, q)\xi^2}{t^{\frac{q-1}{3-q}}} V^{q_1}, \quad t > 0, \quad (6.12)$$

where $B(q, \beta)$ and q_1 are as in Proposition 6.1. Then its inverse q-Fourier transform $U(t, x) = F_q^{-1}[V(t, \xi)](x)$ in G_q^* exists and satisfies equation (6.6) and the initial condition $U(0, x) = \delta(x)$.

Proof. By separation of variables of (6.12) one can verify that its solution is given by

$$V(t, \xi) = e_{q_1}^{-\frac{3-q}{8\beta^{2-q}C_q^{q-1}}\left(\xi t^{\frac{2-q}{3-q}}\right)}.$$

In accordance with Corollary 3.6 the inverse q-Fourier transform for $V(t, \xi)$ exists in G_q^* and, by virtue of Proposition 3.2 and Theorem 3.4, it has the representation

$$U(t, x) = \frac{1}{t^{\frac{1}{3-q}}} G_q\left(\beta(q); \frac{x}{t^{\frac{1}{3-q}}}\right), \tag{6.13}$$

where

$$\beta(q) = \frac{1}{\left[2(3 - q)C_q^{\frac{1}{q-1}}\right]^{\frac{2}{3-q}}}. \tag{6.14}$$

The latter is a solution of equation (6.6). One can verify this by direct substitution.

Moreover, using Plastino–Rocca's Theorem 3.19 with $F_q[f](\xi) \equiv 1$ and equality (3.29), one has

$$U(0, x) = F_q^{-1}[V(0, \xi)](x)$$

$$= \frac{1}{2\pi} \int_{\mathbb{R}} \left[\lim_{\varepsilon \to 0^+} \int_1^2 \delta(q - 1 - \varepsilon)dq\right] e^{ix\xi}d\xi$$

$$= \frac{1}{2\pi} \int_{\mathbb{R}} e^{ix\xi}d\xi = \delta(x),$$

proving the proposition. $\qquad\qquad\qquad\qquad\qquad\qquad\qquad\square$

Notice that, if the initial condition is given in the form $U(0, x) = \delta(x)$ with the Dirac delta function, and $q = 1$, then we obtain (6.3), (6.4) with $\kappa = 1$ and $\hat{\varphi}(\xi) \equiv 1$. Moreover, here $\beta = 1/4$, and hence $B(\beta, 1) = 4\beta = 1$.

Remark 6.1.

(1) The solution (6.13) corresponds to the solution of the nonlinear Fokker–Planck equation obtained in Plastino and Plastino (1995) and Tsallis and Bukman (1996) from an ansatz.

(2) The q-Fourier transform method used for solving the Cauchy problem (6.6), (6.7) can be extended for more general cases as well. For instance, the Fokker–Planck type equation associated with a process X_t with constant drift $\tau = \mu \neq 0$, due to a term $-2i\mu\sqrt{q_n}\, Y_{\mu,q_n}(\xi)$ in equation (3.90), has an additional drift term on the right-hand side of equation (6.38). We note also that with more routine calculations the method can be extended to the case of time dependent drift and diffusion coefficients.

6.3. Observations of q-Gaussians and Related Functions in Physics

6.3.1. *Cold atoms*

It was predicted in 2003 (Lutz, 2003) that the distribution of velocities of cold atoms in dissipative optical lattices would be q-Gaussians instead of the classical Maxwellian distribution. This prediction was made on grounds of the following in-homogeneous linear Fokker–Planck equation (Rayleigh equation):

$$\frac{\partial W}{\partial t} = -\frac{\partial}{\partial p}[K(p)W] + \frac{\partial}{\partial p}\left[D(p)\frac{\partial W}{\partial p}\right], \qquad (6.15)$$

where the drift coefficient $K(p)$ and the diffusion coefficient $D(p)$ satisfy the condition

$$\frac{K(p)}{D(p)} = -\frac{\beta}{1 - \beta(1-q)U(p)}\frac{\partial U(p)}{\partial p},$$

with $U(p) = p^2$, and some (physically meaningful) constants q and β. Its stationary-state solution is shown to be the following q-Gaussian:

$$W(p) = Z_q^{-1}[1 - (1-q)\beta U(p)]^{1/(1-q)}, \qquad (6.16)$$

where $U(p)$ plays the role of a potential, Z_q is a normalizing factor, and the index is given by $q = 1 + 44E_R/U_0 > 1$, where E_R is the recoil energy and U_0 is the potential depth. It was verified in 2006, both in quantum Monte Carlo simulations and in the laboratory with Cs atoms (Duglas *et al.*, 2006) (see also Lutz and Renzoni, 2013), that the prediction was indeed correct.

Further observations of q-Gaussians are available in the literature, for instance in various trapped ions (DeVoe, 2009).

6.3.2. *Anomalous diffusion and granular matter*

On empirical grounds, the nowadays so-called *Porous Medium Equation* was proposed in 1937 (Muskat, 1937). This nonlinear partial differential equation may be written as follows

$$\frac{\partial P(x,t)}{\partial t} = D \frac{\partial^2 [P(x,t)]^\nu}{\partial x^2} \quad (\nu \in \mathcal{R}), \tag{6.17}$$

where D is a generalized diffusion coefficient; for $\nu = 1$ we recover Fourier's celebrated Heat Equation. Plastino and Plastino (1995) showed that, in the presence of an external quadratic confining potential, the solutions can be q-Gaussians with $q = 2 - \nu$. More precisely, this form, which optimizes S_q under appropriate constraints, emerges as the solution when $P(x,0) = \delta(x)$, where $\delta(x)$ denotes Dirac's delta. One year later, this line led to the scaling law (Tsallis and Bukman, 1996)

$$\alpha = \frac{2}{3-q}, \tag{6.18}$$

where α is defined through the $x^2 \sim t^\alpha$ scaling, which, for $q = 1$, recovers the Einstein 1905 scaling $\langle x^2 \rangle \propto t$. The plausibility of the predicted scaling (6.18) was experimentally verified first with the motion of *Hydra* cells (Upadhyaya *et al.*, 2001), and then with defect turbulence in fluids (Daniels *et al.*, 2004). Later on, this scaling was numerically verified in long-range-interacting Hamiltonians (Rapisarda and Pluchino, 2005). Finally, high-precision experiments in granular matter provided strong evidence of the validity of the scaling (6.18) along a wide variation of the experimental parameter (Combe *et al.*, 2015; Viallon-Galiner *et al.*, 2018).

6.3.3. *High-energy physics*

A large number of applications are available in high-energy physics, concerning experiments at the Large Hadron Collider (LHC)/CERN (ALICE, CMS, ATLAS, LHCb Collaborations) and the Relativistic Heavy Ion Collider (RHIC)/Brookhaven (STAR, PHENIX Collaborations) (Bediaga *et al.*, 2000; Adare *et al.*, 2011; Alice, 2017; Wong *et al.*, 2015; Marques *et al.*, 2015; Alice, 2017a,b,c; Rybczynski *et al.*, 2015; Wilk and Wlodarczyk, 2015), Pierre Auger Cosmic Ray Observatory (Tsallis *et al.*, 2003), among others. In such experiments, distributions of the transverse momenta of the ejected hadronic jets are measured along up to impressive fourteen experimental decades. They very well verify, within small log-periodic oscillations, the q-exponential form. The distributions of transverse momenta p_T of the hadronic jets are, along the same fourteen decades, well fitted (Wong *et al.*, 2015) by

$$\frac{dN}{2\pi y(p_T)^{d-1}dp_T}\bigg|_{y=0} = A_q\, e_q^{-E_T/T}, \tag{6.19}$$

where dimensionality $d = 2$, A_q is a normalizing constant, y is the rapidity, $E_T = \sqrt{m^2 + \vec{p}_T^2}$ is the effective transverse energy, and $(q, T) \simeq (1.15, 0.15\,\text{GeV})$.

Many applications also exist in high-energy physics observations out of the Earth (Komatsu and Kumira, 2013, 2014; Yalcin and Beck, 2018; Komatsu, 2017; Beck, 2016; Hou *et al.*, 2017; Kohler, 2017; Bertulani *et al.*, 2018). Let us illustrate with results concerning an observatory out of the Earth (Yalcin and Beck, 2018), which focuses on fluxes of cosmic rays, more precisely matter (electrons) and antimatter (positrons). The corresponding distributions are given by

$$P_{\pm}(E) = A_{\pm}\, E^{d-1}\Big[\big(1 + (q-1)E/T\big)^{\frac{-1}{q-1}} + C_{\pm}\big(1 + (\hat{q}-1)E/\hat{T}\big)^{\frac{-1}{\hat{q}-1}}\Big] \tag{6.20}$$

with dimensionality $d = 3$, $+$ and $-$ correspond, respectively, to electrons and positrons, E is the energy, $A_{\pm}$ are normalizing factors,

$C_\pm$ characterize crossovers, $q = 13/11$, $\hat{q} = 1/(2-q) = 11/9$, $\hat{T} = (11/9)T \simeq 198\,\mathrm{MeV}$, $(A_+, C_+) \simeq (750, 0.0053)$, and $(A_-, C_-) \simeq (100, 0.040)$.

6.3.4. *Solar wind and plasma physics*

Uncountable studies are available in the literature using q-statistics to approach systems in plasma physics such as Burlaga and Vinas (2005), Liu and Goree (2008), Lourek (2016), Bacha *et al.* (2017), Merriche and Tribeche (2017), Livadiotis (2018), Oliveira and Galvao (2018) and Casas *et al.* (2019).

A particularly interesting case is that of the solar wind, which attracts special interest in NASA, due to the data sent to Earth from the Voyager 1 and Voyager 2 space shuttles. Let us briefly review this example. It was conjectured in 2004 (Tsallis, 2004a), the existence of a set of q-indices, namely the q-triplet $(q_{\mathrm{sen}}, q_{\mathrm{rel}}, q_{\mathrm{stat}})$ (where sen, rel and stat stand for *sensitivity, relaxation* and *stationary state*), characterizing basic properties of a given complex system under specific (typically off equilibrium) circumstances. Simple connections were expected between these three indices. Observational values for them were offered in 2005 by NASA researchers (Burlaga and Vinas, 2005), which were $(q_{\mathrm{sen}}, q_{\mathrm{rel}}, q_{\mathrm{stat}}) = (-0.6 \pm 0.2, 3.8 \pm 0.3, 1.75 \pm 0.06)$. These values were conjectured (Tsallis *et al.*, 2005b) to be compatible with $(-1/2, 4, 7/4)$ and with the successive application of the additive and multiplicative dualities, respectively $q \to 2-q$ and $q \to 1/q$. It was finally suggested the following relations:

$$q_{\mathrm{rel}} = 2 - \frac{1}{q_{\mathrm{sen}}} \qquad (6.21)$$

and

$$q_{\mathrm{stat}} = 2 - \frac{1}{q_{\mathrm{rel}}}. \qquad (6.22)$$

Some years later (Baella, 2008; Tsallis, 2009a), the q-triplet was transformed, through $\epsilon \equiv 1 - q$, into the ϵ-triplet $(\epsilon_{\mathrm{sen}}, \epsilon_{\mathrm{rel}}, \epsilon_{\mathrm{stat}}) =$

$(3/2, -3, -3/4)$, and then shown to satisfy

$$\epsilon_{\text{stat}} = \frac{\epsilon_{\text{sen}} + \epsilon_{\text{rel}}}{2},$$

$$\epsilon_{\text{sen}} = \sqrt{\epsilon_{\text{stat}}\,\epsilon_{\text{rel}}},$$

$$\epsilon_{\text{rel}}^{-1} = \frac{\epsilon_{\text{sen}}^{-1} + \epsilon_{\text{stat}}^{-1}}{2}. \tag{6.23}$$

The physical–mathematical interpretation of this remarkable set of relations, in which on the right-hand side one observes the arithmetic, geometric, and harmonic means, remains until today as a highly intriguing open question. These connections and similar ones are further discussed in Gazeau *et al.* (2019).

6.3.5. *Long-range-interacting many-body classical Hamiltonian systems*

Classical many-body Hamiltonian systems including two-body long-range interactions have deserved along recent decades special attention, mainly because they violate the basic assumptions under which BG statistical mechanics is valid (Antoneodo and Tsallis, 1998; Campa *et al.*, 2001; Antoni and Ruffo, 1995; Pluchino *et al.*, 2007; Pluchino and Rapisarda, 2007; Chavanis, 2010; Cirto *et al.*, 2014, 2015, 2018; Nobre and Tsallis, 2003; Rodriguez *et al.*, 2019; Christodoulidi *et al.*, 2014, 2016; Bagchi and Tsallis, 2016, 2018). It is typically (but not exclusively) assumed a d-dimensional system and a two-body attractive interaction whose potential decays as $1/r^{\alpha}$ with $\alpha \geq 0$, r being the dimensionless distance between the two bodies localized at the sites of an hypercubic lattice. Therefore, for a linear chain we have $r = 1, 2, 3 \ldots$; for a square lattice we have $r = 1, \sqrt{2}, 2, \ldots$; for a cubic lattice we have $r = 1, \sqrt{2}, \sqrt{3}, 2, \ldots$. Consequently, the $\alpha \to \infty$ limit corresponds to interactions only between first-neighbors, and $\alpha = 0$ corresponds to a mean-field Hamiltonian, every site being coupled to every other with the same strength. The integral

$$\int_{1}^{\infty} r^{d-\alpha-1}\,dr$$

is proportional to the potential energy per particle. It diverges for $0 \le \alpha/d \le 1$ and converges for $\alpha/d > 1$. This is the criterion which separates long-range from short-range interactions. Hamiltonians which follow this scenario include the XY model (planar rotators), Heisenberg model (three-dimensional rotators and the Fermi–Pasta–Ulam–Tsingou system. They have all three been numerically investigated during recent years. The XY model constitutes a paradigmatic one. For fixed total energy and $\alpha/d < 1$, it exhibits a ferro-paramagnetic phase transition. For energies slightly below the critical value and rather generic initial conditions it evolves, along time, from a quasi-stationary state (QSS1) to a second quasi-stationary state (QSS2). In both of them the one-particle distribution of momenta is non-Maxwellian but instead given by Q-Gaussians. For example, for $\alpha/d = 0.9$, at the QSS2, it is obtained

$$P(\bar{p}) \propto e_{q_p}^{-\beta_q \bar{p}^2/2}, \tag{6.24}$$

with $q_p \simeq 1.58$, $\bar{p}$ being a conveniently rescaled momentum.

6.3.6. *Many-body classical systems with overdamping*

Non-conservative systems, typically associated with memory effects, may also exhibit connections with non-extensive statistical mechanics. Such is the case of some classes of overdamped many-body systems including two-body repulsive interactions, e.g., Type II superconductors (Andrade *et al.*, 2010; Vieira *et al.*, 2016; Curado *et al.*, 2014; Nobre *et al.*, 2015). A paradigmatic case is as follows. If we consider the nonlinear Fokker–Planck equation

$$\frac{\partial P}{\partial t} = -\frac{\partial[A(x)P]}{\partial x} + D\nu \frac{\partial}{\partial x}\Big[P^{\nu-1}\frac{\partial P}{\partial x}\Big], \tag{6.25}$$

where $A(x) = -\alpha x$ $(\alpha > 0)$ is an external confining force, and $(D, \nu) \in \mathcal{R}^2$. The solution is given by

$$P(x,t) = B(t)[1 + \beta(t)(1 - \nu)x^2]_+^{1/(\nu-1)}, \tag{6.26}$$

where $B(t)$ is a normalizing factor, $1/\beta(t)$ characterizes the width of the distribution, and $[y]_+ = y$ for $y > 0$ and zero otherwise.

This solution precisely extremizes the entropy

$$S_\nu(t) = k \frac{1 - \int_{-\infty}^{\infty} [P(x,t)]^\nu}{\nu - 1}, \qquad (6.27)$$

where $k = D/\gamma$, γ being a positive constant. The stationary state is given by

$$P(x,\infty) = B^*[1 - \beta^*(\nu - 1)x^2]^{1/(\nu-1)}, \qquad (6.28)$$

where B^* and β^* are positive constants.

This approach was numerically verified with success for a confined ensemble of N superconducting vortices in the presence of overdamping (i.e., assuming negligible Newtonian inter-particle forces) and a repulsive vortex–vortex interaction of the type of a modified Bessel function, decaying exponentially with distance, a situation which physically corresponds to $\nu = 0$.

6.3.7. *Quantum systems*

There are various kinds of d-dimensional Hamiltonian systems with strong quantum entanglement such that S_{BG} is not extensive, which violates thermodynamics. More precisely, they yield $S_{\mathrm{BG}}(L) \propto L^{d-1}$ for $d > 1$ and $S_{\mathrm{BG}}(L) \propto \ln L$ for $d = 1$, instead of $S_{\mathrm{BG}}(L) \propto L^d$, $\forall d$, where L is the linear size of the system. This anomaly is usually referred to as the *area-law*. One-dimensional such examples are discussed in Caruso and Tsallis (2008), Saguia and Sarandy (2010), Carrasco *et al.* (2016) and Souza *et al.* (2019). An important class of this type is characterised by the so-called central charge $c \geq 0$ (e.g., $c = 1/2$ and $c = 1$, respectively, correspond to the $T = 0$ quantum critical phenomenon occurring in the Ising and XY ferromagnetic chains, respectively). Thermodynamical extensivity was reestablished in Caruso and Tsallis (2008) through the non-additive entropy $S_q(L) \propto L$ with

$$q = \frac{\sqrt{9 + c^2} - 3}{c} \in [0, 1]. \qquad (6.29)$$

Certain classes of nonlinear quantum systems also exhibit interesting connections with q-statistics (Nobre *et al.*, 2011; Costa *et al.*,

2013; Nobre and Plastino, 2017; Plastino and Wedemann, 2017). For instance, the following nonlinear equation for $q \geq 1$

$$i\hbar \frac{\partial}{\partial t} \left[\frac{\Phi(\vec{x}, t)}{\Phi_0} \right] = -\frac{1}{2-q} \frac{\hbar^2}{2m} \nabla^2 \left[\frac{\Phi(\vec{x}, t)}{\Phi_0} \right]^{2-q}. \qquad (6.30)$$

This equation recovers the usual (linear) Schroedinger equation for $q = 1$. The free-particle solution of this equation is given by

$$\Phi(\vec{x}, t) = \Phi_0 \exp_q \left[i(\vec{k} \cdot \vec{x} - \omega t) \right], \qquad (6.31)$$

usually referred to as the q-plane wave.

6.3.8. *Low-dimensional dissipative and conservative maps*

Dissipative and conservative nonlinear maps whose largest Lyapunov exponent is positive exhibit strong chaos and are adequately associated with the additive BG entropy. In contrast, when this Lyapunov exponent is zero, the sensitivity to the initial conditions increases subexponentially with time, very frequently like a power law. In this case, the nonadditive entropy S_q plays a crucial role (Lyra and Tsallis, 1998; Boldovin and Robledo, 2004; Mayoral and Robledo, 2005; Ruiz *et al.*, 2012; Bountis and Skokos, 2012; Tirnakli and Borges, 2016). In what follows, we briefly review two paradigmatic cases, the dissipative logistic map at the so-called Feigenbaum point, and the area-preserving standard map in the neighborhood of vanishing nonlinear coupling.

The logistic map can be defined as follows:

$$x_{t+1} = 1 - ax_t^2, \quad 0 \leq a \leq 2, \quad x_t \in [-1, 1], \ t = 0, 1, 2, \ldots. \qquad (6.32)$$

At the Feigenbaum point $a = a_c = 1.40115518909\ldots$ the Lyapunov exponent vanishes, and we have the sensitivity to the initial conditions

$$\xi \equiv \lim_{\Delta x_0 \to 0} \frac{\Delta x_t}{\Delta x_0} = e_{q_{sen}}^{\lambda_{q_{sen}} t}, \qquad (6.33)$$

with $q_{\text{sen}} = 0.244487701341282066198\ldots$ and $\lambda_{q_{\text{sen}}} > 0$; sen stands for *sensitivity*. Consistently with a Pesin-like identity we verify a *linear* entropy production per unit time

$$S_{q_{\text{sen}}} \sim \lambda_{q_{\text{sen}}} t \quad (t \to \infty). \tag{6.34}$$

Also the sum of a large number of successive iterations of x_t approaches, after centering and scaling, the q-Gaussian attractor $P(x) \propto e_{q_{\text{stat}}}^{-\beta_{q_{\text{stat}}} x^2}$ with $q_{\text{stat}} = 1.65 \pm 0.05$ and $\beta_{q_{\text{stat}}} > 0$.

Let us finally mention a basic relaxation property. We start at $t = 0$ with a large number of initial conditions within a small window belonging to a partition of the space in $W \gg 1$ equally spaced windows and evolve in time. The entropy $S_{q_{\text{sen}}}(t)$ approaches from above its value (of the order of $\ln_{q_{\text{sen}}} W$) at the stationary state as follows:

$$S_{q_{\text{sen}}}(t) - S_{q_{\text{sen}}}(\infty) \propto e_{q_{\text{rel}}}^{-t/\tau_{q_{\text{rel}}}}, \tag{6.35}$$

with $q_{\text{rel}} = 2.249784109\ldots$ and $\tau_{q_{\text{rel}}} > 0$. The q-triplet $(q_{\text{sen}}, q_{\text{stat}}, q_{\text{rel}})$ is an important one and has been discussed in Gazeau *et al.* (2019).

Let us focus now on the standard map (or kicked rotor map), which is conservative for all the values of its nonlinear constant K:

$$p_{i+1} = p_i - K \sin x_i \pmod{2\pi} \quad (K \geq 0)$$

$$x_{i+1} = x_i + p_{i+1} \pmod{2\pi}. \tag{6.36}$$

If we sum a large number of successive values of x_t we approach, after centering and scaling, the probability distribution of the attractor. If K is very large, the Lyapunov exponents over almost all points of the phase space are positive, and the attractor is a Gaussian, which corresponds to the BG scenario. But if K is close to zero, almost all these Lyapunov exponents are nearly vanishing. The attractor has been numerically shown to be a q-Gaussian with $q \simeq 2$ (Tirnakli and Borges, 2016). For $K = 0$ it has been recently proved to be precisely $q = 2$ (Bountis *et al.*, 2020).

6.4. Applications in Economics

In order to study price fluctuations in stock markets, a stochastic process

$$X_t = \frac{\ln S(t + t_0)}{\ln S(t_0)},$$

representing log-returns was introduced in Borland (1998). Here $S(t)$ is the price of a share at time t. X_t solves the stochastic differential equation

$$dX_t = \tau dt + \sigma d\Omega_t,$$

where τ and σ are the drift and volatility coefficients respectively, and Ω_t is a solution to the Îto stochastic differential equation

$$d\Omega_t = [P(\Omega_t)]^{\frac{1-q}{2}} dB_t, \quad t > t_0. \tag{6.37}$$

In this equation B_t is a Brownian motion, and P is a q-Gaussian distribution function. The corresponding Fokker–Planck type equation in the case $\tau = 0$, $\sigma = 1$ reads

$$\frac{\partial V(x, t | x', t')}{\partial t} = ([V(x, t | x', t')]^{2-q})_{xx}, \tag{6.38}$$

which can easily be reduced to the form (6.6). Indeed, setting

$$V(t, x | x', t') = (2 - q)^{1/(q-1)} U(t, x)$$

for fixed (x', t'), we have the equation

$$(2 - q)^{\frac{1}{q-1}} \frac{\partial U}{\partial t} = \left((2 - q)^{\frac{2-q}{q-1}} U^{2-q} \right)_{xx}$$

$$= (2 - q)^{\frac{2-q}{q-1}} \left((2 - q) U^{1-q} U_x \right)_x$$

$$= (2 - q)^{\frac{1}{q-1}} (U^{1-q} U_x)_x, \quad t > 0, \ x \in \mathbb{R},$$

which coincides with equation (6.6).

From the financial applications point of view it is important to know the properties of the stochastic process X_t, since it can be considered as a q-alternative to the Brownian motion. One can effortlessly verify that if $U(t,x)$ is a solution to equation (6.6) for $t > 0$ with an initial condition $U(0,x) = f(x)$, then a solution $V(t,x)$, $t > t'$ to the same equation (6.6) considered for $t > t'$ with an initial condition $V(t',x) = f(x)$ can be represented in the form

$$V(t,x) = U(t - t', x), \quad t > t', \quad x \in \mathbb{R}.$$

It follows that X_t has stationary increments.

Further details on this and other applications are available at Borland (2002, 2002b), Kwapien and Drozdz (2012), Ruiz and Marcos (2018), Borland (2017), Xu and Beck (2017) and Tsallis (2017) to quote but a few.

6.5. Application to Global Optimization

Global optimization consists in numerically finding a global minimum of a given (not necessarily convex) cost function, defined in a continuous D-dimensional space. Such algorithms have a plethora of useful applications. A well-known classical procedure, currently referred to as the *Boltzmann machine*, is the so-called *Simulated Annealing*, which visits phase space with a Gaussian distribution and uses the Boltzmann factor to accept or not the possible new position in phase space.

Inspired by q-statistics, an algorithm was introduced (Tsallis and Stariolo, 1994, 1996), named *Generalized Simulated Annealing* (GSA), which recovers the Boltzmann machine as a particular case.

Like the Boltzmann machine, GSA consists of two algorithms that are to be used with alternation. These are the *Visiting algorithm* and the *Acceptance algorithm*. The visiting algorithm is based on a exploration of phase space using a q_V-Gaussian (instead of using a Gaussian distribution), and the acceptance algorithm is based on a q_A-exponential weight (instead of the Monte Carlo Boltzmann weight). Therefore a GSA machine is characterized by the pair (q_V, q_A). The choice $(1,1)$ is the Boltzmann machine. In practice, the most performant values have been shown to be $q_V > 1$, slightly below the maximal admissible value for the D-dimensional problem (for $D = 1$ the maximal admissible value is $q_V = 3$, and a performant value is $q_V \simeq 2.7$; for D dimensions, the maximal admissible value is $q_V = (D+2)/D)$, and $q_A < 1$.

Part of the simulated annealing procedure consists in the *Cooling algorithm*, which determines how the *effective temperature T* is to be decreased with time, so that the global minimum is eventually attained within the desired precision. A quick cooling is of course computationally desirable. But not too quick, otherwise the rate of success of ultimately arriving to the real global minimum decreases sensibly. The optimal cooling procedure appears to be given by Tsallis and Stariolo (1994, 1996)

$$\frac{T(t)}{T(1)} = \frac{2^{q_V-1} - 1}{(1+t)^{q_V-1} - 1} = \frac{\ln_{q_V}(1/2)}{\ln_{q_V}[1/(t+1)]}, \quad t = 1, 2, 3, \ldots,$$

(6.39)

where $T(1)$ is the initial high temperature imposed onto the system. We verify that, for $q_V = 1$, we have

$$\frac{T(t)}{T(1)} = \frac{\ln 2}{\ln(1+t)}, \quad t = 1, 2, 3, \ldots,$$
(6.40)

and that, for $q_V = 2$, we have

$$\frac{T(t)}{T(1)} = \frac{2}{t}, \quad t = 1, 2, 3, \ldots.$$
(6.41)

For the $D = 1$ upper limit, we have $q_V = 3$, hence

$$\frac{T(t)}{T(1)} = \frac{3}{(1+t)^2 - 1}, \quad t = 1, 2, 3, \ldots.$$
(6.42)

We see therefore a strong influence of q_V on the cooling allowed speed, which can ultimately benefit (decrease) quite strongly the necessary computational time.

6.6. Applications to Complex Networks

The so-called scale-free networks are deeply related to q-statistics (Brito *et al.*, 2016, 2019; Nunes *et al.*, 2017; Thurner and Tsallis,

2005; Emmerich *et al.*, 2014). A connection which relates that random geometrical problem with a quite generic thermal q-statistical one is exhibited in Oliveira *et al.* (2021). In this model, the d-dimensional geographical position of the arriving new node is characterized by an isotropic location distribution

$$p(r) \propto 1/r^{d+\alpha_G}, \tag{6.43}$$

where $r \geq 1$ is the Euclidean distance to the center of mass of the pre-existing graph, and $\alpha_G > 0$. The newly arrived node is attached to one of the pre-existing cluster with a probability

$$\Pi_{ij} \propto \epsilon_i/d_{ij}^{\alpha_A}, \tag{6.44}$$

where d_{ij} is the Euclidean distance between the pre-existing node i and the newly arrived j, $\alpha_A \geq 0$, the "energy"

$$\epsilon_i = \sum_{j=1}^{k_j} w_{ij}/2, \tag{6.45}$$

with w_{ij} randomly obtained from

$$p(w) = \frac{\eta}{w_o \, \Gamma(1/\eta)} e^{-(w/w_0)^\eta}, \quad w_0 > 0, \ \eta > 0. \tag{6.46}$$

It is numerically verified that, for increasingly large values of the total number N of nodes, the energy distribution is given by

$$p_q(\epsilon) = e_q^{-\beta \epsilon}/Z_q \tag{6.47}$$

for all $(\alpha_G, \alpha_A, \eta, w_0, d)$, Z_q being a normalizing factor and q depending only on α_A/d. The value of q characterizes the nonadditive entropic universality class.

6.7. Additional Notes

In this chapter, we considered a few illustrative applications of the abstract mathematical results obtained in previous chapters and their observations in natural processes. This list can be lengthily continued. For example, many applications have emerged along the years in areas such as signal and image processing, and very specifically for medical and biological utility, facial recognition, civil engineering, and others. For these and other applications we refer the reader to the following sources: Hagiwara *et al.* (2017), Tsigelny *et al.* (2008), Sotolongo *et al.* (2010), Bogachev *et al.* (2014, 2017), Mohanalin *et al.* (2010), Capurro et al (1999) and Acharya *et al.* (2018b).

Chapter 7

Some Open Problems

Various relevant aspects of the mathematical foundations of q-statistics have been updated and presented in this book. Naturally, others remain to be developed. Let us mention here a non-exhaustive list of them.

7.1. q-Central Limit Theorem for $q < 1$

The q-generalization of the Fourier transform and of its inverse have been developed in detail within this book for $q \geq 1$. This enables various interesting applications, very particularly the q-central limit theorem, which provides sufficient (but not necessary) conditions for q-Gaussians with $q > 1$ being attractors in the space of distributions of probabilities. This constitutes an important ingredient for understanding some of the many such long-tailed distributions found in nature. However, the analogous discussion for the compact support distributions corresponding to $q < 1$, which also emerge in nature, is still to be done.

7.2. Large Deviation Theory for $q \neq 1$

The standard large deviation theory is a mathematical theory connected with a variety of systems, including those described by the celebrated BG factor in physics. It essentially states the conditions for having, for a large number N of (nearly independent in some sense) random variables, the probability $P(x < x_{\mathrm{peak}}; N)$ decreasing *exponentially* with N, where x is the ratio of favourable events below the value corresponding to the peak of the distribution when $N \to \infty$. More precisely, we expect $P(x < x_{\text{peak}}; N) \sim e^{-r(x) N}$, where the *rate function* $r(x) \geq 0$ with $\lim_{x \to x_{\text{peak}}} r(x) = 0$; $r(x)$ can be identified with a BG relative entropy per random variable. In other words, $\ln P(x < x_{\text{peak}}; N) \sim -r(x) N$ is *extensive* in the thermodynamical sense. This is very welcome since $[r(x) N]$ corresponds to the *total* BG entropy of the system.

Let us focus now on the physical counterpart of the same concept for a classical many-body Hamiltonian $\mathcal{H}_N(\{x_i\}.\{p_i\})$ involving *short-range* interactions between the N particles, whose coordinates and momenta are respectively denoted by $(\{x_i\}, \{p_i\})$. The BG weight at thermal equilibrium is given by

$$P(\{x_i\}, \{p_i\}) \propto e^{-\beta \mathcal{H}_N(\{x_i\},\{p_i\})},$$

where $\beta \equiv 1/kT$ is the inverse temperature. It follows that the quantity

$$\ln P(\{x_i\}, \{p_i\}) \sim -[\beta \, \mathcal{H}_N(\{x_i\}, \{p_i\})/N] \, N$$

is extensive ($N \to \infty$) as well, analogously with the large deviation theory result.

It happens, however, that ubiquitous problems in physics exhibit Q-Gaussian instead of Gaussian attractors. It is therefore expected that, for a wide class of systems, N random variables are strongly correlated in such a way that $P(x < x_{\text{peak}}; N)$ asymptotically vanishes when $N \to \infty$ like a power-law, more precisely like a q-exponential for all values of N not exceedingly small. It is expected that $Q = f(q)$ with a function f such that $f(1) = 1$. In other words, we expect

$$P(x < x_{\text{peak}}; N) \sim e_q^{-r_q(x) N},$$

where the *q-rate function* $r_q(x) \geq 0$ with $\lim_{x \to x_{\text{peak}}} r_q(x) = 0$ (with $r_1(x) = r(x)$); $r_q(x)$ can possibly be identified with some kind of relative q-entropy per random variable. The plausibility of this conjecture has already been numerically illustrated with a probabilistic model (Ruiz and Tsallis, 2012, 2013; Touchette, 2013), with successive iterations of the standard map, and also with the

Coherent Noise Model for earthquakes and the Ehrenfest dog-flea model (Tirnakli *et al.*, 2021). Therefore, the q-generalization of the usual Large Deviation Theory becomes a sort of must. Notice that, assuming that the conjecture is valid for some classes of complex systems, it implies that $\ln_q P(x < x_{\text{peak}}; N) \sim -[r_q(x) N]$ is *extensive* ($N \to \infty$) in the thermodynamical sense, $\forall q$, which once again is consistent with an *extensive* total entropy even outside the BG theory.

We may illustrate its physical counterpart for d-dimensional many-body Hamiltonians involving *long-range* interactions, inversely proportional to $1/r^\alpha$ ($\alpha \geq 0$). Within nonextensive statistical mechanics we expect that

$$P(\{x_i\}, \{p_i\}) \propto e_q^{-\beta_q \mathcal{H}_N(\{x_i\}, \{p_i\})}.$$

Hence

$$\ln_q P(\{x_i\}, \{p_i\}) \sim -[\beta_q \tilde{N}] \, [\mathcal{H}_N(\{x_i\}, \{p_i\})/N\tilde{N}] \, N, \ N \to \infty,$$

where $\tilde{N} \equiv \frac{N^{1-\alpha/d}-1}{1-\alpha/d}$ (Tsallis, 2009a). The quantity $\tilde{N}$ has the limit $1/(\alpha/d - 1)$ for $\alpha/d > 1$ when $N \to \infty$, and behaves as $\ln N$ for $\alpha/d = 1$, and as $\frac{N^{1-\alpha/d}}{1-\alpha/d}$ for $0 \leq \alpha/d < 1$, for large N. The quantity $[\beta_q \tilde{N}] \, [\mathcal{H}_N(\{x_i\}, \{p_i\})/N\tilde{N}]$ is intensive Tsallis (2009a), consequently $\ln_q P(\{x_i\}, \{p_i\})$ is expected to be *extensive*, $\forall q$. This generic extensivity plays a most important role in thermodynamics.

The examples that are available in the literature are all for $q \geq 1$. It remains an open problem what happens for $q < 1$. In any case, it is clear that the rigorous generalization of the standard Large Deviation Theory beyond the $q = 1$ world would be of extreme value.

7.3. (q, α)-Stable Distributions: Open Problems and Some Conjectures

In Chapter 5, two classifications of the q-generalized α-stable distributions were obtained. Both classifications of (Q, α)-stable distributions, where $Q = 2q-1$, discussed in Section 5.8 are restricted to the region $Q = \{1 \leq Q < 3, 0 < \alpha \leq 2\}$. This limitation is caused by the tool used for these classifications, namely, Q-Fourier transform is defined for $Q \geq 1$. However, at least two facts strongly suggest some conjectures regarding the region

$$R_{Q,\alpha} \equiv \{\max\{-1, 1 - 2/\alpha\} < Q < 1, 0 < \alpha \leq 2\},$$

corresponding to the rectangle at the left of the vertical green line (the classic Lévy line) in Fig. 5.7. Theses facts are the positivity of

$$\mu_{q,\alpha} = \begin{cases} \dfrac{q}{2}\sigma_{2q-1}^2[f] \; \nu_{2q-1}[f], & \text{if } \alpha = 2; \\[2ex] \dfrac{2^{2-\alpha}(1 + \alpha(q-1))C_f}{2-q} \displaystyle\int_0^\infty \dfrac{-\Psi_q(y)}{y^{\alpha+1}} dy, & \text{if } (q,\alpha) \in \mathcal{Q}_2. \end{cases}$$

defined in (3.64) (see Proposition 3.15 and Remark 3.9) for $q > \max\{0, 1 - 1/\alpha\}$ (or, equivalently, $Q > \max\{-1, 1 - 2/\alpha\}$) and the continuous extensions of curves in the family C_{qL}. In this region one can see three borderlines: F-II, F-III and F-IV.

- **Conjecture 1.** The borderline F-II separates the regions where the random variables have *finite* and *infinite* Q-variances. More precisely, the random variables corresponding to (Q, α) on and above the line F-II have a *finite* Q-variance and, consequently, the paper Umarov *et al.* (2008) applies. Moreover, as seen in the figure, the q^L-attractors corresponding to the points on the line F-II are the classic Gaussians, because $q^L = 1$ for these (Q, α). It follows from this fact that q^L-Gaussians corresponding to points above F-II have compact support (the blue region in the figure), and q^L-Gaussians corresponding to points on this line and below might have infinite support (though $q < 1$). This requires a proof.
- **Conjecture 2.** The borderline F-III separates the points (Q, α) whose q^L-attractors have *finite* or *infinite* usual variances. More precisely, the points (Q, α) above this line identify attractors (in terms of q^L-Gaussians) with *finite* variance, and the points on this line and below identify attractors with *infinite* variance, which requires a proof.
- **Conjecture 3.** The borderline F-IV with the equation $Q + 2\alpha - 1 = 0$ and joining the points $(1, 0)$ and $(-1, 1)$ is related to attractors in terms of q^*-Gaussians. It follows from (5.57) that

for (Q, α) lying on the line F-IV, the index $q^* = -\infty$. Thus the horizontal lines corresponding to $\alpha < 1$ can presumably be continued only up to the line F-IV with $q^* \in (-\infty, 3 - \frac{4\alpha}{Q+2\alpha-1})$ (see the dashed horizontal brown line in the figure). This requires a proof. If $\alpha \to 0$, the Q-interval becomes narrower, but q^*-interval becomes larger tending to $(-\infty, 3)$.

It is clear that in order to investigate the problems set in these conjectures, one needs to study (Q, α)-stable random variables for $Q < 1$. As was noted in Section 4.6.1, the Vignat–Plastino central limit theorem is established for $q < 1$ as well. But this theorem corresponds to $\alpha = 2$. Extending this theorem for $0 < \alpha < 2$ may help to confirm or refute the above three conjectures. Another possibility is to extend the q-Fourier transform for $q < 1$. Such an attempt was considered in Nelson and Umarov (2008).

The question of how to numerically exhibit a specific (q, α)-stable distribution given its parameters q, α, and its width is highly interesting. Indeed, the attractors for $(q, \alpha) = (q, 2)$ are the q-Gaussians, whose explicit expression is available. The $(1, \alpha)$ attractors are Lévy distributions, whose Fourier transform is explicit and given by a stretched exponential. The numerical construction of these attractors is unique and can be given with any desired finite precision through its inverse Fourier transform (Tsallis and Arenas, 2014). But what about the generic (q, α) attractors? We know that in a log-log representation, they exhibit two successive straight-line regions, an intermediate one with a specific slope dictated by (q, α), and an asymptotic one with a different slope, also dictated by (q, α). But we still do not know how to numerically calculate the entire distribution given generic (q, α) values (for $(q, \alpha) \neq (q, 2)$ and also $\neq (1, \alpha)$), and its characteristic width. Undoubtedly, this is an open practical problem which demands to be properly tackled.

7.4. Asymmetric and Multivariate (q, α)-Stable Distributions

In modeling physical processes, one often needs to use asymmetric or multivariate stable random vectors. In case of strongly dependent states the existing approaches, based on the statistical independence, do not work. Therefore, the study of asymmetric and multivariate versions of the (q, α)-stable distributions would be highly welcome. To our best knowledge, this has never been undertaken.

7.5. Connection with the Möbius Structure of q-Triplets

The various properties of a given complex system are typically characterized by a set of indices q, which are possibly connected through relatively simple relationships. These relations would leave the possibility of only a few among the indices q to be independent, and dictated by the universality class of the system itself and the conditions within which it is placed. These sets of indices appear to follow some sort of Möbius structure (Tsallis *et al.*, 2005a; Tsallis, 2017a,b; Gazeau *et al.*, 2019). An interesting question which remains open is what might be the precise connections of the results that emerged in the present book with that Möbius structure of q-triplets and related sets of indices q.

7.6. Additional Notes and Final Words

Boltzmann–Gibbs statistical mechanics constitutes the heart of a formidable theoretical body with far-reaching branches. Its results can *in principle* be reobtained from using theory of probabilities into mechanics (classical, quantum, relativistic) and electromagnetism. This procedure, sometimes referred to as from *first principles*, is intractable in most of the cases. For instance, if we are dealing with a real classical gas, it requires to solve the Newton equations of motion of say the Avogadro number ($\simeq 6 \times 10^{23}$) of coupled particles. If the system is quantum, the situation is even more dramatic. Indeed, it would require to solve the Schroedinger partial derivative equation for the same number of variables. What the genius of Boltzmann and of Gibbs did, by introducing statistical mechanics, was an unbelievable mathematical shortcut which consists in introducing the BG entropic functional S_{BG}. The optimization of this functional with simple appropriate constraints enables, through by far much simpler operations, the calculation of virtually all the relevant quantities, in particular at the stationary state of the system. Furthermore, in the $N \to \infty$ limit, contact is established with thermodynamics and its Legendre mathematical structure. Nowadays, one century and a half after the proposal of BG statistical mechanics in the 19th century, it still remains as an open problem the mathematical proof, starting from Newton's equations of motion, of the BG weight for a generic classical many-body Hamiltonian system with short-range interactions.

Moreover, the BG theory has very many other connections, for instance with Langevin stochastic differential equations, Fokker–Planck equations, master equations. It also connects with the central limit theorem and the large deviation theory in theory of probabilities, with the ergodic theory, the Liouville equation, the von Neumann equation, the Boltzmann kinetic and the Vlasov equations, the variational method based on the Bogoliubov inequality, cumulant expansions, high- and low-temperature series, random geometrical problems (percolation, network of random resistors, self-avoiding random walks), graph theory, renormalization group theory, and so on.

Nevertheless, in spite of its admirable operational power, the BG theory fails when some of the above simplifying hypothesis are not satisfied. For instance when the maximal Lyapunov exponent of a nonlinear dynamical system vanishes. Consequently, its generalization becomes desirable.

The philosophy of nonextensive statistical mechanics consists in replacing, whenever appropriate, the additive BG entropic functional S_{BG} by a non-additive one, namely, S_q, where the entropic index q is in principle to be obtained from microscopic dynamics, unless its determination turns out to be mathematically intractable. This path is naturally associated with typically nonlinear q-generalizations of the branches and equations listed above. Along these lines, we may mention the attempts for the properly q-generalized variational method (Plastino and Tsallis, 1993; Lenzi *et al.*, 1998; Mendes *et al.*, 2004), cumulant expansion (Rodriguez and Tsallis, 2010), master equation (Curado and Nobre, 2003; Nobre *et al.*, 2004), probabilistic models (Rodriguez *et al.*, 2008) having Q-Gaussians as $N \to \infty$

limiting distributions (whose robustness still remains to be proved), alternative and equivalent versions yielding nonextensive statistical mechanics (Ferri *et al.*, 2005), cross-overs related with simultaneous $N \to \infty$ and *time* $\to \infty$ limits (Christodoulidi *et al.*, 2014). Last but not least, a closer look to the hypotheses of the theorem in Braun and Hepp (1977) might reveal the necessity of some kind of generalization. Indeed, to prove this theorem one goes from the Liouville-theorem phase-space to the standard Vlasov equation by integrating many integrals on the phase-space. If the interactions are short-ranged, this surely is legitimate, but if long-ranged interactions are present, it is not obvious that some zero Lebesgue measure ingredients could not require a more general proof, possibly opening the door to some kind of generalized Vlasov equation. Revisiting all the above mentioned results through a mathematically rigorous perspective would be very welcome.

We have discussed, in the present book, various crucial issues. Further discussions on the mathematical foundations of nonextensive statistical mechanics remain as an endless fascinating task.

Bibliography

Abramowitz M., Stegun I. (1964) *Handbook of Mathematical Functions with Formulas, Graphs, and Mathematical Tables.* Dover, New York City.

Acharya U.R., Hagiwara Y., Koh J.E.W, Oh S.L., Tan J.H., Adam M., Tan R.S. (2018) Entropies for automated detection of coronary artery disease using ECG signals: A review, *Biocybern. Biomed. Eng.* **38**(2), 373.

Acharya U.R., Hagiwara Y., Deshpande S.N., Suren S., Koh J.E.W., Oh S.L., Arunkumar N., Ciaccio E.J., Lim C.M. (2018) Characterization of focal EEG signals: A review, *Future Gen. Comput. Syst.* **91**, 290.

Aczel J., Daroczy Z. (1975) On measures of information and their characterization, in *Mathematics in Science and Engineering* (ed. R. Bellman). Academic Press, New York.

Adams W.F. (2009) *The Life and Times of the Central Limit Theorem.* American Mathematical Society.

Adare A. *et al.* (2011) (PHENIX Collaboration), Measurement of neutral mesons in $p + p$ collisions at $\sqrt{s} = 200\ GeV$ and scaling properties of hadron production, *Phys. Rev. D* **83**, 052004.

ALICE Collaboration, Enhanced production of multi-strange hadrons in high-multiplicity proton-proton collisions, *Nature Physics* **13**, 535 (June 2017).

ALICE Collaboration, Production of π^0 and η mesons up to high transverse momentum in pp collisions at 2.76 TeV, *Eur. Phys. J. C* **77**, 339 (2017).

ALICE Collaboration, Production of $\Sigma(1385)^{\pm}$ and $\Xi(1530)^0$ in p-Pb collisions at $\sqrt{s_{NN}} = 5.02\ TeV$, *Eur. Phys. J. C* **77**, 389 (2017).

ALICE Collaboration, $K^*(892)^0$ and $\Phi(1020)$ meson production at high transverse momentum in pp and Pb-Pb collisions at $\sqrt{s_{NN}} = 2.76\ TeV$, *Phys. Rev. C* **95**, 064606 (2017).

Amari S., Ohara A. (2011) Geometry of q-exponential family of probability distributions. *Entropy* **13**, 1170–1185.

298   *Mathematical Foundations of Nonextensive Statistical Mechanics*

Ananos G.F.J., Tsallis C. (2004) Ensemble averages and nonextensivity at the edge of chaos of one-dimensional maps, *Phys. Rev. Lett.* **93**, 020601.

Andrade Jr. J.S., da Silva G.F.T., Moreira A.A., Nobre F.D., Curado E.M.F. (2010) Thermostatistics of overdamped motion of interacting particles, *Phys. Rev. Lett.* **105**, 260601.

Andrew F.F., Mallows C.L. (1974) Scale mixtures of normal distributions, *J. Royal Stat. Soc. Series B.* **36**, 99–102.

Anteneodo C. (2005) Non-extensive random walks, *Physica A* **358**, 289.

Anteneodo C., Tsallis C. (1998) Breakdown of the exponential sensitivity to the initial conditions: Role of the range of the interaction, *Phys. Rev. Lett.* **80**, 5313.

Antoni M., Ruffo S. (1995) Clustering and relaxation in Hamiltonian long-range dynamics, *Phys. Rev. E* **52**, 2361.

Azzalini A. (1985) A class of of disctributions which includes the normal ones. *Scand J. Statist*, **12**, 171–178.

Azzalini A. (2014) Skew Normal and Related Families. Cambridge University Press, New York.

Bacha M., Gougam L.A., Tribeche M. (2017) Ion-acoustic rogue waves in magnetized solar wind plasma with nonextensive electrons, *Physica A* **466**, 199.

Baella N.O. (2008) private communication.

Bagchi D., Tsallis C. (2016) Sensitivity to initial conditions of d-dimensional long-range-interacting quartic Fermi-Pasta-Ulam model: Universal scaling, *Phys. Rev. E* **93** (5), 062213.

Bagchi D., Tsallis C. (2017) Long-ranged Fermi-Pasta-Ulam systems in thermal contact: Crossover from q-statistics to Boltzmann–Gibbs statistics, *Phys. Lett. A* **381**, 1123–1128.

Bagchi D., Tsallis C. (2018) Fermi-Pasta-Ulam-Tsingou problems: Passage from Boltzmann to q-statistics, *Physica A* **491**, 869.

Baldovin F., Stella A. (2007) Central limit theorem for anomalous scaling due to correlations, *Phys. Rev. E*, **75**, 02010.

Baldovin F., Robledo A. (2002) Universal renormalization-group dynamics at the onset of chaos in logistic maps and nonextensive statistical mechanics, *Phys. Rev. E* **66**, 045104.

Baldovin F., Robledo A. (2004) Nonextensive Pesin identity. Exact renormalization group analytical results for the dynamics at the edge of chaos of the logistic map, *Phys. Rev. E* **69**, 045202.

Beck C. (2001) Dynamical foundations of nonextensive statistical mechanics. *Phys. Rev. Lett.* **87**, 180601.

Beck C. (2016) Cosmological flux noise and measured noise power spectra in SQUIDs, *Scientific Reports* **6**, 28275.

Beck C., Cohen E.G.D. (2003) Superstatistics, *Physica A*, **322**, 267–275.

Beck C., Schloegel F. (1993) *Thermodynamics of Chaotic Systems: An Introduction.* Cambridge University Press, Cambridge.

Bediaga I., Curado E.M.F., Miranda J. (2000) A nonextensive thermodynamical equilibrium approach in e^+e^- → hadrons, *Physica A* **286**, 156.

Bertulani C.A., Shubhchintak, Mukhamedzhanov A.M. (2018) Cosmological Lithium problems, *EPJ Web Conf.* **184**, 01002.

Billingsley P. (1995) *Probability and Measure.* John Wiley and Sons, New York.

Blitvić, N. (2012) The (q, t)-Gaussian process, *J. Func. Anal.* **263**, 3270–3305.

Bogachev M.I., Kayumov A.R., Bunde A. (2014) Universal internucleotide statistics in full genomes: A footprint of the DNA structure and packaging?, *PLoS ONE* **9**(12), e112534.

Bogachev M.I., Markelov O.A., Kayumov A.R., Bunde A. (2017) Superstatistical model of bacterial DNA architecture, *Scientific Reports* **7**, 43034.

Bohner M., Guseinov G.Sh. (2010) The h-Laplace and q-Laplace transforms, *J. Math. Anal. Appl.* **365**, 75–92.

Bologna M., Tsallis C., Grigolini P. (2000) Anomalous diffusion associated with nonlinear fractional derivative Fokker-Planck-like equation: Exact time-dependent solutions, *Phys. Rev. E* **62**, 2213.

Boltzmann L. (1872) Weitere Studien über das Wärmegleichgewicht unter Gas molekülen [Further studies on thermal equilibrium between gas molecules], *Wien, Ber.* **66**, 275.

Boltzmann L. (1877) Uber die Beziehung eines allgemeine mechanischen Satzes zum zweiten Haupsatze der Warmetheorie, Sitzungsberichte, K. Akademie der Wissenschaften in Wien, *Math.-Naturwissenschaften* **75**, 67–73; English translation (On the Relation of a General Mechanical Theorem to the Second Law of Thermodynamics) in *Kinetic Theory*, Vol. 2: *Irreversible Processes* (ed. S. Brush), Pergamon Press, Oxford, 1966, 188–193.

Boon J.P., Tsallis C. (2005) Nonextensive statistical mechanics: New trends, new perspectives, *Europhys News* **36** (6).

Borges E.P. (1998) A q-generalization of circular and hyperbolic functions, *J. Phys. A: Math. Gen.* **31**, 5281–5288.

Borges E.P. (2004) A possible deformed algebra and calculus inspired in nonextensive thermostatistics, *Physica A* **340**, 95–101.

Borges E.P., Tsallis C., Ananos G.F.J., Oliveira P.M.C. (2002) Nonequilibrium probabilistic dynamics at the logistic map edge of chaos, *Phys. Rev. Lett.* **89**, 254103.

Borland L. (1998) Microscopic dynamics of the nonlinear Fokker-Plank equations: A phenomenological model, *Phys. Rev. E* **57**, 6634–6642.

Borland L. (2002) Closed form option pricing formulas based on a non-Gaussian stock price model with statistical feedback, *Phys. Rev. Lett.* **89**, 098701.

Borland L. (2002) A theory of non-gaussian option pricing, *Quantitative Finance* **2**, 415.

Borland L. (2017) Financial market models, in *Complexity and Synergetics*. Springer, Heidelberg, pp. 257.

Bountis A., Skokos H. (2012) *Complex Hamiltonian Dynamics*, Springer Series in Synergetics. Springer, Berlin.

Bountis A., Veerman J.J.P., Vivaldi F. (2020) Cauchy distributions for the integrable standard map, *Phys. Lett. A* **384**, 126659.

Bozejko M., Kummerer B., Speicher R. (1997) q-Gaussian processes: Noncommutative and classical aspects, *Commun. Math. Phys.* **185**, 129–154.

Bozejko M., Speicher R. (1991) An example of a generalized brownian motion, *Commun. Math. Phys.* **137**, 519–531.

Bozejko M., Speicher R. (1992) An example of a generalized brownian motion II, *Quant. Prob. Related Topics* **VII**, 67–77.

Bradley R.C. (2003) Introduction to strong mixing conditions, V I,II, Technical report, Department of Mathematics, Indiana University, Bloomington (Custom Publishing of IU).

Braun W., Hepp K. (1977) The Vlasov dynamics and its fluctuations in the IfN limit of interacting classical particles, *Commun. Math. Phys.* **56**, 125–146.

Brito S.G.A., da Silva L.R., Tsallis C. (2016) Role of dimensionality in complex networks, *Sci. Reports* **6**, 27992.

Brito S.G.A., Nunes T.C., L.R. da Silva L.R., Tsallis C. (2019) Scaling properties of d-dimensional complex networks, *Phys. Rev. E* **99**, 012305.

Budini A. (2015) Extended q-Gaussian and q-exponential distributions from gamma random variables, *Phys. Rev. E* **91**, 052113.

Burlaga L.F., Ness N.F., Acuna M.H. (2007) Magnetic fields in the heliosheath and distant heliosphere: Voyager 1 and 2 observations during 2005 and 2006, *Astrophys. J.* **668**, 1246–1258.

Burlaga L.F., Vinas A.F. (2005) Triangle for the entropic index q of nonextensive statistical mechanics observed by Voyager 1 in the distant heliosphere, *Physica A* **356**, 375.

Campa A., Giansanti A., Moroni D., Tsallis C. (2001) Classical spin systems with long-range interactions: Universal reduction of mixing, *Phys. Lett. A* **286**, 251.

Capurro A., Diambra L., Lorenzo D., Macadar O., Martins M.T., Mostaccio C., Plastino A., Perez J., Rofman E., Torres M.E., Velluti J. (1999) Human dynamics: The analysis of EEG signals with Tsallis information measure, *Physica A* **265**, 235.

Carati A., Galgani L., Gangemi F., Gangemi R. (2019) Relaxation times and ergodicity properties in a realistic ionic-crystal model, and the modern form of the FPU problem, *Physica A* **532**, 121911.

Carrillo J.A., Toscani G. (2000) Asymptotic L^1-decay of solutions of the porous medium equation to self-similarity, *Indiana Univ. Math. J.* **49**, 113.

Carrasco J.A., Finkel F., Gonzalez-Lopez A., Rodriguez M.A., Tempesta P. (2016) Generalized isotropic Lipkin-Meshkov-Glick models: Ground state entanglement and quantum entropies, *J. Statist. Mech.* 033114.

Caruso F., Pluchino A., Latora V., Vinciguerra S., Rapisarda A. (2007) Analysis of self-organized criticality in the Olami-Feder-Christensen model and in real earthquakes, *Phys. Rev. E* **75**, 055101.

Caruso F., Tsallis C. (2008) Nonadditive entropy reconciles the area law in quantum systems with classical thermodynamics, *Phys. Rev. E* **78**, 021102.

Casas A.A., Nobre F.D., Curado E.M.F. (2019) New type of equilibrium distribution for a system of charges In a spherically-symmetric electric field, *EPL* **126**, 10005.

Celikoglu A., Tirnakli U. (2006) Sensitivity function and entropy increase rates for z-logistic map family at the edge of chaos, *Physica A* **372**, 238–242.

Chavanis P.H., Campa A. (2010) Inhomogeneous Tsallis distributions in the HMF model, *Eur. Phys. J. B* **76**, 581.

Chow Y.S., Teicher H. (1978) *Probability Theory: Independence, Interchangeability, Martingales.* Springer, New York.

Christodoulidi H., Bountis T., Tsallis C., Drossos L. (2016) Dynamics and statistics of the Fermi–Pasta–Ulam β–model with different ranges of particle interactions, *J. Statist. Mech.*, 123206.

Christodoulidi H., Tsallis C., Bountis T. (2014) Fermi-Pasta-Ulam model with long-range interactions: Dynamics and thermostatistics, *EPL* **108**, 40006.

Cieśliński J.L. (2011) Improved q-exponential and q-trigonometric functions, *Appl. Math. Lett.* **24**, 2110–2114.

Cieśliński J.L. (2012) New definitions of exponential, hyperbolic and trigonometric functions on time scales, *J. Math. Anal. Appl.* **388**, 8–22.

Cirto L.J.L., Assis V.R.V., Tsallis C. (2014) Influence of the interaction range on the thermostatistics of a classical many-body system, *Physica A* **393**, 286.

Cirto L.J.L., Lima L.S., Nobre F.D. (2015) Controlling the range of interactions in the classical inertial ferromagnetic Heisenberg model: Analysis of metastable states, *JSTAT* P04012.

Cirto L.J.L., Rodriguez A., Nobre F.D., Tsallis C. (2018) Validity and failure of the Boltzmann weight, *Europhys. Lett.* **123**, 30003.

Combe G., Richefeu V., Stasiak M., Atman A.P.F. (2015) Experimental validation of nonextensive scaling law in confined granular media, *Phys. Rev. Lett.* **115**, 238301.

Conway J. (1978) Functions of one complex variable, in *Graduate Texts in Mathematics,* Vol. **11**, 2nd ed. Springer.

Costa Filho R.N., Alencar G., Skagerstam B.-S., Andrade Jr. J.S. (2013) Morse potential derived from first principles, *Europhys. Lett.* **101**, 10009.

Curado E.M.F., Nobre F.D. (2003) Derivation of nonlinear Fokker-Planck equations by means of approximations to the master equation, *Phys. Rev. E* **67**, 021107.

Curado E.M.F., Souza A.M.C., Nobre F.D., Andrade R.F.S. (2014) Carnot cycle for interacting particles in the absence of thermal noise, *Phys. Rev. E* **89**, 022117.

Curado E.M.F., Tsallis C (1991) Generalized statistical mechanics: Connection with thermodynamics, *J. Phys. A* **24**, L69. [Corrigenda: **24**, 3187 (1991) and **25**, 1019, 1992].

Daniels K.E., Beck C., Bodenschatz E. (2004) Defect turbulence and generalized statistical mechanics, in *Anomalous Distributions, Nonlinear Dynamics and Nonextensivity* (eds. H.L. Swinney and C. Tsallis); *Physica D* **193**, 208.

Daroczy Z. (1970) Generalized information measures, *Inf. Control* **16**, 36–51.

Dehling H., Denker M., Philipp W. (1986) Central limit theorem for mixing sequences of random variables under minimal condition, *Ann. Probab.*, **14** (4), 1359.

Dehling H.G., Mikosch T., Sorensen M. eds. (2002) *Empirical Process Techniques for Dependent Data.* Birkhäser, Boston-Basel-Berlin.

R.G. DeVoe. (2009) Power-law distributions for a trapped ion interacting with a classical buffer gas, *Phys. Rev. Lett.* **102**, 063001.

Diaz R., Pariguan E. (2009) On the Gaussian q-distribution, *J. Math. Anal. Appl.*, **358**, 1–9.

Diniz P.R.B., Murta L.O., Brum D.G., de Araujo D.B., Santos A.C. (2010) Brain tissue segmentation using q-entropy in multiple sclerosis magnetic resonance images, *Braz J. Med. Biol. Res.* **43**, 77.

Dudley R.M. (1999) *Uniform Central Limit Theorems.* Cambridge University Press, New York.

Douglas P., Bergamini S., Renzoni F. (2006) Tunable Tsallis distributions in dissipative optical lattices, *Phys. Rev. Lett.* **96**, 110601.

Doukhan P. (1994) Mixing properties and examples, *Lecture Notes Statist.* **85**.

Durrett R. (2005) *Probability: Theory and Examples.* Thomson, New York.

Eckman J.-P., Ruelle D. (1985) Ergodic theory of chaos and strange attractors, *Rev. Mod. Phys.* **57**, 617.

Einstein A. (1910) Theorie der Opaleszenz von homogenen Flüssigkeiten und Flüssigkeitsgemischen in der Nähe des kritischen Zustandes, *Ann. Phys. Leipzig*, **33**, 1275.

Emmerich T., Bunde A., Havlin S. (2014) Structural and functional properties of spatially embedded scale-free networks, *Phys. Rev. E* **89**, 062806.

Ernst T. (2012) *A Comprehensive Treatment of q-Calculus*, Birkhäuser, Basel.

Ernst T. (2003) A method for q-calculus, *J. Nonlin. Math. Phys.* **10**(4), 487–525.

Euler L. (1748) *Introductio in Analysin Infinitorum.* T. 1, Chapter XVI, Lausanne, p. 259.

Feinsilver Ph. (1989) Elements of q-harmonic analysis, *J. Math. Anal. Appl.* **141**, 509–526.

Feller W. (1945) Fundamental limit theorems in probability, *Bull. Amer. Math. Soc*, **51**, 800–832.

Feller W. (1966) *An Introduction to Probability Theory and its Applications II.* John Wiley and Sons, Inc, New York and London and Sydney.

Ferri G.L., Martinez S., Plastino A. (2005) Equivalence of the four versions of Tsallis's statistics, *J. Statist. Mecha.: Theory Exp.*, PO4009+14.

Fischer H. (2011) *A History of the Central Limit Theorem. From Classical to Modern Probability Theory.* Springer, Berlin.

Fitouhi A., Bouzffour F. (2012) The q-cosine Fourier transform and the q-heat equation, *Ramanujan J.* **28**, 443–461.

Frank T.D. (2005) *Nonlinear Fokker-Planck Equations: Fundamentals and Applications*, Series Synergetics Springer, Berlin.

Gazeau J.-P., C. Tsallis. (2019) Möbius transforms, cycles and q-triplets in statistical mechanics, *Entropy* **21**, 1155.

Gell-Mann M., Tsallis C. eds. (2004), *Nonextensive Entropy — Interdisciplinary Applications*, Oxford University Press, New York.

Gibbs J.W. (1902) *Elementary Principles in Statistical Mechanics — Developed with Especial Reference to the Rational Foundation of Thermodynamics.* C. Scribner's Sons, New York, (Yale University Press, New Haven, 1948; OX Bow Press, Woodbridge, Connecticut, 1981).

Gnedenko B.V., Kolmogorov A.N. (1954) *Limit Distributions for Sums of Independent Random Variables*. Addison-Wesley, Reading.

Gneiting T. (1997) Normal scale mixtures and dual probability densities, *J. Statist. Comput. Simul.* **59**, 375–384.

Graf U. (2010) *Introduction to Hyperfunctions and Their Integral Transformations*. Birkhäuser, Basel.

Hagiwara Y., Sudarshan V.K., Leong S.S., Vijaynanthan A., Ng K.H. (2017) Application of entropies for automated diagnosis of abnormalities in ultrasound images: A review, *J. Mech. Med. Biol.* **17** (7), 1740012.

Hanel R., Thurner S., Tsallis C. (2009) Limit distributions of scale-invariant probabilistic models of correlated random variables with the q-Gaussian as an explicit example, *Eur. Phys. J. B* **72**, 263–268.

Havrda J., Chavrat F. (1967) Concept of structural α-entropy, *Kybernetika* **3**, 30–35.

Hilhorst H.J. (2009) Central limit theorems for correlated variables: Some critical remarks, *Braz. J. Phys.* **39**(2A), 371–379.

Hilhorst H.J. (2010) Note on a q-modified central limit theorem, *J. Stat. Mecha.: Theory Exp.* **2010**(10), 10023.

Hilhorst H.J., Schehr G. (2007) A note on q-Gaussians and non-Gaussians in statistical mechanics, *J. Stat. Mech.*, 06003.

Hou S.Q., He J.J., Parikh A., Kahl D., Bertulani C.A., Kajino T., Mathews G.J., Zhao G. (2017) Non-extensive statistics solution to the cosmological Lithium problem, *Astrophys. J.* **834**, 165.

Huang K. (1987) *Statistical Mechanics*, 2nd edn. John Wiley and Sons, New York.

Jackson F.H. (1908) On q-functions and a certain difference operator, *Trans. Roy Soc. Edin.* 46, 253–281.

Jacod J., Shiryaev A.N. (2003) *Limit Theorems for Stochastic Processes*, Springer, New York.

Jauregui M., Tsallis C. (2011) q-Generalization of the inverse Fourier transform, *Phys. Lett. A*, **375**, 2085–2088.

Jauregui M., Tsallis C., Curado E.M.F. (2011) q-Moments remove the degeneracy associated with the inversion of the q-Fourier transform, *J. Statist Mech.: Theory Exp.*, 10016.

Jiang X., Hahn M. (2003) Central limit theorems for exchangeable random variables when limits are mixture of normals, *J. Theoret. Probab.* **16**, 543–571.

Jiang X., Hahn M., Umarov S. (2010) On q-Gaussians and exchangeability, *J. Phys. A: Math. Theor.* **43**(6), 165208.

Jona-Lasinio G. (1975) The renormalization group: A probabilistic view, *Nuovo Cimento* **26 B** 99.

Jona-Lasino G. (2001) Renormalization group and probability theory, *Phys. Rep.* **352**, 439–458.

Kallenberg O. (2002) *Foundations of Modern Probability.* Springer-Verlag.

Kalogeropoulos N. (2018) The τ_q-Fourier transform: Covariance and uniqueness, *Modern Phys. Lett. B* **32** (4), 1850149.

Kaniadakis G. (2002). Statistical mechanics in the context of special relativity, *Phys. Rev. E*, 66, 056125.

Kaniadakis G., Lissia M., Scarfone A.M. (2005) Two-parameter deformations of logarithm, exponential, and entropy: A consistent framework for generalized statistical mechanics, *Physics Rev. E*, **71**, 046128.

Kaniadakis G., Scarfone A.M. (2002) A new one-parameter deformation of the exponential function, *Physica A* **305**, 69–75.

Kasaki M., Koike K.-i. (2017) On skew q-Gaussian distribution. *International J. Stat Syst.* **12**(4), 773–789.

Keilson J., Steutel F.W. (1974) Mixtures of distributions, moment inequalities and measures of exponentiality and normality, *Ann. Probab.* **2**, 112–130.

Koelink E., Van Assche W. (2009) Leonhard Euler and q-analogue of the logarithm, *Proc. Am. Math. Soc.* **137**, 1663–1676.

Kohler S. (2017) Fixing the Big Bang theory's Lithium problem, NOVA — research highlights, *J. Am. Astronom. Soc.* (15 February 2017 Features).

Komatsu N. (2017) Cosmological model from the holographic equipartition law with a modified Renyi entropy, *Eur. Phys. J. C* **77**, 229.

Komatsu N., Kimura S. (2013) Entropic cosmology for a generalized black-hole entropy, *Phys. Rev. D* **88**, 083534.

Komatsu N., Kimura S. (2014) Evolution of the universe in entropic cosmologies via different formulations, *Phys. Rev. D* **89**, 123501.

Korn G.A., Korn T.M. (1968) Mathematical Handbook for Scientists and Engineers, 2nd edn., McGraw-Hill, NY.

Kwapien J., Drozdz S. (2012) Physical approach to complex systems, *Phys. Rep.* **515**, 115.

Lay D., Lay S., McDonald J. (2014) *Linear Algebra and Its Applications.* Pearson, London.

Lima H.S., Tsallis C. (2020) Exploring the neighborhood of q-exponentials, Entropy **22**, 1402.

Lenzi E.K., Malacarne L.C., Mendes R.C. (1998) Perturbation and variational methods in nonextensive Tsallis statistics, *Phys. Rev. Lett.* **80**(2), 218–221.

Linhard L., Nielsen V. (1971) Studies in statistical mechanics, *Det Kongelige Danske Videnskabernes Selskab Matematisk-fysiske Meddelelser* (Denmark) **38** (9), 1–42.

Liu B., Goree J. (2008) Superdiffusion and non-Gaussian statistics in a driven-dissipative 2D dusty plasma, *Phys. Rev. Lett.* **100**, 055003.

Livadiotis G. (2018) Thermodynamic origin of kappa distributions, EPL/Perspective **122**, 50001.

Lourek I., Tribeche M. (2016) On the role of the κ-deformed Kaniadakis distribution in nonlinear plasma waves, *Physica A* **441**, 215.

Lu D. (2009) q-Difference equation and the Cauchy operator identities, *J. Math. Anal. Appl.* **359**, 265–274.

Lucena L.C., da Silva L.R., Tsallis C. (1995) Departure from Boltzmann-Gibbs statistics makes the hydrogen-atom specific heat a computable quantity, *Phys. Rev. E* **51**, 6247.

Lutz E. (2003) Anomalous diffusion and Tsallis statistics in an optical lattice, *Phys. Rev. A* **67**, 051402(R).

Lutz E., Renzoni F. (2013) Beyond Boltzmann-Gibbs statistical mechanics in optical lattices, *Nat. Phys.* **9**, 615–619.

Lyra M.L., Tsallis C. (1998) Nonextensivity and multifractality in low-dimensional dissipative systems, *Phys. Rev. Lett.* **80**, 53.

Mandelbrot B. (1997) *Fractals and Scaling in Finance*. Springer.

Marques L., Cleymans J., Deppman A. (2015) Description of high-energy pp collisions using Tsallis thermodynamics: Transverse momentum and rapidity distributions, *Phys. Rev. D* **91**, 054025.

Marsh J.A., Fuentes M.A., Moyano L.G., Tsallis C. (2006) Influence of global correlations on central limit theorems and entropic extensivity, *Physica A*, **372**, 183–202.

Marsh J., Earl S. (2005) New solutions to scale-invariant phase-space occupancy for the generalized entropy S_q, *Phys. Lett. A* **349**, 146–152.

Mayoral E., Robledo A. (2004) Multifractality and nonextensivity at the edge of chaos of unimodal maps, *Physica A* **340**, 219.

Mayoral E., Robledo A. (2005) Tsallis' q index and Mori's q phase transitions at edge of chaos, *Phys. Rev. E* **72**, 026209.

McCulloch J. H. (1996) Financial applications of stable distributions, in *Handbook of Statistics*, Volume 14, (eds. G. S. Maddala and C. R. Rao), North-Holland, New York.

Meerschaert M.M., Scheffler H.-P. (2001) Limit distributions for sums of independent random vectors, *Heavy Tails in Theory and Practice*. John Wiley and Sons, Inc, New York.

Mello P.A., Shapiro B. (1988) Existence of a limiting distribution for disordered electronic conductors, *Phys. Rev. B* **37**, 5860.

Mello P.A., Tomsovic S. (1992) Scattering approach to quantum electronic transport, *Phys. Rev. B* **46**, 15963.

Mendes R.C., Lopes C.A., Lenzi E.K., Malacarne L.C., (2004) Variational methods in nonextensive Tsallis statistics: A comparative study, *Physica A* **344**, 562–567.

Mendes R.S., Tsallis C. (2001) Renormalization group approach to nonextensive statistical mechanics, *Phys. Lett. A* **285**, 273–278.

Merriche A., Tribeche M. (2017) Electron-acoustic rogue waves in a plasma with Tribeche-Tsallis-Cairns distributed electrons, *Ann. Phys.* **376**, 436.

Metzler R., Klafter J. (2000) The random walk's guide to anomalous diffusion: A fractional dynamics approach, *Phys. Rep.* **339**, 1, 1–77.

Mittnik S., Rachev S. (1993) Modeling asset returns with alternative stable distributions, *Econom. Rev.* **12**, 261–330.

Mohanalin J., Beenamol, Kalra P.K., Kumar N. (2010) A novel automatic microcalcification detection technique using Tsallis entropy and a type II fuzzy index, *Comput. Math. Appl.* **60** (8), 2426.

Montroll E. W., Shlesinger M.F. (1982) On $1/f$ noise and other distributions with long tails, *Proc. Nat. Acad. Sci. USA* **79**, 3380–3383.

Montroll E. W., Bendler J.T. (1984) On Lévy (or stable) distributions and the Williams-Watt model of dielectric relaxation, *J. Statist. Phys.* **34**, 129–162.

Moyano L.G., Tsallis C., Gell-Mann M. (2006) Numerical indications of a q-generalized central limit theorem, *Europhys. Lett.* **73**, 813–819.

Muskat M. (1937) *The Flow of Homogeneous Fluids Through Porous Media.* McGraw Hill, New York.

Naudts J. (2010) The q-exponential family in statistical physics, *J. Phys.: Conf. Ser.* **201**, 012003.

Nelson K., Umarov S. (2008) The relationship between Tsallis statistics, the Fourier transform, and nonlinear coupling, arXiv:0811.3777 [cs.IT].

von Neumann J. (1927) Thermodynamik quantenmechanischer Gesamtheiten, Nachrichten von der Gesellschaft der Wissenschaften zu Gottingen S. 273.

NEXT-biblio (2021) http://tsallis.cat.cbpf.br/biblio.htm

Nivanen L., Le Mehaute A., Wang Q.A. (2003) Generalized algebra within a nonextensive statistics, *Rep. Math. Phys.* **52**, 437–444.

Nobre F.D., Curado E.M.F., Rowlands G. (2004) A procedure for obtaining general nonlinear Fokker–Planck equations, *Physica A* **334**, 109–118.

Nobre F.D., Tsallis C. (2003) Classical infinite-range-interaction Heisenberg ferromagnetic model: Metastability and sensitivity to initial conditions, *Phys. Rev. E* **68**, 036115.

Nobre F.D., Rego-Monteiro M.A., Tsallis C. (2011) Nonlinear generalizations of relativistic and quantum equations with a common type of solution, *Phys. Rev. Lett.* **106**, 140601.

Nobre F.D., Plastino A.R. (2017) A family of nonlinear Schrodinger equations admitting q-plane wave, *Phys. Lett. A* **381**, 2457.

Nobre F.D., Curado E.M.F., Souza A.M.C., Andrade R.F.S. (2015) Consistent thermodynamic framework for interacting particles by neglecting thermal noise, *Phys. Rev. E* **91**, 022135.

Nolan J. (2002) Stable distributions. Birkhäuser.

Nunes T.C., Brito S., da Silva L.R., Tsallis C. (2017) Role of dimensionality in preferential attachment growth in the Bianconi-Barabasi model, *J. Statist. Mech.*, 093402.

de Oliveira R.M., Brito S., da Silva L.R., Tsallis C. (2021) Connecting complex networks to nonadditive entropies, *Sci. Rep.* **11**, 1130.

Oliveira D.S., Galvao R.M.O. (2018) Non-extensive transport equations in magnetized plasmas for non-Maxwellian distribution functions, *Phys. Plasmas* **25**, 102308.

Ol'shanetskii M.A., Rogov V-B.K. (1999) The q-Fourier transform of q-generalized functions, *Sbornik: Mathematics*, **190**(5), 717.

Otto F. (2001) The geometry of dissipative evolution equations: The porous medium equation, *Comm. Partial Diff. Eq.* **26**, 101.

Peligrad M. (1986) Recent advances in the central theorem and its weak invariance principle for mixing sequences of random variables (a survey), in *Dependence in Probability and Statistics*, (eds. E. Eberlein and M.S. Taqqu), Progress in Probability and Statistics Vol. **11**. Birkhäuser, Boston, p. 193.

Penrose O. (1970) *Foundations of Statistical Mechanics: A Deductive Treatment*. Pergamon, Oxford.

Pierce R.D. (1996) RCS characterization using the alpha-stable distribution, in *Proceedings of the National Radar Conference,* IEEE Press, pp. 154–159.

Plastino A.R., Plastino A. (1995) Non-extensive statistical mechanics and generalized Fokker-Planck equation, *Physica A* **222**, 347.

Plastino A., Rocca M.C. (2011) A direct proof of Jauregui-Tsallis' conjecture, *J. Math. Phys.* **52**(10), 103503.

Plastino A., Rocca M.C. (2012) Inversion of Umarov-Tsallis-Steinberg's q-Fourier transform and the complex-plane generalization, *Physica A* **391**, 4740–4747.

Plastino A., Rocca M.C. (2013) Reflections on the q-Fourier transform and the q-Gaussian function, *Physica A: Statist. Mech. Appl.*, **392**(18), 3952–3961.

Plastino A.R., Tsallis C. (1993) Variational method in generalized statistical mechanics, *J. Physics A: Math. Gen.* **26**, L893–L896.

Plastino A.R., Wedemann R.S. (2017) Nonlinear wave equations related to nonextensive thermostatistics, *Entropy* **19**, 60.

Pluchino A., Rapisarda A., Tsallis C. (2007) Nonergodicity and central limit behavior in long-range Hamiltonians, *Europhys. Lett.* **80**, 26002.

Pluchino A., Rapisarda A. (2007) Nonergodicity and central limit behavior for systems with long-range interactions, *SPIE* **2**, 6802-32.

Prato D., Tsallis C. (1999) Nonextensive foundation of Levy distributions, *Phys. Rev. E* **60**, 2398–2401.

Queiros S.M.D., Tsallis C. (2007) Nonextensive statistical mechanics and central limit theorems II — Convolution of q-independent random variables, in *Complexity, Metastability and Nonextensivity*, (eds. S. Abe, H.J. Herrmann, P. Quarati, A. Rapisarda and C. Tsallis). American Institute of Physics Conference Proceedings 965, Catania, pp. 21–33.

Rapisarda A., Pluchino A. (2005) Nonextensive thermodynamics and glassy behavior, *Europhys. News* **36**, 202. [Europhysics News Special Issue Nonextensive Statistical Mechanics: New Trends, New Perspectives, eds. J.P. Boon and C. Tsallis (November/December 2005].

Reed M., Simon B. (1972) *Methods of Modern Mathematical Physics. I: Functional Analysis*, 1st ed. Academic Press Inc, Cambridge, USA.

Renyi A. (1961) On measures of information and entropy, in *Proceedings of the Fourth Berkeley Symposium*, Vol. 1 University of California Press, Berkeley, Los Angeles, pp. 547.

Renyi A. (1970) *Probability Theory*, North-Holland, Amsterdam.

Rio E. (2000) Theorie asymptotique des processus aleatoires faiblement dependants, *Mathematiques et Applications*, Vol. 31, Springer, Berlin.

Risken H. (1989) *The Fokker-Planck Equation. Methods of Solution and Applications.* 2nd edn. Springer, Berlin.

Rodriguez A., Schwammle V., Tsallis C. (2008) Strictly and asymptotically scale-invariant probabilistic models of N correlated binary random variables having q–Gaussians as $N \to \infty$ limiting distributions, *J. Statist. Mech.* 09006.

Rodriguez A., Tsallis C. (2014) Connection between Dirihlet distributions and a scale-invariant probabilistic model based on Leibnitz-like piramids, *J. Statist. Mech.: Theory Exp.* **2014**, 12027.

Rodriguez A., Tsallis C. (2010) A generalization of the cumulant expansion. Application to a scale-invariant probabilistic model, *J. Math. Phys.* **51**, 073301.

Rodriguez A., Nobre F.D., Tsallis C. (2019) d-Dimensional classical Heisenberg model with arbitrarily-ranged interactions: Lyapunov exponents and distributions of momenta and energies, *Entropy* **21**, 31.

Ruelle D. (1986) Resonances of Chaotic dynamical systems, *Phys. Rev. Lett.* **56**, 405.

Ruiz G., Tsallis C. (2009) Nonextensivity at the edge of chaos of a new universality class of one-dimensional unimodal dissipative maps, *Eur. Phys. J. B* **67**, 577–584.

Ruiz G., Tirnakli U., Borges E.P., Tsallis C. (2017a) Statistical characterization of the standard map, *J. Statist. Mech.*, 063403.

Ruiz G., Tirnakli U., Borges E.P., Tsallis C. (2017b) Statistical characterization of discrete conservative systems: The web map, *Phys. Rev. E* **96**, 042158.

Ruiz G., Bountis T., Tsallis C. (2012) Time-evolving statistics of chaotic orbits of conservative maps in the context of the central limit theorem, *Int. J. Bifurcation Chaos* **22** (9), 1250208.

Ruiz G., de Marcos A.F. (2018) Evidence for criticality in financial data, *Eur. Phys. J. B* **91**, 1.

Ruiz G., Tsallis C. (2012) Towards a large deviation theory for strongly correlated systems, *Phys. Lett. A* **376**, 2451–2454.

Ruiz G., Tsallis C. (2013) Reply to Comment on "Towards a large deviation theory for strongly correlated systems", *Phys. Lett. A* **377**, 491–495.

Rybczynski M., Wilk G., Wlodarczyk Z. (2015) System size dependence of the log-periodic oscillations of transverse momentum spectra, European *Phys. J. Web Conf.* **90**, 01002.

Saguia A., Sarandy M.S. (2010) Nonadditive entropy for random quantum spin-S chains, *Phys. Lett. A* **374**, 3384.

Samorodnitsky G., Taqqu M.S. (1994) *Stable Non-Gaussian Random Processes*. Chapman and Hall. New York.

Sato M. (1959) Theory of hyperfunctions, I, *J. Faculty Sci. University of Tokyo. Sect. 1, Mathematics, Astronomy, Physics, Chemistry* **8**(1), 139–193.

Sato M. (1960) Theory of hyperfunctions, II, *Faculty Sci., University of Tokyo. Sect. 1, Mathematics, Astronomy, Physics, Chemistry,* **8**(2), 387–437.

Scarfone A.M. (2017) κ-Deformed Fourier transform, *Physica A: Statist. Mech. Appl.* **480**, 63–78.

Schmitt F.G., Seuront L. (2001) Multifractal random walk in copepod behavior, *Physica A* **301**, 375–396.

Shannon C.E. (1948) *Bell system Tech. J.* **27**, 379 and 623 (1948); A mathematical theory of communication, *Bell System Technical Journal* **27**, 379–423 and 623–656 (1948); and *The Mathematical Theory of Communication*, University of Illinois Press, Urbana, 1949.

Sharma B.D., Mittal D.P. (1975) New non-additive measures of entropy for discrete probability distributions, *J. Math. Sci.* **10**, 28.

Shiryaev A.N. (2016) *Probability. Graduate Texts in Math.* Springer, New York.

Sornette D. (2001) Critical Phenomena in Natural Sciences. Springer, Berlin.

Sotolongo-Grau O., Rodriguez-Perez D., Antoranz J.C., Sotolongo-Costa O. (2010) Tissue radiation response with maximum Tsallis entropy, *Phys. Rev. Lett.* **105**, 158105.

Souza A.M.C., Rapcan P., Tsallis C. (2020) Area-law-like systems with entangled states can preserve ergodicity, *Eur. Phys. J. Special Topics.* Vol. 229, pp. 759–772.

Souza A.M.C., Tsallis C. (1997) Student's t and r-distributions: Unified derivation from an entropic variational principle, *Physica A* **236**, 52–57.

Thistleton W.J., Marsh J.A., Nelson K.P., Tsallis C. (2009) q-Gaussian approximants mimic non-extensive statistical-mechanical expectation for many-body probabilistic model with long-range correlations, *Cent. Eur. J. Phys.* **7**, 387–394.

Thurner S., Tsallis C. (2005) Nonextensive aspects of self-organized scale-free gas-like networks, *Europhys. Lett.* **72**, 197.

Tirnakli U., Borges E.P. (2016) The standard map: From Boltzmann-Gibbs statistics to Tsallis statistics, *Sci. Rep.* **6**, 23644.

Tirnakli U., Jensen H.J., Tsallis C. (2011) Restricted random walk model as a new testing ground for the applicability of q-statistics, *Europhys. Lett.* **96**, 40008.

Tirnakli U., Tsallis C., Ay N. (2020) Approaching a large deviation theory for complex systems, arXiV: 2010.09508.

Touchette H. (2009) The large deviation approach to statistical mechanics, *Phys. Rep.* **478**, 1.

Touchette H. (2013) Comment on "Towards a large deviation theory for strongly correlated systems, *Phys. Lett. A* **377** (5), 436–438.

Tsallis C. (1988) Possible generalization of Boltzmann-Gibbs statistics, *J. Stat. Phys.* **52**, 479–487.

Tsallis C. (1994) What are the numbers that experiments provide? *Quimica Nova* **17**(6), 468–471.

Tsallis C. (2004) Dynamical scenario for nonextensive statistical mechanics, in *News and Expectations in Thermostatistics*; *Physica A* **340**, 1–10.

Tsallis C. (2004) Some thoughts on theoretical physics, *Physica A* **344**, 718.

Tsallis C. (2004) What should a statistical mechanics satisfy to reflect nature?, in *Anomalous Distributions, Nonlinear Dynamics and Nonextensivity* (eds. H.L. Swinney and C. Tsallis); *Physica D* **193**, 3.

Tsallis C. (2005) Nonextensive statistical mechanics, anomalous diffusion and central limit theorems, *Milan J. Math.* **73**, 145.

Tsallis C. (2006) On the extensivity of the entropy S_q, the q-generalized central limit theorem and the q-triplet, in *Proceedings of the International Conference on Complexity and Nonextensivity: New Trends in*

Statistical Mechanics (Yukawa Institute for Theoretical Physics, Kyoto, 14–18 March 2005), eds. S. Abe, M. Sakagami and N. Suzuki, Prog. Theor. Phys. Supplement, Vol. 162, p. 1.

Tsallis C. (2009) *Introduction to Nonextensive Statistical Mechanics — Approaching a Complex World.* Springer, New York.

Tsallis C. (2009) Entropy, *Encyclopedia of Complexity and Systems Science.* Springer, Berlin.

Tsallis C. (2017) Economics and finance: q-statistical features galore, in (Special Issue) *Entropic Applications in Economics and Finance* (eds. M. Stutzer and S. Bekiros). *Entropy* **19**, 457.

Tsallis C. (2017) Generalization of the possible algebraic basis of q-triplets, *Eur. Phys. J. Special Topics* **226**, 455–466.

Tsallis C. (2017) Statistical mechanics for complex systems: On the structure of q-triplets, in *Physical and Mathematical Aspects of Symmetries, Proceedings of the 31st International Colloquium in Group Theoretical Methods in Physics*, (eds. Duarte, S., Gazeau, J.-P., Faci, S., Micklitz, T., Scherer, R., and Toppan, F.) Springer, pp. 51–60,

Tsallis C., Anjos J.C., Borges E.P. (2003) Fluxes of cosmic rays: A delicately balanced stationary state, *Phys. Lett. A* **310**, 372.

Tsallis C., Anteneodo C., Borland L., Osorio R. (2003) Nonextensive statistical mechanics and economics, *Physica A* **324**, 89.

Tsallis C., Arenas (2014) Nonextensive statistical mechanics and high energy physics, *Euro. Phys. J. J* **71**, 00132.

Tsallis C., Bukman D.J. (1996) Anomalous diffusion in the presence of external forces: Exact time-dependent solutions and their thermostatistical basis, *Phys. Rev. E* **54**, R2197.

Tsallis C., Gell-Mann M., Sato Y. (2005) Asymptotically scale-invariant occupancy of phase space makes the entropy S_q extensive, *Proc. Natl. Acad. Sci. USA* **102**, 15377.

Tsallis C., Gell-Mann M., Sato Y. (2005) Extensivity and entropy production, *Europhys News* **36**, 186.

Tsallis C., Haubold H.J. (2015) Boltzmann-Gibbs entropy is sufficient but not necessary for the likelihood factorization required by Einstein. *EPLA* **110**, 30005 [doi: 10.1209/0295-5075/110/30005].

Tsallis C., Mendes R.S., Plastino A.R. (1998) The role of constraints within generalized nonextensive statistics, *Physica A* **261**, 534.

Tsallis C., Queiros S.M.D. (2007) Nonextensive statistical mechanics and central limit theorems I — Convolution of independent random variables and q-product, in *Complexity, Metastability and Nonextensivity* (eds. S. Abe, H.J. Herrmann, P. Quarati, A. Rapisarda and C. Tsallis). American Institute of Physics Conference Proceedings Vol. 965, pp. 8–20.

Tsallis C., Plastino A.R., Alvarez-Estrada R.F. (2009) Escort mean values and the characterization of power-law-decaying probability densities, *J. Math. Phys.* **50** 043303.

Tsallis C., Plastino A.R., Zheng W.-M. (1997) Power-law sensitivity to initial conditions — New entropic representation, *Chaos, Solitons Fractals* **8**, 885–891.

Tsallis C., Stariolo D.A. (1994) Generalized simulated annealing, *Notas de Fisica/CBPF* 026.

Tsallis C., Stariolo D.A. (1996) Generalized simulated annealing, *Physica A* **233**, 395.

Tsigelny I.F., Kouznetsova V.L., Sweeney D.E., Wu W., Bush K.T., Nigam S.K. (2008) Analysis of metagene portraits reveal distinct transitions during kidney organogenesis, *Sci. Signal.* **1**, (49).

Uchaykin V.V., Zolotarev V.M. (1999) Chance and stability, *Stable distributions and their applications.* VSP, Utrecht.

Umarov S. (2015) *Introduction to Fractional and Pseudo-Differential Equations with Singular Symbols.* Springer, Berlin.

Umarov S., Gorenflo R. (2005) On multi-dimensional symmetric random walk models approximating fractional diffusion processes, *Fract. Calculus Appl. Anal.* **8**, 73–88,

Umarov S., Hahn M., Kobayashi K. (2018) *Beyond the Triangle: Brownian Motion, Ito Calculus, and Fokker-Planck Equation – Fractional Generalizations.* World Scientific, New Jersey.

Umarov S., Queiros S.M-D. (2010) Functional-differential equations for F_q-transforms of q-Gaussians, *Physica A: Math. Theor.* **43**(9), 095202.

Umarov S., Steinberg S. (2006) Random walk models associated with distributed fractional order differential equations, *IMS Lecture Notes — Monograph Series*, **51**, 117–127.

Umarov S., Tsallis C. (2008) On a representation of inverse Fq-transform, *Phys. Lett. A.* **372,** 29, 4874–4876.

Umarov S., Tsallis C. (2007) On multivariate generalizations of the q-central limit theorem consistent with nonextensive statistical mechanics, in *Complexity, Metastability and Nonextensivity*, (eds. S. Abe, H.J. Herrmann, P. Quarati, A. Rapisarda and C. Tsallis) American Institute of Physics Conference Proceedings Vol. 965, pp. 34–42.

Umarov S., Tsallis C. (2016) The limit distribution in the q-CLT for $q \geq 1$ is unique and can not have a compact support, *J. Phys. A: Math. Theoret.* **49**, 415204.

Umarov S., Tsallis C., Gell-Mann M., Steinberg S. (2010) Generalization of symmetric α-stable Lévy distributions for $q > 1$, *J. Math. Phys.* **51**(3), 033502.

Umarov S., Tsallis C., Steinberg S. (2008) On a q-central limit theorem consistent with nonextensive statistical mechanics, *Milan J. Math.* **76**, 307–328.

Upadhyaya A., Rieu J.-P., Glazier J.A., Sawada Y. (2001) Anomalous diffusion and non-Gaussian velocity distribution of Hydra cells in cellular aggregates, *Physica A* **293**, 549.

Vajda I. (1968) Axioms of α-entropy of generalized probability distribution, *Kybernetika* **4**, 105–112 [in Czeck].

Van Leeuwen H., Maassen H. (1998) A q-deformation of the Gauss distribution, *J. Math. Phys.* **36** (9), 4773.

Vázquez J.L. (2007) The Porous Medium Equation: Mathematical Theory, Oxford Mathematical Monographs. Oxford University Press, New York.

Viallon-Galiner L., Combe G., Richefeu F., Atman A.P.F. (2018) Emergence of shear bands in confined granular systems: Singularity of the q-statistics, *Entropy* **20**, 862.

Vieira C.M., Carmona H.A., Andrade Jr. J.S., Moreira A.A. (2016) General continuum approach for dissipative systems of repulsive particles, *Phys. Rev. E* **93**, 060103(R).

Vignat C., Plastino A. (2005) The p-sphere and the geometric substratum of power-law probability distributions, *Phys. Lett. A.* **343**, 411–416.

Vignat C., Plastino A. (2006) Poincare's observation and the origin of Tsallis generalized canonical distributions, *Physica A.* **365**, 167.

Vignat C., Plastino A. (2007) Scale invariance and related properties of q-Gaussian systems, *Phys. Lett. A.* **365**, 370.

Vignat C., Plastino A. (2007) Central limit theorem and deformed exponentials, *J. Phys. A: Math. Theor.* **40**, F969–F978.

Wehrl A. (1978) General properties of entropy, *Rev. Mod. Phys.* **50**, 221.

Wilk G., Wlodarczyk Z. (2000) Interpretation of the nonextensive parameter q in some applications of Tsallis statistics and Lévy distributions, *Phys. Rev. Lett. A* **84** (13), 2770–2773.

Wilk G., Wlodarczyk Z. (2015) Tsallis distribution decorated with log-periodic oscillation, *Entropy* **17** (1), 384–400.

Wikipedia. Chi distribution.

Wong C.Y., Wilk G., Cirto L.J.L., Tsallis C. (2015) From QCD-based hard-scattering to nonextensive statistical mechanical descriptions of transverse momentum spectra in high-energy pp and $p\bar{p}$ collisions, *Phys. Rev. D* **91**, 114027.

Xu D., Beck C. (2017) Symbolic dynamics techniques for complex systems: Application to share price dynamics, *Europhys. Lett.* **118**(3), 30001.

Yalcin G.C., Beck C. (2018) Generalized statistical mechanics of cosmic rays: Application to positron-electron spectral indices, *Sci. Rep.* **8**, 1764.

Yoshihara K. (1992) Weakly dependent stochastic sequences and their applications, V.1. Summation theory for weakly dependent sequences Sanseido, Tokyo.

Zand J., Tirnakli U., Jensen H.J. (2015) On the relevance of q-distribution functions: The return time distribution of restricted random walker, *J. Phys. A: Math. Theor.* **48**, 425004.

Zaslavsky G. (2002) Chaos, fractional kinetics, and anomalous transport, *Phys. Rep.*, **371**, 461–580.

Zhang R. (2014) On asymptotics of the q-exponential and q-gamma functions, *J. Math. Anal. Appl.* **411**, 522–529.

Index